Neuromorphic Circuits
for Nanoscale Devices

RIVER PUBLISHERS SERIES IN BIOMEDICAL ENGINEERING

Series Editor:

DINESH KANT KUMAR
RMIT University
Australia

Indexing: All books published in this series are submitted to the Web of Science Book Citation Index (BkCI), to SCOPUS, to CrossRef and to Google Scholar for evaluation and indexing.

The "River Publishers Series in Biomedical Engineering" is a series of comprehensive academic and professional books which focus on the engineering and mathematics in medicine and biology. The series presents innovative experimental science and technological development in the biomedical field as well as clinical application of new developments.

Books published in the series include research monographs, edited volumes, handbooks and textbooks. The books provide professionals, researchers, educators, and advanced students in the field with an invaluable insight into the latest research and developments.

Topics covered in the series include, but are by no means restricted to the following:

- Biomedical engineering
- Biomedical physics and applied biophysics
- Bio-informatics
- Bio-metrics
- Bio-signals
- Medical Imaging

For a list of other books in this series, visit www.riverpublishers.com

Neuromorphic Circuits for Nanoscale Devices

Pinaki Mazumder

Yalcin Yilmaz

Idongesit Ebong

Woo Hyung Lee

University of Michigan, Ann Arbor, USA

River Publishers

Routledge
Taylor & Francis Group

LONDON AND NEW YORK

Published 2019 by River Publishers

River Publishers

Alsbjergvej 10, 9260 Gistrup, Denmark

www.riverpublishers.com

Distributed exclusively by Routledge

4 Park Square, Milton Park, Abingdon, Oxon OX14 4RN

605 Third Avenue, New York, NY 10158

First published in paperback 2024

Neuromorphic Circuits for Nanoscale Devices / by Pinaki Mazumder, Yalcin Yilmaz, Idongesit Ebong, Woo Hyung Lee.

Routledge is an imprint of the Taylor & Francis Group, an informa business

Publisher's Note
The publisher has gone to great lengths to ensure the quality of this reprint but points out that some imperfections in the original copies may be apparent.

While every effort is made to provide dependable information, the publisher, authors, and editors cannot be held responsible for any errors or omissions.

ISBN: 978-87-7022-060-6 (hbk)
ISBN: 978-87-7004-363-2 (pbk)
ISBN: 978-1-003-33891-8 (ebk)

DOI: 10.1201/9781003338918

Contents

Preface

In 1987 when I was wrapping up my doctoral thesis at the University of Illinois, I had a rare opportunity to listen to John Hopfield of California Institute of Technology describing his groundbreaking research in neural networks to spellbound students in the Loomis Laboratory of Physics at Urbana-Champaign. He pedagogically described how to design and fabricate a recurrent neural network chip to rapidly solve the benchmark Traveling Salesman Problem (TSP), which is provably NP-complete in the sense that no physical computer could solve the problem in asymptotically bounded polynomial time as number of cities in the TSP increases rapidly.

John Hopfield's seminal work established that if the "objective function" of a combinatorial algorithm can be expressed in quadratic form, the synaptic links in a recurrent artificial neural network could be accordingly programmed to reduce (locally minimize) the value of the objective function through massive interactions between the constituent neurons. Hopfield's neural network consists of laterally connected neurons that can be randomly initialized and then the network will iteratively reduce the intrinsic Lyapunov energy function of the network to reach a local minima state. Notably, the Lyapunov function decreases in a monotone fashion under the dynamics of the recurrent neural networks where neurons are not provided with self-feedback.

Soon after I joined the University of Michigan as an assistant professor, I developed, at first, an analog neural network with asynchronous state updates working with one of my doctoral students, and then digital neural chip with synchronous state updates working with another student. These neural circuits were designed to repair VLSI chips by formulating the repair problem in terms of finding the node cover, edge cover, or pair matching in a bipartite graph. In our graph formalism, one set of vertices in the bipartite graph represented the faulty circuit elements, and the other set of vertices represented the spare circuit elements. In order to restructure a faulty VLSI chip into a fault-free operational chip, the spare circuit elements were automatically invoked through programmable switching elements after identifying the faulty circuit elements through embedded built-in self-testing circuitry.

Most importantly, like the TSP problem, the 2-D array repair can be shown as an NP-complete problem because the repair algorithm seeks the optimal number of spare rows and spare columns that can be assigned to bypass faulty components such as memory cells, word-line and bit-line drivers, sense amplifiers, etc. located inside the memory array. Therefore, simple digital circuits comprising counters and other blocks woefully fail to solve such self-repair problems. Notably, one cannot use external digital computers to determine how to repair the embedded arrays, as input and output pins of the VLSI chip cannot be deployed to connect them with deeply embedded circuit blocks.

In 1989 and 1992, I received two NSF grants to expand the neuromorphic self-healing design styles to a wider class of embedded VLSI modules such as memory array, processors array, programmable logic array, and so on. However, this approach to improving VLSI chip yield by built-in self-testing and self-repair was a bit ahead of its time as the state-of-the-art microprocessors in early 1990's had only a few hundred thousands of MOS transistors and sub-micron CMOS technology was robust. Therefore, after developing the neural-net based self-healing VLSI chip design methodology for various types of VLSI circuit blocks, I stopped working on CMOS neural networks. I was not particularly interested in pursing applications of neural networks for other types of engineering problems, as I wanted to remain focused in VLSI research. Our research on neuromorphic VLSI circuits is published in our new book, titled: "Learning in Energy-Efficient Neuromorphic Computing Algorithm and Architecture Co-Design Neuromorphic Circuits for Nanoscale Devices", *John Wiley & Sons*, UK, 2019.

On the other hand, in late 1980's, there were mounting concerns among technology prognosticators about the impending red brick wall ending the shrinking era in CMOS. Therefore, to promote several types of emerging technologies that might push the frontier of VLSI technology, in early 1990's Defense Advanced Research Projects Agency (DARPA) in USA had initiated the Ultra Electronics: Ultra Dense, Ultra Fast Computing Components Research Program, and Ministry of International Trade & Industry (MITI) in Japan had launched the Quantum Functional Devices (QFD) project. Early successes with a plethora of innovative non-CMOS technologies in both research programs led to the launching of National Nanotechnology Initiative (NNI), which is a U.S. Government research and development (R&D)

initiative involving 20 departments and independent agencies to bring about revolution in nanotechnology to impact the industry and society at large.

During the period of 1995 and 2012, my research group had at first focused on quantum physics based device and circuit modeling for quantum tunneling devices, and then extensively worked on cellular neural network (CNN) circuits for image and video processing by using 1-D (resonant tunneling diodes), 2-D (nanowires) and 3-D (quantum dots) constrained quantum devices. Subsequently, we developed learning based neural network circuits by using resistive synaptic devices (commonly known as memristors) and CMOS neurons. We also developed analog voltage programmable nanocomputing architectures by hybridizing quantum tunneling and memristive devices together in the computing nodes.

The bulk of this book contains thesis work of three of my doctoral students who are listed as coauthors on the cover of the book. In addition, I requested some of my visiting research scientists to contribute to our collaborative work on CNN to broaden the scope of the book. The book is generally organized on the following way:

Chapter 1 through Chapter 4 describe resistive RAM memory i.e., memristor crossbar memory from physics, device modeling, circuit simulation, architectures and performance evaluation perspectives. Chapter 5 and Chapter 6 describe learning based neuromorphic design using memristors. Specifically, spiking timing dependent plasticity (STDP) and Q-learning algorithms are implemented on memristor substrates. Chapter 7 through Chapter 10 describes various types of quantum tunneling devices and how they can be utilized to design ultra-fast and low-power memristor based image processing, video motion processing, and color-image processing algorithms. Additionally, in order to achieve multiple functionalities on a single array of quantum dots or boxes, the processor array is combined with programmable memristors that alter the spatio-temporal characteristics of the processor array. Therefore, this hybrid design can be construed as a novel way to implement analog voltage programmable nanoscale computers as well as various types of spatiotemporal filtering systems. Chapter 11 provides the design of memristor based cellular neural networks (CNN) that Prof. Duan Sukai's group in South West University, Chongquin, China started at Michigan when he visited my research group in 2011 and 2012, while Chapter 12 provides a more rigorous analysis of memristor based CNN that Prof. Yongbi Yu of University of Electronics, Science and Technology in China (UESTC) started while he visited my research group in 2013 and 2014.

In order to set up these disparate research topics using memristors, quantum tunneling and spin torque nanomagnetic devices. I requested Prof. Steve Kang, Prof. Kamran Eshraghian and Dr. Jason Eshraghian to write a pedagogical introduction to the book for readers who are not fully conversant of these nascent technologies as well as their applications in neuromorphic computing. Chapter 1 provides an excellent review of these technologies from their basic operational principles so that the book can be adopted for advanced courses on naoscale neuromporphic circuits and architectures.

In Chapter 2, W. H. Lee and I describe the first crossbar memory technology that we developed using the silver-amorphous oxide-silicon based memristor structure. To teach and train senior undergrad and grad students, details of crossbar memory architectures, the chapter provides appropriate analytical models and calculations, including modeling of static power consumption for scalable crossbar design.

In Chapter 3, I. Ebong and I introduce the practical issues of crossbar memory design that was overlooked in the previous chapter that does not account for degeneration of cell behavior if multiple Read operations are made. Specifically, a procedure to program and erase a single-level cell (SLC) memristor memory is presented. The procedure is proven to have an adaptive scheme that stems from the device properties and makes accessing the memristor memory more reliable.

In Chapter 4, Y. Yilmaz and I address the design of reliable architectures of multi-level cell (MLC) based memristors that utilize a reduced constraint read-monitored-write scheme. Additionally, we describe a novel read technique that can successfully distinguish resistive states under the existence of resistance drift due to read/write disturbances in the array is presented. Again, to teach students, we provide derivations of analytical relations to set forth a design methodology in selecting peripheral device parameters.

In Chapter 5, I. Ebong and I describe the design of a spiking timing dependent plasticity (STDP) based Winner-Takes-All (WTA) neural networks architecture for the position detection of an object on a 2-dimensional gridded structure. We show that the analog approach to STDP implementation with memristors is superior to a digital-only approach.

In Chapter 6, I. Ebong and I present an attempt to bridge higher level learning to a memristor crossbar, therefore paving the way to realizing self-configurable circuits. The approach, or training methodology, is compared to Q-Learning in order to re-emphasize that reliably using memristors may require not knowing the precise resistance of each device, but instead working with relative magnitudes of one device to another.

In Chapter 7, S. Li, I. Ebong and I present a resonant tunneling diode (RTD)-based CNN architecture by elaborately describing its operation through driving-point-plot analysis, stability and settling time study, and circuit simulation. A comparative study between different CNN implementations reveals that the RTD-based CNN can be designed superior to conventional CMOS technologies in terms of integration density, operating speed, and functionality.

In Chapter 8, W. H. Lee and I introduce a novel approach to color image processing that utilizes multi-peak resonant tunneling diodes for encoding color information in quantized states of the diodes. The multi-peak resonant tunneling diodes (MPRTDs) are organized as a two-dimensional array of vertical pillars which are locally connected by programmable passive and active elements with a view to realizing a wide variety of color image processing functions such as quantization, color extraction, image smoothing, edge detection and line detection. In order to process color information in the input images, two different methods for color representation schemes have been used: one using color mapping and the other using direct RGB representation.

In Chapter 9, W. H. Lee and I demonstrate the design of a nanoscale velocity-tuned filter that employs resonant tunneling diodes to perform temporal filtering to track moving and stationary objects. The new velocity-tuned filter is not only amenable for nanocomputing, but also superior to other approaches in terms of area, power and speed. We show through analytical modeling that the proposed nanoarchitecture for velocity-tuned filter is asymptotically stable in the specific region.

In Chapter 10, Y. Yilmaz and I present an innovative architecture for image processing by a programmable artificial retina comprising quantum dots and variable resistance devices. We have designed an analog programmable resistive grid-based architecture mimicking the cellular connections of a biological retina in the most basic level, which is capable of performing various real time image processing tasks such as edge and line detections. The unit cell structure employs 3-D confined resonant tunneling diodes called quantum dots for signal amplification and latching, and these dots are interconnected between neighboring cells through non-volatile continuously variable resistive elements.

In Chapter 11, Prof. Duan's group presents a compact CNN model based on memristors along with its performance analysis and applications. In the new CNN design, the memristor bridge circuit acts as the synaptic circuit element and substitutes the complex multiplication circuit used in traditional CNN architectures. Additionally, the negative differential resistance (NDR) and nonlinear I–V characteristics of the memristor have been leveraged to replace the linear resistor in conventional CNNs. The proposed CNN design has several merits, for example, high-density, non-volatility, and programmability of synaptic weights. The proposed memristor-based CNN design operations for implementing several image processing functions are illustrated through simulation and contrasted with conventional CNNs. Monte-Carol simulation has been used to demonstrate the behavior of the proposed CNN due to variations in memristor synaptic weights.

Finally, in Chapter 12, Prof. Yu's group describes the memristor-based WTA neural network and the memristor-based recurrent neural network. At first, they explain the theoretical principles for the design of two memristive neural networks and perform the dynamic analysis to study the behaviors of these neural networks. Based on this theoretical analysis, they apply the WTA neural network to employ to skin diseases classifier with improved simulation results.

Thank you.

Pinaki Mazumder   September 14th, 2019

Address:

4765 BBB Building
Division of Electrical and Computer Engineering
Department of Electrical Engineering and Computer Science
University of Michigan, Ann Arbor, MI 48109-2122
Ph: 734-763-2107
E-mail: mazum@eecs.umich.edu, pinakimazum@gmail.com
Website: http://www.eecs.umich.edu/~mazum

Acknowledgement

First, I would like to thank several of my senior colleagues who had encouraged me to carry on my research in neural computing during the past three decades after I published my first paper on self-healing of VLSI memories in1989 by adopting the concept of the Hopfield network. Specifically, I would like to thank Prof. Leon O. Chua and Prof. Ernest S. Kuh of the University of California at Berkeley, Prof. Steve M. Kang, Prof. Kent W. Fuchs, and Prof. Janak H. Patel of the University of Illinois at Urbana-Champaign, Jacob A. Abraham of University of Texas at Austin, Prof. Supriyo Bandyopadhyay of Virginia Commonwealth University, Prof. Sudhakar M. Reddy of University of Iowa, and Prof. Tamas Roska and Prof. Csurgay Arpad of Technical University of Budapest, Hungary.

Second, I would like to thank my colleagues at the National Science Foundation where I served in the Directorate of Computer and Information Science and Engineering (CISE) as a Program Director of Emerging Models and Technologies program from January 2007 to December 2008, and then I served in the Engineering Directorate (ED) as a Program Director of Adaptive Intelligent Systems program from January 2008 to December 2009. Specifically, I would like to thank Dr. Robert Grafton and Dr. Sankar Basu of Computing and Communications Foundation Division of CISE, and Dr. Radhakrisnan Baheti, Dr. Paul Werbos and Dr. Jenshen Lin of Electrical Communications and Cyber Systems (ECCS) Division of ED for providing me with research funds over the past so many years to conduct research on learning-based systems that enabled me to delve deep into CMOS chip design for brain-like computing.

Besides the National Science Foundation, I received funding to develop nanoscale neuromorphic circuits from Dr. Gernot Pomrenke of Air Force Office of Scientific Research. He had lead the prestigious Ultra Electronics Program at the Defense Advanced Research Projects Agency (DARPA) to foster emerging technologies and their unique applications. After the Ultra Program ended in 1999, Dr. Lawrence Cooper of the Office of Naval Research (ONR) supported my work that enabled my research group to design several types of neural networks by using quantum tunneling devices such as resonant tunneling diodes, nanowires and quantum dots. After Larry retired from

ONR in 2004, Dr. Chaggan Baatar took over as the Program Manager of Larry's program and continued to support my research. I am also thankful to Dr. Dwight Woolard of the Army Research Office and Dr. Todd Hilton of DARPA for providing me with funding to develop nanoscale neuromorphic circuitry using resistive memories or memristors. Special thanks go to Prof. Kyounghoon Young of Korea Advanced Institute of Science and Technology (KAIST) and Prof. Kwang Seo Seok of Seoul National University for providing me funding through a collaborative research sponsored by Korean Government under its Tera and Nano Devices initiative.

Dedication

Western classical music had flourished in the Seventeenth and the Eighteenth Centuries because many anonymous wealthy patrons who admired the music of maestros like Ludwig von Beethoven, had secretly financed their creative endeavors so that they could exclusively devote to composing music without being overly burdened by teaching pupils to eke out a living.

Modern engineering research similarly requires tons of research dollars to pay graduate students, research scientists, and even to buy out investigators' myriad responsibilities in their universities, to fabricate proof-of-concept integrated chips using in-house and external foundry facilities, to conduct testing, measurements, and validation of the invention, and finally to disseminate research results through publishing in archival journals and presenting the work to the research community in international conferences and workshops. We, the research investigators, are deeply grateful to our program managers who not only provide research funding to allow our shows to go on, but also inspire and challenge us to embark on voyages in uncharted ocean (of knowledge) in radar-less vessels. Ultimately, in our research career, el viaje es la recompense, the journey is the reward.

This book is dedicated to those unheralded patrons of research in the following government agencies, who had provided research funding to us.

United States of America (NSF, DARPA, AFOSR and ARO),
Australia (Department of Foreign Affairs and Trade, and
   Commonwealth Government),
China (National Nature Science Foundation), and
South Korea (Australia-Korea Foundation and Korean Government
   TND Program)

Thank you on behalf of all authors of this book.

Pinaki Mazumder
University of Michigan

List of Contributors

Idongesit Ebong, *University of Michigan, Ann Arbor, MI, USA;*
E-mail: idong@umich.edu

Jason Eshraghian, *Department of Electrical Engineering and Computer Science, University of Michigan, Ann Arbor, MI, USA;*
E-mail: jeshraghian@gmail.com

Jiagui Wu, *South West University, China*

Kamran Eshraghian, *iDataMap Corporation, Eastwood, SA, Australia*

Lefei Men, *School of Information and Software Engineering, University of Electronic Science and Technology of China (UESTC), Chengdu, Sichuan, China*

Nyima Tashi, *School of Information Science and Technology, Tibet University, Lhasa, China*

Pinaki Mazumder, *University of Michigan, Ann Arbor, MI, USA;*
E-mail: pinakimazum@gmail.com

Qingqing Hu, *School of Information and Software Engineering, University of Electronic Science and Technology of China (UESTC), Chengdu, Sichuan, China*

Qishui Zhong, *School of Aeronautics and Astronautics, UESTC, Chengdu, Sichuan, China*

Shouming Zhong, *School of Mathematical Science, UESTC, Chengdu, Sichuan, China*

Shukai Duan, *South West University, China; E-mail: duansk@swu.edu.cn*

Sing-Rong Li,

Sung-Mo Kang, *Baskin School of Engineering, University of California, Santa Cruz, CA, USA*

Wenbo Song, *South West University, China*

Woo Hyung Lee, *University of Michigan, Ann Arbor, MI, USA;*
E-mail: leewh99@gmail.com

Xiaofang Hu, *South West University, China*

Xingwen Liu, *College of Electrical and Information Engineering,*
Southwest Minzu University, Chengdu, Sichuan, China

Yalcin Yilmaz, *University of Michigan, Ann Arbor, MI, USA;*
E-mail: yalciny@umich.edu

Yongbin Yu, *School of Information and Software Engineering, University of*
Electronic Science and Technology of China (UESTC), Chengdu, Sichuan,
China; E-mail: ybyu@uestc.edu.cn

List of Figures

List of Tables

List of Abbreviations

BRS	Background Resistance Sweep
CMOS	Complementary Metal Oxide Semiconductor
CNN	Cellular Neural/Nonlinear Network
CPI	Cycle Per Instruction
DLC	Diode Leakage Current
DRAM	Dynamic Random Access Memory
DSP	Digital Signal Processing
EE	Edge Extraction
FET	Field Effect Transistor
FPGA	Field Programmable Gate Array
HDD	Hard Disk Drive
HF	Hole Filling
HSS	High State Simulation
IOR	Inhibition Of Return
IPC	Instruction Per Cycle
KCL	Kirchoff's Current Law
LIF	Leaky-Integrate-and Fire
LTD	Long Term Depression
LTP	Long Term Potentiation
M-CNN	Memristor-based Cellular Neural/Non-linear network
MMD	Magnetic Memory Devices
MMOST	Memristor-MOS Technology
MOBILE	MONOstable-BIstable Logic Element
MPRTD	Multi-Peak Resonant Tunneling Diodes
MRAM	Magnetic Random Access Memory
MRAM	Magnetoresistive Random-Access Memory
MRNN	Memristor-based Recurrent Neural Network
MRS	Minimum Resistance Sweep
MUX	Multiplexer
NDR	Negative Differential Resistance

NMOS	N-type Metal Oxide Semiconductor
ODE	Ordinary Differential Equations
PCRAM	Phase-Change Random Access Memory
PCRD	Phase-Change Memory Devices
PDR	Positive Differential Resistance
PE	Processing Elements
PMOS	P-type Metal Oxide Semiconductor
PVCR	Peak-to-Valley Current Ratio
RC	Read Circuitry
RGB	Red Green Blue
RP	Reverse Polarity
RRAM	Resistive Random-Access Memory
RTD	Resonant Tunneling Diode
SA	Sense Amplifier
SDD	Solid State Drive
SRAM	Static Random Access Memory
STDP	Spike-Timing-Dependent-Plasticity
STT-RAM	Spin-Transfer Torque Random-Access Memory
VLSI	Very Large Scale Integration
VP	Peak Voltage
VTF	Velocity Tuned Filter
WTA	Winner-Take-All
XOR	Exclusive Or

1

Introduction

Jason Eshraghian, Sung-Mo Kang
and Kamran Eshraghian

1.1 Discovery

It has been a long and winding road for the memristor to reach where it is today, and the journey only continues to accelerate as an increasing number of researchers dedicate their focus to understanding how the memristor can be used as the next major technological disruption. It started off as an inkling of an idea. A concept conjured out of the need for completeness; Leon Chua noticed that something was missing in the world around us. The resistor existed to form a relationship between voltage and current. Inductors formed the link between magnetic flux and current. And the capacitor for charge and voltage. There was a void to fill in linking charge and flux, and Chua hypothesized that someday, the memristor may very well be the answer [1].

Unlike the resistor, there was no law of physics demanding a memristor's existence, but there was no law prohibiting it either. Dmitri Mendeleev made the astute observation the physical and chemical properties of elements had some relation to their atomic mass in a 'periodic' way. After arranging all of them, he noticed there were gaps in horizontal rows. Instead of seeing this as a problem, he treated it as an opportunity for new discoveries. By calculating the atomic mass of the missing elements, he could predict their properties, and no surprises – he was correct. When gallium was discovered in 1875, its properties were eerily close to Mendeleev's prediction. While nature is not obliged to complete the table, blank spots are great places to search for a new chemical element. Or a new circuit element, such as the memristor.

As it turns out, a solid-state device that can link charge and flux would also be able to have an electronically controllable resistance. This is of course where the memristor received its name, being a device that can store 'memory' in the form of its resistance. And in this introduction, we will

1

describe exactly how this is possible. It is important to understand how the discovery of an idealized memristor is contentious. Plenty of emulators have been presented [2–5], but these require an internal supply. Solid-state devices capable of storing memory in the form of resistance have also been fabricated which, under very specific circumstances, can even be approximated to behave as ideal memristors. But the possibility of using flux as a mechanism to vary charge (and vice versa) in the manifestation of an ideal memristor remains an open question.

Regardless, a good starting point for comprehending this controversial and often misunderstood device is the moment Chua first conceptualized the memristor back in 1971 [1]. This work serves to clarify how an ideal memristor behaves, and how the concept extends to the 'memristive device' or the 'memristive system' which was introduced a few years later in 1976 [6]. Readers will understand that a memristive device substitutes magnetic flux (or charge) for some other device-dependent feature, be it ion conduction, phase transitions, or electron spin direction. In doing so, memristive devices are still able to retain the features of ideal memristors that make them so useful.

In continuing the history lesson, the generalized notion of a memristive device was finally linked to the titanium dioxide (TiO_2) solid-state circuit element invented by Hewlett-Packard in 2008 [7], which was the catalyst leading to the explosion of memristor research. Many circuit designers and physicists have since shifted their focus to using the memristor as a possible solution in the race to overcoming the limitations of process scaling in CMOS.

The nanoscale advantage of memristors has been used in logic circuits, their dense integration exploited in memory storage [8], their probabilistic nature in security [9], their reconfigurability in analog and digital circuits [10–12], and a combination of these properties in neuromorphic computing [13]. We are on the cusp of mass-producing a highly disruptive emerging technology. However, the convoluted road of discovery of the memristor and the nonlinearity of its behavior has caused much confusion and misunderstanding of what is classified as a memristor and how it behaves. As a result of this, devices and mechanisms, which exhibit the pinched hysteresis fingerprint of the memristor have in the past been misclassified. From neuronal activity, modeled by the Hodgkin-Huxley equations [14], to the quantum-physics based resonant tunnel diode under certain operating regions [15]. In this introduction, we will attempt to clear up some of these misconceptions, whilst laying down a strong foundation of truly understanding what the memristor is. Starting with the theoretical ideal memristor, through to the more recent solid-state discoveries and their implementation in neuromorphic computing.

1.2 The Missing Memristor

1.2.1 Definitions

So, what does a device that links charge and flux look like? Thinking about the question in this form may be why it took almost 4 decades to realize the memristor after it was conceptualized. As we will see, the TiO$_2$ device developed by Hewlett-Packard Labs displays no tangible connection with flux or other magnetic phenomena [7]. Chua answers that charge and flux, q and φ, should only be considered mathematically – not as physical mechanisms. Before we consider these solid-state memristive devices, we will return to the basics and have an in-depth look at ideal memristors. From there, it will be easy to understand why memristive devices still retain the advantages of their ideal counterparts, regardless of a flux-based mechanism.

A memristor is characterized by the following criteria:

- passivity
- two-terminal device
- pinched hysteresis loop in the V-I plane under bipolar periodic input
- with a zero crossing at the origin.

If current is the rate at which charge flows, then charge is the time integral of an electric current. Magnetic flux can be defined using a similar relationship, but with voltage. These definitions can be used to more intuitively describe a memristor in terms of voltage and current instead of q and φ.

The simplest form of the memristor equation is just a charge-dependent (or alternatively, flux-dependent) Ohm's Law:

$$v = M(q)i, \tag{1.1}$$
$$i = v/M(\varphi), \tag{1.2}$$

where linear resistance R is replaced with a charge- or flux-dependent resistance M, also known as memristance. In the charge-dependent case, memristance varies based on the accumulated (or historical) amount of charge that has passed through the memristor. Therefore, in the absence of any current, the amount of accumulated charge remains constant, and M does not change. The memristor is therefore a good candidate as a passive non-volatile memory. This is to say that memory is stored using the value of memristance M, and still remains even after power is turned off.

Now let us assume that we have a flux-controlled ideal memristor and express it in an equivalent but slightly different form. If we apply

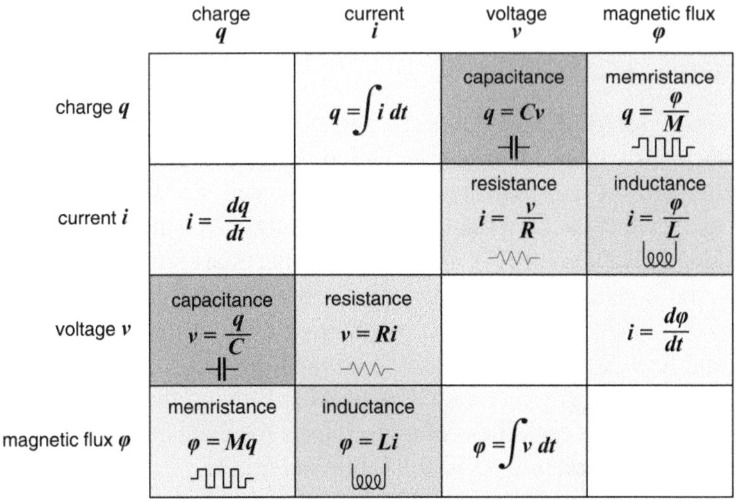

	charge q	current i	voltage v	magnetic flux φ
charge q		$q = \int i\, dt$	capacitance $q = Cv$	memristance $q = \dfrac{\varphi}{M}$
current i	$i = \dfrac{dq}{dt}$		resistance $i = \dfrac{v}{R}$	inductance $i = \dfrac{\varphi}{L}$
voltage v	capacitance $v = \dfrac{q}{C}$	resistance $v = Ri$		$i = \dfrac{d\varphi}{dt}$
magnetic flux φ	memristance $\varphi = Mq$	inductance $\varphi = Li$	$\varphi = \int v\, dt$	

Figure 1.1 Matrix of fundamental circuit elements and variable relationships.

a time-varying voltage (such as an AC supply, or periodic pulses), then we are varying flux and current over time:

$$i(t) = G(\varphi(t))v(t). \tag{1.3}$$

Conductance G (sometimes referred to as memductance) is the inverse of resistance, *1/R*, and is shown to be a function of time-varying flux and, in practice, controlled by voltage. The charge-controlled memristor can be similarly derived:

$$v(t) = i(t)/G(q(t)). \tag{1.4}$$

From Figure 1.1, we know that voltage is the time derivative of flux, and charge is the time derivative of current. Given that conductance is the inverse of resistance, $G = I/V$, it follows that:

$$G = \frac{dq}{d\varphi}. \tag{1.5}$$

Equations (1.3–1.5) are the form in which Chua characterized memristors in [1], though there are still some further conditions to satisfy before one can assert this is a true mathematical representation of an ideal memristor. The rich dynamics of memristors are also not quite clear from these equations, so we need to delve further. The best way to understand the complex nonlinear dynamics of the memristor is to track the motion of the state variable on the

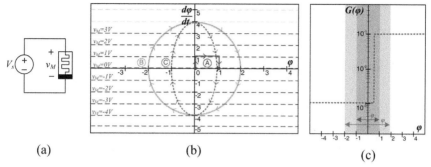

(a) (b) (c)

Figure 1.2 (a) Memristor connected to a voltage supply (b) Phase-plane of an ideal memristor plotting φ vs. $d\varphi/dt$ with the dynamic route map of a number of different driving inputs: Ⓐ shows the flux response to a pulse input, Ⓑ shows the response to an AC input (see Equation (1.6)), and Ⓒ also depicts the AC response, but at a higher input frequency (see Equation (1.7)). (c) Idealized flux-dependent conductance φ vs. $G(\varphi)$. In this case, we assume low conductance $G_L = 10^{-3}$ S, and high conductance $G_H = 10^{-1}$ S, and a switching threshold of $\varphi_T = 0.5$. The shaded regions depict the x-range across which Ⓑ and Ⓒ vary, which is a function of input frequency.

phase plane in Figure 1.2(b). The state in a flux-controlled memristor is flux, which we know to be $d\varphi/dt = V$.

Figure 1.2 shows the dynamic route map (DRM) of an ideal memristor on its phase plane. When no voltage is applied, $V = d\varphi/dt = 0$, it means flux does not change, and that any point that lies along the x-axis must therefore remain stationary.

1.2.2 DC Response of an Ideal Memristor

Now let us apply a voltage pulse $V = 1\,V$ for exactly $t = 1\,s$ duration across the terminals of the memristor. The rectangular motion path Ⓐ from Figure 1.2 shows the evolution of flux during this process. This is validated by knowing that the time integral of such a voltage pulse would be 1.

If we track φ across the Figure 1.2(c) conductance-state curve, then it becomes easy to see how the memristor is able to switch. For the step response, while the DC pulse is applied, φ linearly increases at a rate of $V_s = d\varphi/dt = 1$ unit per second[1]. It starts at the low level of conductance,

---

[1]We note that the charge-flux relation is purely theoretical. While the SI unit of magnetic flux is the *Weber* (Wb), we leave it dimensionless here to avoid misleading the reader as such a device has not been discovered. Flux is often substituted for some other physical mechanism. We will explore this in more depth in Section 1.3.

and after V_s is switched off, φ ceases to change and remains at the higher level of conductance. Therefore, this memristor exhibits non-volatile memory.

1.2.3 AC Response of an Ideal Memristor

What might happen when we apply an AC signal? As an example, let us use a voltage source of $V_s = 4\sin(2t)V$. The effect this sinusoidal input has on flux is depicted by the circular motion path Ⓑ in Figure 1.2, following the directions of the arrowheads. Flux φ can be analytically solved:

$$\varphi = \int \frac{d\varphi}{dt} dt$$

$$= \int V_S \, dt = \int 4\sin(2t)dt$$

$$= -2\cos(2t) + \varphi_0 \tag{1.6}$$

We can assume the integral constant term $\varphi_0 = 0$. Plotting V_s v. φ shows the circular trajectory on the phase-plane of the memristor. At initial time $t = 0$, substitution into (1.6) gives initial flux conditions of $\varphi(t = 0) = -2$. It is easy to see that φ crosses the y-axis of Figure 1.2(b) every time V_s reaches its maximum or minimum value. The range of values for which flux cycles between are $-2 \leq \varphi(t) \leq 2$.

Again, we can track φ on the conductance plot of Figure 1.2(c). This shows the route that conductance takes in order to periodically switch. Under AC excitation, we apply a continuously varying range of voltages so it is useful to find the current response in terms of voltage, which can also be calculated as graphed in Figure 1.3.

From the $d\varphi/dt$ v. φ phase plane in Figure 1.2(b), and the $G(x)$ v. φ curve in Figure 1.2(c), we've been able to derive $v_M(t)$, $\varphi(t)$, $i(t)$ and $q(t)$. We then use the time-varying $\varphi(t)$ and $q(t)$ curves to find a relationship between charge and flux in Figure 1.3(f), which is able to completely characterize an ideal memristor.

We do the same with $v_M(t)$ and $i(t)$ to find a relationship between voltage and current in Figure 1.3(g). The two slopes in Figures 1.3(f–g) represent the high and low conductance levels. Therefore, a *pinched hysteresis loop* exists in the voltage-current plane of an ideal flux-controlled memristor [16].

An equivalent process can be undertaken to find the pinched hysteresis loop of a charge-controlled memristor.

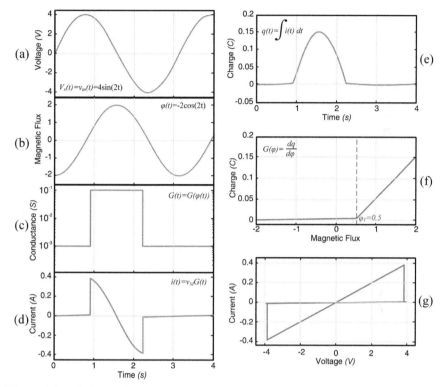

Figure 1.3 Ideal memristor response. (a) Voltage vs. Time (b) Magnetic Flux vs. Time, (c) Conductance vs. Time (d) Current vs. Time – note that current still flows in the high resistance state, it is just ~100 orders of magnitude less than the peak value, (e) Charge vs. Time (f) Monotonically increasing Magnetic Flux vs. Charge curve (g) Pinched hysteresis loop in the Voltage-Current plane.

1.2.4 AC Response of an Ideal Memristor: Higher Frequencies

It is interesting to see what happens when we increase the frequency of the driving voltage. We can try increasing it by twice the value of the previous example to give an input of $V_s = 4\sin(4t)\,V$, while holding the amplitude constant. Flux φ is again analytically solved in the same way as (1.6):

$$\varphi = \int V_S\,dt$$
$$= -\cos(4t) + \varphi_0 \tag{1.7}$$

This time, flux oscillates between $-1 \leq \varphi \leq 1$. The upper and lower limits have decreased from ± 2 as they were in (1.6). With reference to the phase-plane in Figure 1.2(b), the motion of flux is shown in the motion path ©. Intuitively, this makes sense. An increase in frequency corresponds to an increase in the rate of change of voltage, which leaves less time for flux to vary at every iteration of voltage. Therefore, the motion path starts to 'squeeze' inwards.

The motion path will continue to squeeze with an increasing frequency, and as frequency approaches ∞, the bounds of flux will tend to $-\frac{1}{\infty} \leq \varphi \leq \frac{1}{\infty}$. This range is so tiny that we can make the approximation $d\varphi/dt = 0$ for all time. If one were able to measure the change of flux, it would appear to remain constant.

Mapping this constant flux value over to the same $G(\varphi)$ v. φ curve from Figure 1.2(c), shows an important result. Conductance stays constant, despite a changing voltage. With G now a constant, Equation (1.3) suddenly becomes a linear Ohm's Law equation. This alludes to a significant finding: *under sufficiently high frequencies of AC excitation, a memristor will behave as a linear resistor*.

This does not change the φ v. q curve in Figure 1.3(f). We simply look to the value of φ and calculate the conductance of that instantaneous point using (1.5) – equivalent to the charge-flux slope at the relevant point. The key takeaway is that regardless of frequency, the φ v. q curve can still fully characterize an ideal memristor. So once φ stops crossing the conductance threshold, the V-I curve will become linear with a slope equivalent to either of the conductance values in Figure 1.2(c), depending on the initial condition of φ.

In a case where there are two distinguishable conductance (or resistance) states, the memristor can behave as a resistive switch or as memory, which is stored in the form of the flux-mechanism, and readable by calculating resistance (often performed by finding the current response).

1.2.5 Some Further Observations

There are still some issues to address, as the above ideal memristor model does not perfectly depict how a memristor will behave in the real world. We can consider these issues by addressing a number of conditions the model must fulfill, in order to be a memristor.

As long as the time-invariant φ v. q curve is: (i) nonlinear, (ii) continuous, and (iii) strictly monotonically increasing, then the criteria for an ideal memristor will be satisfied.

We will now explore what the implications of these three criteria are, and why they are necessary in order to have a memristor.

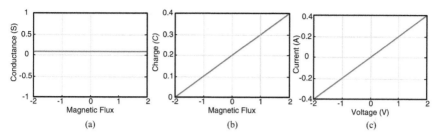

Figure 1.4 A linear flux-charge curve corresponding to resistance. (a) Single-valued conductance function (b) Resulting linear flux vs. charge (b) voltage vs. current curve obeying Ohm's Law.

1.2.5.1 Requirement 1: Nonlinearity

This requires the charge-flux relationship to not be linear, as is the case in Figure 1.3(f). If Requirement 1 was left unsatisfied and φ v. q was linear, then $G(\varphi)$ v. φ would be a single-valued function. That is to say, the memristor would have a constant conductance (and constant resistance) regardless of driving frequency. Requirement 1 is necessary as otherwise the device would simply be a resistor. This can be followed in Figure 1.4.

1.2.5.2 Requirement 2: Continuous φ *v.* q

If Requirement 2 was unsatisfied, then there would be an instantaneous jump in charge as φ increases. The relationship in Figure 1.5 implies that for $\varphi(t) = 0$, $\frac{dq}{dt} = \infty$. A model which passes infinite current at every switch would poorly represent how switching occurs in the real world.

One could further narrow this criteria by replacing '*continuous*' for '*continuously differentiable*', in which case the φ-q curve in Figure 1.2(f) would only be piecewise differentiable. That is to say, differentiable through its subdomain, even though it is not differentiable between the pieces. The result of this is a rise and fall time of $t = 0$ for current, or instantaneous switching. In practice, the rise time would not be zero. As we live in an analog world, there would always be some finite but small time-interval with a minimum energy requirement for a physical device to switch completely from a low to a high resistance.

The abrupt jumps in Figures 1.2(c), (e) and (g) are due to the piecewise linear nature of the assumed φ v. q curve. This can be fixed by ensuring $G(\varphi)$ is a continuous function (or continuously differentiable function for non-zero rise and fall times). Therefore, we can accurately model finite switching time by smoothing $G(\varphi)$ in Figure 1.3(b) with the curve having a continuous

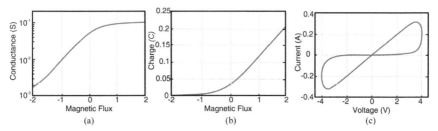

Figure 1.5   Ideal memristor response where $G(\varphi)$ is a continuous function (a) Magnetic Flux vs. Conductance represented with a logistic function (b) resulting Magnetic Flux vs. Charge curve (c) pinched hysteresis loop in the voltage-current plane. Note the transitions are smoother than the previous case in Figure 1.2(g), thus modeling a more physically viable switching mechanism.

derivative as shown below. This can be done by modeling the relationship between magnetic flux and conductance using a logistic function.

While Figure 1.2(b) is fixed under all circumstances by virtue of the flux-voltage relationship, the $G(\varphi)\, v.\, \varphi$ relationship would be device-dependent and need not be a step function as we have used in Figure 1.2(c).

It would be an interesting exercise to see the effect of increasing frequency on this more realistic form of the ideal memristor. In such a case, as voltage frequency increases, the range of the periodic flux will again decrease and converge to zero. However, the two conductance curves will gradually converge to behave as a linear resistor. The difference between this case and the step-function $G(\varphi)$ curve in Figure 1.2(c) is that the latter instantaneously collapsed into a linear resistor, once flux ceased to cross the threshold.

1.2.5.3 Requirement 3: Strictly monotonically increasing φ v. q

This third criteria of a strictly monotonically increasing φ-q curve serves to remove negative resistances. If one applies a DC voltage across the terminals of a memristor and violated this requirement, there would be a point where, as flux increases, the amount of charge flowing through the device would decrease. A decrease in charge (with corresponding increase in flux) means a negative current, which implies current is flowing in the opposite direction to our voltage source in Figure 1.2(a).

Instantaneous resistance can be calculated by $R = \frac{d\varphi}{dq}$ and the moment the charge-flux curve starts to decrease, $dq < 0$ which implies $R < 0$. As we know, memristors are passive devices, and a negative resistance implies some active internal power supply is providing this injection of current in the opposite direction. As long as $G(\varphi) = 0$, then the memristor will satisfy the passivity criterion, and this can be ensured with a monotonically increasing φ-q curve.

Equivalently, this means that the conductance curve cannot go below the x-axis.

1.2.6 Summary

Thus far, we have considered how the concept of the memristor came about. We have looked at its DC response, AC response, how its behaviour alters with changing frequency, and how an appropriate charge-flux relationship will ensure a passive, nonlinear two-terminal memristor.

And, while the ideal memristor is exactly what its name suggests – an idealization – what may be perceived as a purely theoretical exercise actually bears a strong likeness to the solid-state memristive devices that have thus far been discovered, even without any flux-dependence. Therefore, let us see how this links back to the real world.

1.3 Memristive Devices and Systems

1.3.1 Definitions

Without magnetic flux, what does memory exactly store in a physical memristor? Can it still be called *'the fourth fundamental circuit element'* if flux is no longer the physical mechanism by which we switch the device? These are some of the questions that will be explored in this section, as we consider what brought on the generalization of the memristor.

In 1976, a few years after the postulation of the memristor, Chua and Kang broadened the memristor definition to further classify *'memristive devices and systems'* as no longer strictly reliant on flux to store memory [6]. This generalization meant one could class a few pre-existing systems as a memristive device, such as the thermistor [17], the ionic systems as described by the Hodgkin-Huxley membrane model [14], and the discharge tube [18].

Of particular interest, is that the Hewlett-Packard TiO_2 memristor discovered in 2008, is by this definition a memristive device. The tongue-in-cheek title of their seminal *Nature* paper *'The missing memristor found'*[2] was the catalyst that led to researchers referring to 'memristors' and 'memristive devices' interchangeably, which naturally resulted in some confusion [7].

---

[2]The title back-referencing Chua's original 1971 paper title, *'Memristor – the missing circuit element'*.

Chua addressed this confusion by dubbing these memristive devices as 'generic' and 'extended' memristors [19], depending on their behavior. And we will come to see that the solid-state memristors thus far been discovered will typically fall into the categories of either generic or extended memristors.

Regardless of nomenclature, it turns out that all three categories of memristors (ideal, generic and extended) exhibit the sort of nonlinearity that is drastically revolutionizing the way many researchers are designing their circuits. As we will see, these three subclasses of memristors all exhibit indistinguishable advantages insofar as they are electronically-controlled resistors, and store memory in the form of their resistance. By analogy, when a pn-junction is in reverse bias, the size of the depletion region becomes controllable as a pseudo-dielectric layer. Varying this reverse bias voltage allows one to vary the size of the dielectric layer, which gives rise to a variation in measurable capacitance – this device is a varactor diode. In the absence of a pair of parallel metal plates, we can still treat this device as an electronically-controllable capacitor, regardless of the physical mechanism used to drive it. It has correspondingly found its place in a range of useful applications, such as voltage-controlled oscillators and RF filters.

Memristors are not wholly dissimilar. A range of physical mechanisms can be used to realize a non-volatile electronically-variable resistive memory. Also, whether they are metal-oxide based or dependent on phase transitions, they can all be mathematically characterized as a memristive system, and we cannot disregard their usefulness by futile arguments of whether or not the memristor is *'fundamental'* – if it can achieve the same ends as a flux-mechanism of switching, then an argument on the fundamentality of generic and extended memristors would only serve philosophical ends, and would have little impact on the possible circuits that could be developed.

A time-invariant[3] voltage-controlled memristive system is described by:

<u>Generic Memristor</u>   <u>Extended Memristor</u>

$$\frac{dx}{dt} = f(x, v)$$

$$i = G(x)v \qquad (1.8)$$

$$\frac{dx}{dt} = f(x, v)$$

$$i = G(x, v)v \qquad (1.9)$$

---

[3]Time-variance is briefly touched upon in Chua and Kang's 1976 paper. Here, we will follow suit and keep things invariant for simplicity.

The state-dependent Ohm's Law for the generic case is the exact same as that in (1.3), where φ is now substituted for a state x. The dynamical state equation for dx/dt is a little different now, as it has an additional dependence on the state x. If we translate all our prior observations of the q-φ ideal model over to q-x, then these generic and extended cases will be able to broadly characterize all electronically-controlled resistive switches. We can show this via the following steps:

1. understanding the *resistive switching mechanisms* in various functional materials
2. analyzing the *electrical models* that exist for these resistive switching mechanisms
3. performing a comparison to the definition of *memristive device* as per (1.8) and (1.9)
4. performing a comparison to the definition of the *ideal memristor* as per (1.3).

These will be performed in the following subsections, concurrently with consideration of the physical description for the underlying mechanisms behind resistive switching.

1.3.2 Resistive Switching Mechanisms

When resistive random-access memory (ReRAM) was first introduced, the mechanism was not well understood, and a whole range of resistive switches were broadly classified as ReRAM without specifying the physical nature of their switching mechanisms. This has led to all non-volatile resistive switching materials as adjusted by an electric field being grouped under the banner of ReRAM, with the exception of a few particularized categories, including phase-change chalcogenides and magnetoresistive memories.

A number of theories exist in describing the switching mechanism, and while empirical evidence has made a strong case for understanding the primary mechanisms in different materials, the difficulty in localized imaging has led to uncertainty in characterization. Resistive switching can be a product of ionic transport, joule-heating induced phase change, formation of conductive filaments, and contact interfaces. And differentiating these effects is a difficult and non-trivial exercise.

In describing some of these mechanisms below, we note that it is unlikely for these effects to exist independently from one another.

1.3.3 Transport of Mobile Ions

As previously alluded to, under the lead of R. Stanley Williams, the team at Hewlett Packard Labs pursued memristor research with renewed interest upon realizing its connection with their solid-state TiO_2 thin-film device. This was not quite the discovery of the memristor – resistive switching in TiO_2 thin-films in fact predates Chua's postulation of the memristor potentially back to 1965 [20]. This was the discovery of the link between resistive RAM (ReRAM) and the memristor.

HP's original memristive device consists of one layer of insulating TiO_2 and one layer of oxygen deficient TiO_{2-x}, sandwiched between platinum electrodes used to apply an electric field.

The theory proposed in [7] indicates that electrical switching occurs by drift of the positively charged vacancies through the barrier, thus increasing the effective width of the insulating layer as in Figure 1.6(b). This process is reversible by attracting the positive vacancies with a negative field, which increases the width of the insulating TiO_2 layer as in Figure 1.6(c).

The following year, HP published a dynamical model of their TiO_2 device to better understand state-evolution in the presence of an electrical field.[4] Using regression techniques to fit their experimental data, they presented

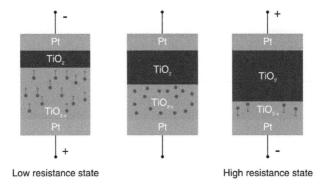

Low resistance state   High resistance state

Figure 1.6   Transport of mobile ions in a TiO_2 memristor. (a) Positive voltage repels oxygen vacancies expanding the conductive oxygen-deficient layer (b) Resistance state persists after current has ceased (c) Negative voltage attracts oxygen vacancies, expanding the insulating TiO_2 layer.

---

[4]A dynamical model was also provided in the original 2008 *Nature* manuscript, though dw/dt had dependence on R_{ON} and R_{OFF} – the high and low resistive states. As we've seen, these two are not necessarily constant values and are themselves a state-dependent term in memristive devices.

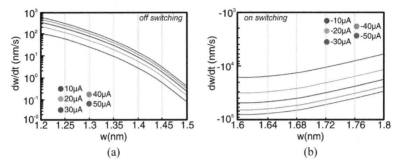

Figure 1.7 TiO_2 memristor model from [20] (a) Dynamical route map based on ion transport for off-switching in (1.10) (b) On-switching in (1.11).

analytical expressions for off-switching ($i > 0$):

$$\frac{dw}{dt} = f_{off}\left(\frac{i}{i_{off}}\right) \exp\left[\exp\left(-\frac{w - a_{off}}{w_c} - \frac{|i|}{b}\right) - \frac{w}{w_c}\right], \qquad (1.10)$$

and for on-switching ($i < 0$):

$$\frac{dw}{dt} = f_{on}\left(\frac{i}{i_{on}}\right) \exp\left[\exp\left(-\frac{w - a_{on}}{w_c} - \frac{|i|}{b}\right) - \frac{w}{w_c}\right]. \qquad (1.11)$$

All fitting parameters $f_{off}, i_{off}, a_{off}, b, w_c, f_{on}, i_{on}, a_{on}$ are given in [21].

Equations in (1.10) and (1.11) are plotted in Figure 1.7. This makes physical sense with respect to what has been described of ion transport. Apply a positive voltage, and the width of the conductive TiO_{2-x} layer expands. Increase the applied voltage and the dynamic route map will shift upwards, corresponding to a faster change in the width w. The rationality behind how this works is quite similar to analysis of the ideal memristor, though just for a different curve shape.

1.3.4 Formation of Conduction Filaments

In their search for ways to further speed up switching times, increase the on/off resistance ratios, and enhance reliability of their memristor, HP continued to experiment with other metal-oxide resistive switches. Their experiments showed that tantalum-oxide was one of the more controllable and reliable options.

Two layers of tantalum oxides with different stoichiometries are stacked – one oxygen rich layer, and one oxygen deficient, with the application of an external voltage enabling the control of exchanging oxygen vacancies between the layers. This differs from ion transport theory in that the oxygen vacancies form a conductive pathway, and it is commonly accepted that the two switching mechanisms are inextricably linked. In fact, most resistive switching phenomena are believed to be a result of the generation of a conducting filament between two electrodes, acting as a circuit breaker/switch. The formation of the conducting filament occurs across a nanometre-scale in the space of nanoseconds [22].

For example, in Figure 1.8, a negative voltage at the oxide-rich layer generates a force that repels negatively charged oxygen ions. This leaves behind a conductive oxygen vacancy. With enough vacancies, this mechanism will form a conducting pathway for electrical transport. Conversely, a positive voltage will suck oxygen atoms back into the conductive path, thus increasing resistance.

The tantalum oxide memristor from [24] was described with the following compact model:

$$\frac{dx}{dt} = \lambda \sin(\eta v) \begin{cases} \frac{1-x}{\tau_0} & v > 0 \\ \frac{x}{\tau_0} & v < 0 \end{cases} \qquad (1.13)$$

$$I = G(x, v) = \left[\frac{\alpha}{v}(1 - e^{-\beta v}) + \frac{x\gamma}{v} \sinh(\delta v) \right] v \qquad (1.14)$$

where the state variable x now represents the overall area fraction of the filament formed due to the increased concentration of oxygen vacancies.

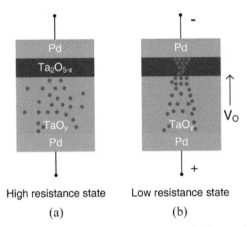

High resistance state   Low resistance state

(a)   (b)

Figure 1.8   An example of creation of a conductive filament in the tantalum oxide memristor from [23]. (a) Scattered oxygen vacancies resulting in a high resistance state (b) formation of a conductive filament with direction of oxygen vacancy motion denoted V_O.

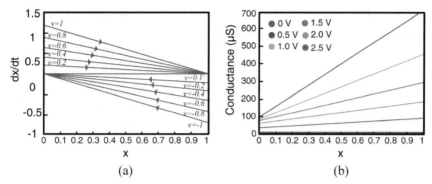

Figure 1.9 Tantalum oxide memristor model from [23] (a) Dynamical route map based on conductive filament formation in (1.13) (b) State-conductance dependence from (1.14). Note that as voltage increases, the slope of the curve decreases for values $v > 0$.

The variables α, β, γ, δ, η, λ and τ are all material dependent properties. The form of these equations precisely matches that of the extended memristor in (1.9). To simplify analysis and understanding, we treat all of these constants as '1' to generate the plots of dx/dt and $G(x)$:

What we see now is a voltage-dependence on conductance. Mathematically, this characteristic sets the 'extended' memristor apart from its 'generic' sibling and is quite straightforward to manage in numerical analysis.

To gain some intuition on how this models conductive filaments, we consider voltage v from (1.13) and (1.14) as applied at the tantalum layer.[5] As shown in Figure 1.9, a positive voltage at the oxygen deficient terminal will increase x due to the increase of filament area. The limited amount of oxygen vacancies means the rate of change of x tapers off and does not continue indefinitely. This change in x can then be tracked along the conductance plot where we notice something interesting: for larger applied voltages, the conductance appears to decrease. The cause of this might be attributed to large positive voltages partially counteracting the filamentary effect by causing a total transfer of oxygen vacancies to the top layer of Figure 1.8(b). The lack of oxygen vacancies in the tantalum oxide (TaO_y) layer causes it to become highly insulating, and the more resistive of the two layers – i.e., the more resistive material has switched.

---

[5] As you may have noticed, different stoichiometries are possible in different structures of layers to generate distinct set of behaviors. Though in all cases, conduction filaments are expected to be the primary resistive switching mechanism.

What is especially useful about this is that we can now precisely adjust conductance across a range of analog values using voltage. We will shortly see how this is extremely valuable in multi-state neuromorphic applications.

1.3.5 Phase Change Transitions

A number of laboratories have used transmission electron microscopes in identifying some other switching mechanism that can occur in tandem with conductive filaments – namely, phase transitions [25–27]. For example, phase transitions in Pt/CuO/Pt switching cells were reported in [28], where the oxygen-deficient conductive filament consists of two different Cu-O phases: CuO and Cu_2O.

There are also solid-state cells that use phase transitions as their primary resistive switching mechanism, and these are made up of a group of elements known as chalcogens. Phase change materials are conventionally alloys of metalloids (Ge: germanium, As: arsenic, Sb: antimony and Te: tellurium) that when alloyed together, result in a chalcogenide. Metalloid alloys have multiple stable states that each come with their own distinct resistance characteristics.

As shown in Figure 1.10, the amorphous phase bears a resemblance to the disordered composition of glass and is highly resistive, while the crystalline phase is more likened to a conductive metal.

To switch between these states, the materials must be heated to a certain temperature, which can be done by applying a voltage above a certain threshold for a given amount of time. As an example, assume the device is initially in the high resistance amorphous state. If a voltage is applied

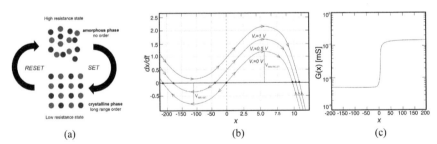

(a)   (b)   (c)

Figure 1.10   Phase change transitions (a) between amorphous and crystalline states (b) Dynamical route map from (1.15) – note the asymmetry between the left and right planes, which is a result of significantly differing switching times (c) State-conductance dependence from (1.16).

across its terminals, then the device dissipates heat as it is resistive, and will therefore also increase in temperature. If it is held at its *'set'* voltage (e.g., $V_{SET} \approx 0.7$ V [29]), the material will become hot enough for its molecules to realign into a crystalline structure. That is, if the temperature is maintained for sufficient time (e.g., $t_{SET} \approx 100$ ns [29]).

To *'reset'* the cell, we apply a much higher voltage (e.g., $V_{RESET} \approx 1.2$ V [29]) to push the temperature high enough to turn the molecular structure back into a molten state, thus melting down the crystalline structure. This voltage must be applied for the duration of its reset switching time (e.g., $t_{RESET} \approx 10$ ns [29]) in order to successfully transition back to the high resistance amorphous phase.

This process is quite different to the devices we have come across so far. Instead of relying on a state threshold to switch conductance levels (for example, $\varphi_T = 0.5$ from Figure 1.2(c)), we now have a voltage threshold instead to enable successful switching. This voltage must not only be applied for a certain amount of time, but by virtue of being a voltage threshold, cannot be cumulatively applied to induce switching. In other words, all our previously examined cases could theoretically have multiple voltage pulses applied to incrementally change the state of x, w or φ. In a resistive switch with an idealized form of phase transition (and in absence of secondary switching mechanisms and imperfections such as temperature-dependent resistance drift and thermal instability) this device will *not* store memory of anything more than a low or high resistance state – an amorphous or crystalline phase.

Does this still fall within the descriptive umbrella of a memristor? Some would argue no, because phase change memory devices do not rely on oxygen vacancies to switch states, as is the case for ion transport and conducting filaments. But based on the actual definition of a generic or extended memristor, there is no reason phase change memory cannot be described by such a pair of dynamical equations as in (1.8) or (1.9). In a case with ideal phase transitions, phase change memory from [29] have been modeled as:

$$\frac{dx}{dt} = \begin{cases} \alpha x(x + \beta) \pm |v_R| & x \leq 0 \\ -\eta x(x - \gamma) \pm |v_R| & x \geq 0 \end{cases} \quad (1.15)$$

$$i(x) = v_R \left(\frac{g_s}{1 + e^{sx}} + g_r \right) \quad (1.16)$$

where $|v_R|$ is added to the dynamical expression for *set* and subtracted for *reset*. The values for α, β, η, γ, s, g_s and g_h are all material-dependent constants, and are calculable based on a set of characteristic phase-plane equations that can model the device. This has also been referred to as a 'bistable memristor' in that there are two points on its dynamical route map that exhibit stability, and the state will always tend towards either of the two points in steady-state [30]. To switch from the left-hand stable point, a voltage slightly higher than $V_{MIN-SET}$ must be applied in order to shift the curve completely to the positive half of dx/dt, which enables the point to move across to the right-hand side of the plot. The time it takes to cross the axis is the switching time for the device in one direction.

There is of course the possibility of enhancing the model for devices which undergo partial phase transitions. Multi-state memories are extremely powerful tools in neuromorphic computing, though its simpler, bistable counterpart has been shown to enable selector-less memories too. It is evident that design trade-offs begin to emerge, once we realize the vast types of memristive devices that can be exploited.

1.3.6 Resonant Tunneling Diodes

The resonant tunnelling diode (RTD) is another device that presents its own series of trade-offs, by using a different resistive switching mechanism. By making use of the quantum tunneling phenomenon it is able to exhibit hysteretic behavior. One common configuration of the RTD is the double barrier quantum well structure. A double barrier give rise to confined energy states, which inhibits electrons from passing through the structure unless they are at a certain energy level. This mechanism makes them candidates for applications in multi-level memory cells and logic, and gives rise to resistive switching characteristics [31].

In a simplified 2D view, the device consists of heavily doped contacts made from a semi-conductor with a relatively small bandgap in regions I and V of Figure 1.11(a). These regions are called the emitter and collector, respectively. When the device is in equilibrium with no external bias, the electrons are probabilistically not at the requisite energy level to overcome the barriers in accordance to classical physics. Nor may the electrons pass through at the resonant energy, according to quantum theory. In the absence of an applied field, there is an even Fermi distribution throughout the RTD which implies that no current flows.

As the voltage applied to region I increases, so does the Fermi level of region I with respect to the other regions: current begins to flow. In a quantum well, the electrons that flow through it may only have discrete energy values.

These energy values can be calculated as per the derivation in [32], and when the electron energy at the emitter contact is equivalent to the discrete permissible states in the well, then a near-unity transmission probability exists for the electron to tunnel through the double barrier at this resonant energy. Thus we have resonant tunneling. The value of current at the voltage that enables resonant tunneling is also known as the peak current I_P, as depicted in Figure 1.11(b). At this point, the number of tunneling electrons per unit area reaches a maximum. In terms of resistive switching, this can be viewed as the *set* process in Figure 1.11(c) as the device is entering a low resistance state [33].

If voltage is increased further beyond this peak, then an interesting effect known as negative differential resistance occurs in the RTD. When the electron energy at the emitter junction increases beyond that of the resonant energy, then the probability of an electron able to tunnel into the quantum well while conserving its transverse momentum rapidly decreases. Accordingly, the tunneling current density sharply drops. This can also be seen by the downwards arrow in Figure 1.11(c), and the minimum point is often referred to as the valley current I_V.

From here, our analysis of resistive switching within RTDs can take one of two directions. In one case, we continue increasing the voltage applied at the emitter. This will increase the electron energy distribution, and electron thermionic emission will enable the charge carriers to overcome the work function of the materials that make up the barriers. Cumulatively, the higher-level resonant energy through the quantum well will also facilitate tunneling through the barriers. A combination of classical and quantum theories thus lead to the continuing increase in current density, depicted in the rising tail-end of the V-I characteristics in Figure 1.11(b).

The second direction our analysis may take is if we decrease the voltage from the valley current instead further increasing voltage. A reversed DC voltage sweep can be applied to bring the cell back down to the high resistance state in the *reset* process in Figure 1.11(c), demonstrated in [33]. One major difference between the all-graphene RTD and the other resistive switches we've considered thus far is that the high resistance state is actually a negative differential resistance.

In summary, if the peak current I_P is driven by peak voltage V_P, valley current I_V is driven by valley voltage V_V, and the peak current is also attainable by applying a higher, secondary voltage V_S from Figure 1.11(b) then the important RTD parameters as presented in [32] are easily derived

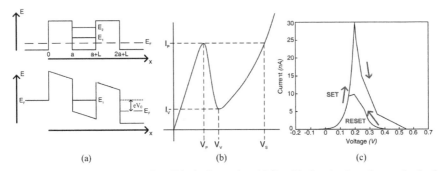

Figure 1.11   Resonant Tunneling Diode Operation (a) Double barrier junction under both equilibrium (top) and under the action of an applied bias (bottom), (b) Current-voltage characteristics [32], (c) Set and reset operation [33].

as below:

$$R_{SET} = \frac{V_P}{I_P} \tag{1.17}$$

$$R_{SET2} = \frac{V_S - V_V}{I_P - I_V} \tag{1.18}$$

$$|R_N| = \frac{V_V - V_P}{I_P - I_V} \tag{1.19}$$

Equation (1.19) can be adapted to the local reset peak to calculate reset resistance R_{RESET} from Figure 1.11(c).

1.3.7 Magnetoresistive Memory, Nanoparticles and Multi-State Devices

Thus far, we have covered ideal memristors, and four modes of physical resistive switching – ionic transport, conductive filaments, phase transitions and quantum tunneling While these are some of the most well-known resistive switching mechanisms, there are still multitudes of other technologies out there each with their own advantages and challenges.

1.3.7.1 Spin-transfer torque

One example of an emerging technology is the spin-transfer torque memristor, which uses the variation of the direction of electron spin to change resistance [34]. When there is one fixed magnetic layer with its polarity in the same direction as the free magnetic layer (in the parallel state, as depicted in Figure 1.12), the STT-memristor is able to pass high currents due to the low

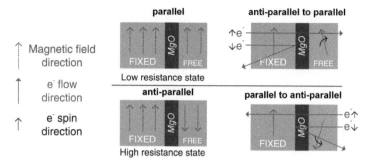

Figure 1.12 Write operation of a spin-transfer torque magnetic tunnel junction memristor. The parallel state exists when the field of the free magnetic layer points in the same direction as that of the fixed layer, or in the high conductance state. The anti-parallel state exists when the field of the free magnetic layer points in the opposite direction to that of the fixed layer, or in the low conduction state. The switching mechanism between the two states is determined by the direction of current flow.

resistance resulting from the identical field directions. When the polarities are in opposing directions, or in anti-parallel, then the device will exhibit a large resistance.

The switching mechanism can be described as follows. For the *set* process (i.e., anti-parallel to parallel), an electron current must be applied from the fixed layer to the free layer. In such a case, only electrons whose spin direction is aligned with the fixed magnet are passed through to the free layer. The polarity of the free layer thus realigns with the electron spin due to a transfer in angular momentum from the electron spin to the free field direction, and the device switches to a low resistance. The electrons whose spin directions oppose that of the fixed layer are reflected at the tunnel junction (in Figure 1.12 using MgO).

Conversely, the *reset* process can be performed using a current in the opposite direction from the free layer to the fixed layer. The off-spin current is reflected back, thus switching the free magnet from parallel to antiparallel using the same mechanism described above. The memristor therefore switches to the high resistive state.

1.3.7.2 Nanoparticles

Another interesting example is colloidal metal-oxide nanoparticles, which have been shown to have good integrability with organic materials [35]. ZnO is distributed in a solution that exhibits large resistance ratios between their On and Off states, which is highly desirable for many applications, especially low-cost memory circuits. It has been proposed that the switching mechanism for ZnO nanoparticles layered between Indium-Tin-Oxide (ITO) and aluminium is due to the alteration of the barrier at the Al/ZnO interface, arising from ionic drift of oxygen vacancies that was explained in Section 1.3.3, coupled with the effect from adsorption and desorption of oxygen molecules.

1.3.7.3 Multi-state memories

We have previously alluded to multi-state memristors, which can come in many shapes and forms. This may arise due to the analog and continuous transition of ion drift, or due to the creation of multiple parallel conductive filaments. Regardless of the mechanism-based theories that exist, characterization of a multi-state memristor can come in an infinite range of forms – a hypothetical case may see many discrete steps in a $G(x)$ curve, or the more likely scenario with a continuous function with a predetermined range of tolerance between states. This latter form has been developed and applied in image processing [36], multi-level ratioed logic gates [37], convolutional filtering [38], and neuromorphic computing applications [13]. As will be elucidated in the next section, brain-inspired neuromorphic processors use architectures where memory and computation are localized together. The classical memory-related constraints, including size, latency and throughput, present major performance bottlenecks in traditional von Neumann architectures. Other interesting schemes for expanding the number of available conductivity states have involved using input frequency modulation [39], or state-coupling [40, 41] to widen the range of functional states available to access from the device (Figure 1.13).

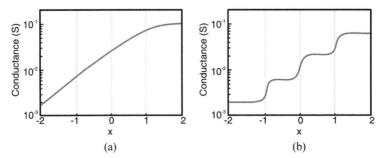

Figure 1.13 Examples of multi-state memories. (a) continuous conductance, (b) theoretical 4-state multi-step conductance.

1.4 Neuromorphic Computing

Memristors have been used in constructing dense memory cells, creating reprogrammable logic [42, 43] and exploiting their stochasticity in security [44]. Memory naturally makes sense as it is a ReRAM device. There has been good success in memory circuits where phase change RAM was implemented in Intel's 3D Xpoint [45] in March 2017. Security applications have also shown fruitful outcomes by exploring circuit-level dynamical characteristics to generate chaotic behaviors [4, 46, 47], and their atomic level uncertainty in switching mechanisms for generating physically unclonable functions (PUFs) [9].

But where it has been suggested that memristors will be most disruptive is in neuromorphic computing. Beyond von Neumann architecture, neuromorphic computing has been coined as a bet to overcome the limits imposed by CMOS scaling. What makes neuromorphics so promising is that they require far less power to process Artificial Intelligence (AI) algorithms, akin to the way our brains use under 20 watts to process information and store memory. The mainstream in digital electronics is focused on taking devices we understand at the micrometer scale and trying to scale it down a thousand times smaller and expecting it to behave the same way. But as we approach sub-5nm feature sizes, the fields of electrons start to perturb the behavior of those around them. It becomes worthwhile to try the opposite strategy: look at what occurs naturally at the nanoscale and trying to build something out of these neurons. To move further along the trajectory of Moore's Law, therefore, we have three general propositions: *P1) the fundamental technology of circuits needs to be altered*, or *P2) the architecture needs to be optimized for our specific needs*, or, alternatively, *P3) a hybrid of the two*.

It is obvious where we are heading with proposition *P1*: the memristor. By exploiting the memristor's ability to behave like an artificial neuron, we can construct architectures that are specifically optimized for certain types of machine learning algorithms such as deep neural networks. With respect to proposition *P2*, the huge rise in the popularity of machine learning has come with it a certain dependence in our daily lives. For example, industries depend on it for fraud detection, health diagnosis and predictive analysis. Neural networks are presently among the best machine learning algorithms available and are inspired by the way neurons in the human brain function. The human brain has an estimated 100 billion neurons each with up to 10,000 synaptic connections and somehow consumes less power than a lightbulb. Neural networks can also process huge amounts of data, with the number of parameters often well in excess of tens of millions, though typical von Neumann computing architectures are not optimized for them. The motivation of neuromorphic computing is to understand the underlying principles of how the brain processes information, and to then translate those principles into hardware in order to generate something useful. A key question in designing a neuromorphic chip is understanding the structure of the algorithms it is likely going to run, and neuromorphic computing is therefore one proposed solution to continuing along the trajectory of Moore's Law.

Thus, combining propositions *P1* and *P2*, proposition *P3* is postulated. As it is possible to compute any function using a neural network[6], accelerators and neuromorphic chips give us an opportunity to keep a generalized, and value-oriented version of Moore's Law alive for this specific, but increasingly important use. Plenty of researchers have discovered that memristors are capable of functioning as synapses and neurons, and so a big task is to scale these into larger networks with novel structures to develop memristor-based VLSI systems that converge to the design specifications of the human brain.

1.4.1 Memristive Synapse

The human brain can perform complex tasks such as unstructured data classification and image recognition through the communication of large numbers of neurons. Neurons communicate with one another at junctions via synapses, which can be either electrical or chemical. At a chemical synapse, an action potential triggers the presynaptic neuron to release neurotransmitters. These molecules bind to receptors on the postsynaptic cell, which will have an effect on the probability of the neuron to fire an action potential. It is accepted that these action potentials give off an all-or-nothing response. Either they happen at full strength, or not at all (Figure 1.14).

The synapses between two neurons, on the other hand, are much more flexible. One way to adjust this strength is by varying the amount of neurotransmitters that are released. In neural networks, this mechanism can be simulated by using memristors and using its reconfigurable conductance to represent the strength, or weight, of the synapse. As just one of many possible examples, a memristor may act as a synapse by forming the junction between pre- and post-synaptic CMOS neurons [48]. When a presynaptic spike is triggered before a postsynaptic spike, or equivalently there is a positive

---

[6]Two caveats come with this statement: 1) we are limited to approximating functions, though to any desired accuracy, and 2) this only applies to continuous functions.

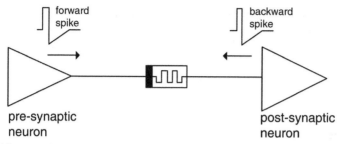

Figure 1.14 Memristor-based synapse for spike time dependent plasticity, with waveforms used as a presynaptic spike (left) and postsynaptic spike (right) propagated from CMOS neurons.

voltage applied across the memristor, then this results in an increase in the state of the memristor. In all of our analyses of memristors thus far, a positive voltage induces an increase in conductance (or at least shifts the state closer towards an increase in conductance).

There will of course be exceptions to this, such as in certain conductive filaments where a high enough voltage will cause a complete transfer of oxygen vacancies from the vacancy-rich material. The effect of a complete transfer of conductive oxygen vacancies causes the device to suddenly switch back to a high resistive state [23]. This point is raised, as not all resistive switches will be suited for all tasks. A good choice of memristor here would be one that can reliably exhibit multi-state memory, such that it can reflect a varying range of conductance values.

In the case where a postsynaptic spike is triggered before a presynaptic spike, the memristor sees a negative voltage, causing a decrease in state, and a drop in conductance. This is represented by

$$\Delta t = t_{pre} - t_{post} \tag{1.20}$$

where Δt is the relative timing between the pre- and post-synaptic firing, defined as the interval from the initial time of the pre-synaptic spike to that of the post-synaptic spike. When $\Delta t > 0$ current through the memristive synapse will increase, causing the conductance also increases, and vice versa. The dependency of memristor conductance, often referred to as 'weight' in this context, forms the basis of temporally correlated learning rules, such as spike-timing-dependent-plasticity (STDP), which is necessary to mimic a biological brain by the neural network.

1.4.2 Memristive Neuron

Beyond synapses, the memristor has also been implemented as a neuron. The picture of the neuron in the previous section is still incomplete – a real biological neuron that responds to sensory input (such as light through the retina) works by accepting many inputs through its dendritic tree. The dendritic tree of a neuron is made up of many branches, called dendrites, all of which pass the input to the body of the neuron where these signals are accumulated. If the sum of inputs exceeds a threshold, then the neuron is more likely to fire. In models that omit stochasticity, the neuron will be guaranteed to fire. This firing is typically in the form of a sudden spike in the potential across the cell membrane.

The biological neuron is still tremendously more complicated, though this captures the essence of the 'artificial neuron model'. The idea is that if the sum of weighted inputs exceeds a given threshold, then the output will be high as determined by a transfer function. Curiously, the logistic function used to characterize the state-conductance relationship in earlier sections is a popular choice of transfer function, and we are aware that multiple inputs are accumulated by virtue of an increasing state or flux in a memristor. Therefore, it is not essential for the conductance of the memristor to continuously evolve – all it needs is thresholding [49].

In this case, it may be desirable to have a state-conductance relationship more likened to that of the discrete step function in Figure 1.2(c). The use of memory retentive states means that initial conditions become relevant, and that an extra *read* step is required to gauge how close the device is to reaching the threshold.

This can be eliminated by using phase change neurons in their bistable form as presented in (1.15). The state will always initialize at a known point of bistability, so the state-threshold can be treated equivalently to a voltage-threshold. Furthermore, a device that can be modeled by a large value of s in the logistic expression from (1.16) will tend towards conductance switching. The different types of memristors that are available mean that neurons and synapses can be implemented in different ways. Selecting the right device with the appropriate threshold, reliability and switching speed to match the learning rule required to process larger operations is a big design decision in optimizing any memristor-based neuromorphic circuit.

1.4.3 Memristive Neural Networks

When many of these neurons and synapses are combined in larger networks, they become capable of performing complex computations in ways that are much more efficient than traditional von Neumann architectures. In fact, a neural network is nothing more than a collection of interconnected artificial neurons which, when assigned correct weights, is a powerful method to analyze data and perform cognitive tasks. The use of memristors in neural networks simplify the hardware mapping of DNN algorithms in a few ways. Firstly, they require less devices per multiply-and-accumulate (MAC) operation than what a purely CMOS chip would need. The MAC operation is self-descriptive from Figure 1.15, where the weighing of the inputs is 'multiply', and the summation corresponds to 'accumulation', and it is the dominant computation during inference in neural networks. Only one memristor is required per multiply process as opposed to potentially over 40 transistors in CMOS for a binary multiplier. Another advantage is that the physical distance between memory (kernel or weight storage) and computation is decreased. This reduces delay and vulnerability to signal degradation, as well as requires less power to store and fetch weights – this particular bottleneck of von Neumann architectures is no longer a problem. The use of memristor-based neuromorphic circuits and accelerators shows promise as they enable computation-in-memory architecture for data-intensive applications.

The most popular implementation for processing machine learning inference in the use of memristors is by structuring them into a dense crossbar architecture [37, 50]. The crossbar consists of two perpendicular nanowire layers which are the top and bottom electrodes, and the memristive material is deposited between the two layers. As a result, a memristor is formed at each junction as shown in the schematic diagram in Figure 1.16.

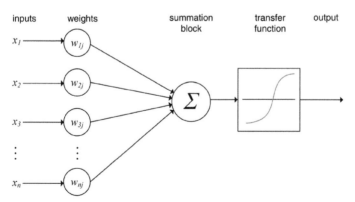

Figure 1.15 Artificial neuron model. Each neuron accepts n inputs, each of which are modulated by a weight. These are then accumulated at the summation block, and passed through a transfer function, which may take on many forms such as the logistic function, a rectified linear unit (ReLU), the hyperbolic tan function amongst many others.

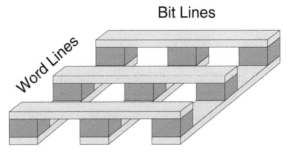

Figure 1.16 3×3 Memristor Crossbar.

The three processes a crossbar performs are reading, writing and training.

1.4.3.1 Crossbar processing: Read

It is straightforward to read from an individual memristor. For instance, say we would like to read from only the top device as per Figure 1.17(b). In such a case, a read voltage is applied at V_1 sufficiently small so as to not perturb the state of the memristor. All other rows (word-lines) have V_2 through to V_m will be grounded, and all column wires (bit-wires) will be fixed at zero. Care must be taken with the bit-lines as grounding them may undesirably direct current through a different branch. Conventionally, they are clamped using a device with asymmetric conductance, like a diode, to ensure all current flows downwards. The measured current at the output can be used to calculate the weight of the memristor using Ohm's Law.

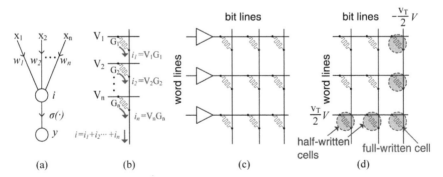

Figure 1.17 Hardware mapping of neural networks (a) artificial neuron model (b) read during multiply-accumulate operation along a single bit-line, (c) vector-matrix multiplier, (d) v/2 write scheme.

This process does not fully utilize the parallelism of crossbars and would require one read process for every memristor in the crossbar – this would be extremely time-consuming. Instead, the more efficient approach is to simultaneously apply input voltages across V_1, V_2 through to V_m. The current response at the base of the bit-line will be the sum-of-products of the voltages and conductance. When this output is passed through a separate sigmoid unit (or other form of mathematical transfer function), then that will represent the full computational process of an artificial neuron. Each memristor contributes to one multiply operation, and with a multi-state memristor, can have a weight set to any of these values. Contrast this to a binary multiplier built in CMOS, which potentially requires over 40 transistors.

But this is only taking advantage of one single bit-line – we can continue to optimize. If this MAC operation is distributed across various bit-lines, then we start to see the possibility of higher precision calculations where each column represents a different bit. Alternatively, if speed is desired over precision, the weights of other bit-lines can be assigned to those of other neurons that accept identical inputs. Thus, we can efficiently process multiple channels in one single cycle.

1.4.3.2 Crossbar processing: Write

During the write process, the weights of synapses are written to the conductance values of the memristors. The multiple types of memristors available means there are multiple modes of writing schemes possible. In the $V/2$ writing scheme depicted in Figure 1.17(d), the bit-line and word-line corresponding to the selected cell will have equal but opposite voltages of half the threshold, so the total potential across the memristor will be enough to switch the device (assuming the voltages are held for the required set or reset time). Assuming all memristors are in their initialized states, only the selected memristor will switch conductance states as all other memristors have insufficient energy (V/2). This can be problematic, as all non-selected memristors will be partially selected (or pseudo-selected) [51]. Unless they are each reset after each cycle, then future writes will result it unintended switching.

One method is to use the idealized form of phase transition memristors which, thanks to their bistable nature, will return back to their original state in the absence of switching. The drawback of this mechanism is the inability to take advantage of multiple states, and it follows that a consequence of bistable memristors is the inability to write more than one bit at a time. A way to circumvent this is to make use of the additional columns as additional bits within the same neuron.

There is a conceptually simple approach to using multi-state metal-oxide devices, and also avoiding pseudo-selection in cells. By implementing the one-transistor-one-memristor (1T1M) scheme in each cell, we can use the transistor as a pass device to block all unintended writing. The transistor is connected to a gate wire, which controls which memristors are accessed by the driving voltages and the bit- and word-lines. This method has been deployed experimentally with 64-state memristors, though comes at the cost of a reduction in volume density of each cell [36]. There is flexibility in that one is not limited to the use of a transistor – diodes have also been demonstrated to enable in 1D1M cells too.

1.4.3.3 Crossbar processing: Training

Now that we know how to read from and write to different types of memristors, the question becomes, what are we writing? Each connection between neurons has a weight associated with it, and these weights must be optimized to generate an output that tells us information about the inputs that could not be interpreted on inspection. In general, training is performed by minimizing a loss function. There are plenty of algorithms that optimize functions, and one of the most common choices for its simplicity is the gradient descent method. Understanding the specifics of the algorithm is an involved process, so to maintain our focus on memristive crossbar application we have not included it here.

The implementation of on-chip training has thus far proven to be a niche application. The ISAAC processor in [37], for example, relies on weights that are pre-trained off-chip. The weights that are calculated offline on a GPU are then mapped to conductance on the crossbar. The advantage of separating the training process from a crossbar is obvious – it saves additional circuitry to perform training, and it ensures uniformity across all devices which is desirable in commercial embedded systems, such as mobile phone processors. However, there is a demand for chips that can perform weight updates with complex data dependencies and continuous on-chip training.

This drawback of the ISAAC processor was addressed in the PipeLayer neural network accelerator, which uses metal-oxide memristors to support the acceleration of both the training (learning) and the inference phases [50]. In inference, data moves forward through the network and in training, data moves backwards to update weights based on error minimization. In PipeLayer, they have improved the efficiency of their processing by using a spike-based scheme instead of voltage-levels for data inputs as was done in ISAAC. While spiked-based processing requires more cycles to provide information, the drawback is offset by removing the overhead that comes with DAC's, and they have further eliminated the need for ADC's by using integrate-and-fire neurons in their design.

These are examples of neuromorphic chips that enable the acceleration of machine learning algorithms. Memristor crossbar arrays can also be used to implement asynchronous STDP learning rules, where memristors are used as synapses in the way described in the previous section. STDP learning in biology occurs asynchronously and online, which means that synaptic weights are updated simultaneously with computation and spike transmission. The work in [52] scales the idea of memristors behaving as a synapse in Figure 1.14, to a larger crossbar structure and depicts the influence of action potential shapes on the resulting STDP memristor weight update function.

1.5 Looking Forward

This introductory chapter serves to provide a fundamental understanding of memristors, memristive devices, and their resistive switching mechanisms. From the quantum effects in spin-transfer memristors and resonant tunneling diodes, to migration of ions and vacancies. These physical mechanisms are the key to unravelling how to harness resistance switching as a means of simultaneously storing memory and processing computation, in the realization of their role in stretching the inevitable limitations that come with Moore's Law.

The nanoscale devices that we explored in this introduction, including RRAM, STTRAM-based memristors, and RTDs, and their general applications in neuromorphic computing, multi-level memory cells and neural network processing will serve as a starting point in building more complex architectures in the chapters to come. Many further considerations in addition to the design considerations, challenges and trade-offs, still remain. These include performance evaluations of crossbar memory storage, binarized digital storage and multi-level architectures, and how learning rules such as STDP can be applied, which will all be considered in further depth in the earlier chapters of this book. The use of relative magnitudes of resistances between compositely connected memristors will be shown to achieve a higher level of learning in crossbar architectures, following which we will shift focus to the use of RTDs for image processing and temporal filtering. To complete our analysis of nanoscale resistive switching applications, we'll turn to the use of memristors in hardware implementation of cellular neural networks.

There is a dominant trend towards using resistive switching in nanoscale devices as a means to process and store information, and a vast range of physical mechanisms in which these can be achieved. But only by furthering our understanding of the likeness of these components to neurons and synapses, may we start applying these emerging memory technologies to the development of ultra-dense, fast and efficient learning mechanisms, as comparable to the biological brain.

Acknowledgments

This work was supported by the Australian Department of Foreign Affairs and Trade, Australia-Korea Foundation under Grant AKF00640, the Commonwealth Government of Australia through the Australian Government Research Training Program Scholarship, and iDataMap Corporation.

References

[1] Chua, L., 1971. "Memristor – the missing circuit element", *IEEE Transactions on Circuit Theory* 18(5), pp. 507–519.

[2] Kim, H., Sah, M.P., Yang, C., Cho, S. and Chua, L.O., 2012. "Memristor emulator for memristor circuit applications", *IEEE Transactions on Circuits and Systems I: Regular Papers*, 59(10), pp. 2422–2431.

[3] Sánchez-López, C., Mendoza-Lopez, J., Carrasco-Aguilar, M.A. and Muñiz-Montero, C., 2014. "A floating analog memristor emulator circuit", *IEEE Transactions on Circuits and Systems II: Express Briefs*, 61(5), pp. 309–313.

[4] Zheng, C., Iu, H.H., Fernando, T., Yu, D., Guo, H. and Eshraghian, J.K., 2018. "Analysis and generation of chaos using compositely connected coupled memristors", *Chaos: An Interdisciplinary Journal of Nonlinear Science*, 28(6), p. 063115

[5] Shin, S., Zheng, L., Weickhardt, G., Cho, S. and Kang, S.M.S., 2013. "Compact circuit model and hardware emulation for floating memristor devices", *IEEE Circuits and Systems Magazine*, 13(2), pp. 42–55.

[6] Chua, L.O. and Kang, S.M., 1976. "Memristive devices and systems", *Proceedings of the IEEE*, 64(2), pp. 209–223.

[7] Strukov, D.B., Snider, G.S., Stewart, D.R. and Williams, R.S., 2008. "The missing memristor found", *Nature,* 453(7191), p. 80.

[8] Eshraghian, K., Cho, K.R., Kavehei, O., Kang, S.K., Abbott, D. and Kang, S.K., 2011. Memristor MOS content addressable memory (MCAM): Hybrid architecture for future high performance search engines. IEEE Transactions on Very Large Scale Integration (VLSI) Systems, 19(8), pp. 1407–1417.

[9] Kim, J., Ahmed, T., Nili, H., Yang, J., Jeong, D.S., Beckett, P., Sriram, S., Ranasinghe, D.C. and Kavehei, O., 2018. A physical unclonable function with redox-based nanoionic resistive memory. IEEE Transactions on Information Forensics and Security, 13(2), pp. 437–448.

[10] Yu, D., Iu, H.H.C., Fernando, T. and Eshraghian, J., 2016. Memristive and memcapacitive astable multivibrators. Oscillator Circuits: Frontiers in Design, Analysis and Applications, 32, p. 51.

[11] Pershin, Y.V. and Di Ventra, M., 2010. Practical approach to programmable analog circuits with memristors. IEEE Transactions on Circuits and Systems I: Regular Papers, 57(8), pp. 1857–1864.

[12] Shin, S., Kim, K. and Kang, S.M., 2011. Memristor applications for programmable analog ICs. IEEE Transactions on Nanotechnology, 10(2), pp. 266–274.

[13] Indiveri, G. and Liu, S.C., 2015. Memory and information processing in neuromorphic systems. Proceedings of the IEEE, 103(8), pp. 1379–1397.

[14] Hodgkin, A.L. and Huxley, A.F., 1952. A quantitative description of membrane current and its application to conduction and excitation in nerve. The Journal of physiology, 117(4), pp. 500–544.

[15] Esaki, L. and Tsu, R., 1970. Superlattice and negative differential conductivity in semiconductors. IBM Journal of Research and Development, 14(1), pp. 61–65.

[16] Chua, L., 2014. "If it's pinched it's a memristor", *Semiconductor Science and Technology*, 29(10), p. 104001.

[17] Sapoff, M. and Oppenheim, R.M., 1963. Theory and application of self-heated thermistors. Proceedings of the IEEE, 51(10), pp. 1292–1305.

[18] Francis, V.J., 1948. Fundamentals of discharge tube circuits.

[19] Chua L., 2015. Everything you wish to know about memristors but are afraid to ask. Radioengineering, 24(2), p. 319.

[20] Argall, F., 1968. Switching phenomena in titanium oxide thin films. Solid-State Electronics, 11(5), pp. 535–541.

[21] Pickett, M.D., Strukov, D.B., Borghetti, J.L., Yang, J.J., Snider, G.S., Stewart, D.R. and Williams, R.S., 2009. Switching dynamics in titanium dioxide memristive devices. Journal of Applied Physics, 106(7), p. 074508.

[22] Jeong, D.S., Thomas, R., Katiyar, R.S., Scott, J.F., Kohlstedt, H., Petraru, A. and Hwang, C.S., 2012. Emerging memories: resistive switching mechanisms and current status. Reports on progress in physics, 75(7), p. 076502.

[23] Yang, Y., Sheridan, P. and Lu, W., 2012. Complementary resistive switching in tantalum oxide-based resistive memory devices. Applied Physics Letters, 100(20), p. 203112.

[24] Zhang, T., Yin, M., Lu, X., Cai, Y., Yang, Y. and Huang, R., 2017. Tolerance of intrinsic device variation in fuzzy restricted Boltzmann machine network based on memristive nano-synapses. Nano Futures, 1(1), p. 015003.

[25] Kwon, D.H., Kim, K.M., Jang, J.H., Jeon, J.M., Lee, M.H., Kim, G.H., Li, X.S., Park, G.S., Lee, B., Han, S. and Kim, M., 2010. Atomic structure of conducting nanofilaments in TiO_2 resistive switching memory. Nature nanotechnology, 5(2), p. 148.

[26] Hwan Kim, G., Ho Lee, J., Yeong Seok, J., Ji Song, S., Ho Yoon, J., Jean Yoon, K., Hwan Lee, M., Min Kim, K., Dong Lee, H., Wook Ryu, S. and Joo Park, T., 2011. Improved endurance of resistive switching TiO_2 thin film by hourglass shaped Magnéli filaments. Applied Physics Letters, 98(26), p. 262901.

[27] Strachan, J.P., Pickett, M.D., Yang, J.J., Aloni, S., David Kilcoyne, A.L., Medeiros-Ribeiro, G. and Stanley Williams, R., 2010. Direct identification of the conducting channels in a functioning memristive device. Advanced materials, 22(32), pp. 3573–3577.

[28] Yajima, T., Fujiwara, K., Nakao, A., Kobayashi, T., Tanaka, T., Sunouchi, K., Suzuki, Y., Takeda, M., Kojima, K., Nakamura, Y. and Taniguchi, K., 2010. Spatial redistribution of oxygen ions in oxide resistance switching device after forming process. Japanese Journal of Applied Physics, 49(6R), p. 060215.

[29] Dong, X., Xu, C., Xie, Y. and Jouppi, N.P., 2012. Nvsim: A circuit-level performance, energy, and area model for emerging nonvolatile memory. IEEE Transactions on Computer-Aided Design of Integrated Circuits and Systems, 31(7), pp. 994–1007.

[30] Ascoli, A., Tetzlaff, R., Chua, L.O., Strachan, J.P. and Williams, R.S., 2016, March. Fading memory effects in a memristor for Cellular Nanoscale Network applications. In Proceedings of the 2016 Conference on Design, Automation & Test in Europe (pp. 421–425). EDA Consortium.

[31] Mazumder, P., Kulkarni, S., Bhattacharya, M., Sun, J.P. and Haddad, G.I., 1998. Digital circuit applications of resonant tunneling devices. Proceedings of the IEEE, 86(4), pp. 664–686.

[32] Sun, J.P., Haddad, G.I., Mazumder, P. and Schulman, J.N., 1998. Resonant tunneling diodes: Models and properties. Proceedings of the IEEE, 86(4), pp. 641–660.

[33] Pan, X. and Skafidas, E., 2016. Resonant tunneling based graphene quantum dot memristors. Nanoscale, 8(48), pp. 20074–20079.

[34] Wang, X., Chen, Y., Xi, H., Li, H. and Dimitrov, D., 2009. Spintronic memristor through spin-torque-induced magnetization motion. IEEE electron device letters, 30(3), pp. 294–297.

[35] Wang, D.T., Dai, Y.W., Xu, J., Chen, L., Sun, Q.Q., Zhou, P., Wang, P.F., Ding, S.J. and Zhang, D.W., 2016. Resistive switching and synaptic behaviors of TaN/Al2O3/ZnO/ITO flexible devices with embedded Ag nanoparticles. IEEE Electron Device Lett, 37(7), pp. 878–881.

[36] Li, C., Hu, M., Li, Y., Jiang, H., Ge, N., Montgomery, E., Zhang, J., Song, W., Dávila, N., Graves, C.E. and Li, Z., 2018. Analogue signal and image processing with large memristor crossbars. Nature Electronics, 1(1), p. 52.

[37] Lee, J., Eshraghian, J.K., Jeong, M., Shan, F., Iu, H.H.C., Cho, K., 2019. Nano-programmable logics based on double-layer anti-facing memristors. Journal of nanoscience and nanotechnology, 19(3), pp. 1295–1300.

[38] Shafiee, A., Nag, A., Muralimanohar, N., Balasubramonian, R., Strachan, J.P., Hu, M., Williams, R.S. and Srikumar, V., 2016. ISAAC: A convolutional neural network accelerator with in-situ analog arithmetic in crossbars. ACM SIGARCH Computer Architecture News, 44(3), pp. 14–26.

[39] Eshraghian, J.K., Kang, S.M., Baek, S.B., Orchard, G., Iu, H.H.C., Lei, W. 2019. Analog weights in ReRAM accelerators. IEEE International Conference on Artificial Intelligence Circuits and Systems. IEEE.

[40] Eshraghian, J.K., Iu, H.H., Fernando, T., Yu, D. and Li, Z., 2016, May. Modelling and characterization of dynamic behavior of coupled memristor circuits. In Circuits and Systems (ISCAS), 2016 IEEE International Symposium on (pp. 690–693). IEEE.

[41] Eshraghian, J.K.J., Iu, H.H. and Eshraghian, K., 2018. Modeling of Coupled Memristive-Based Architectures Applicable to Neural Network Models. In Memristor and Memristive Neural Networks. InTech.

[42] Gao, L., Alibart, F. and Strukov, D.B., 2013. Programmable CMOS/memristor threshold logic. IEEE Transactions on Nanotechnology, 12(2), pp. 115–119.

[43] Cho, S.W., Eshraghian, J.K., Eom, J.S. and Cho, K.R., 2016. Storage logic primitives based on stacked memristor-CMOS technology. Journal of nanoscience and nanotechnology, 16(12), pp. 12726–12731.

[44] Nili, H., Adam, G.C., Hoskins, B., Prezioso, M., Kim, J., Mahmoodi, M.R., Bayat, F.M., Kavehei, O. and Strukov, D.B., 2018. Hardware-intrinsic security primitives enabled by analogue state and nonlinear conductance variations in integrated memristors. Nature Electronics, 1(3), p. 197.

[45] Hady, F.T., Foong, A., Veal, B. and Williams, D., 2017. Platform storage performance with 3D XPoint technology. Proceedings of the IEEE, 105(9), pp. 1822–1833.

[46] Petras, I., 2010. Fractional-order memristor-based Chua's circuit. IEEE Transactions on Circuits and Systems II: Express Briefs, 57(12), pp. 975–979.

[47] Corinto, F., Ascoli, A. and Gilli, M., 2011. Nonlinear dynamics of memristor oscillators. IEEE Transactions on Circuits and Systems I: Regular Papers, 58(6), pp. 1323–1336.

[48] Jo, S.H., Chang, T., Ebong, I., Bhadviya, B.B., Mazumder, P. and Lu, W., 2010. Nanoscale memristor device as synapse in neuromorphic systems. Nano letters, 10(4), pp. 1297–1301.

[49] Eshraghian, J.K., Cho, K., Zheng, C., Nam, M., Iu, H.H.C., Lei, W. and Eshraghian, K., 2018. Neuromorphic Vision Hybrid RRAM-CMOS Architecture. IEEE Transactions on Very Large Scale Integration (VLSI) Systems, 26(12), pp. 2816–2829.

[50] Song, L., Qian, X., Li, H. and Chen, Y., 2017, February. Pipelayer: A pipelined reram-based accelerator for deep learning. In High Performance Computer Architecture (HPCA), 2017 IEEE International Symposium on (pp. 541–552). IEEE.

[51] Eshraghian, J.K., Cho, K.R., Iu, H.H., Fernando, T., Iannella, N., Kang, S.M. and Eshraghian, K., 2017. Maximization of Crossbar Array Memory Using Fundamental Memristor Theory. IEEE Transactions on Circuits and Systems II: Express Briefs, 64(12), pp. 1402–1406.

[52] Serrano-Gotarredona, T., Masquelier, T., Prodromakis, T., Indiveri, G. and Linares-Barranco, B., 2013. STDP and STDP variations with memristors for spiking neuromorphic learning systems. Frontiers in neuroscience, 7, p. 2.

2

Crossbar Memory Simulation and Performance Evaluation

Woo Hyung Lee and Pinaki Mazumder

In this chapter, a crossbar memory cell is electrically modeled, and a specific peripheral circuitry, including the column and row decoders, a sensing circuitry for detecting the difference in the resistance of the cell, and the control circuit for reading and writing are designed. An analytical model on static power dissipation was conducted to suggest an optimal design for power dissipation. To improve the area overhead, MUX logics are introduced to reduce the numbers of sense amplifiers attached to all the bit lines. In addition, scaling issues of the crossbar memory design due to the size mismatch between crossbar memory arrays and CMOS peripheral circuitry are discussed.

2.1 Introduction

2.1.1 Motivation

Scaling of the CMOS technology has encountered a serious setback due to the increasing leakage currents of the CMOS FET devices in the turned-off state and their wide threshold voltage fluctuations due to the process parameter scattering. To overcome the insurmountable limitations of CMOS scaling problems, nanoscale devices such as carbon nanotubes, nanowires and molecular devices are now sought to replace the conventional CMOS devices in ultra-dense digital chips [1–6].

Implementation of these nanoscale devices requires the invention of novel architectures that take advantage of ultra small feature sizes of nanodevices [7]. Innovations for nano-architectures at the circuit level and nanocomputer system architectures at the board level are warranted to reap the benefits of nanoscale CMOS technology. Also, defect-tolerant architectures are required since the architectures become severely unreliable due to high defect densities and process variation at nanoscale. Crossbar structures [1–6], in which an active material is sandwiched between two sets of conducting nanowires crossing each other, show promising properties to address these characteristics at both the nano-architecture and the system-architecture levels.

The crossbar structure offers many advantages as memory devices can be realized since each cell can be realized as a two-terminal device formed by the two crossed nanowires trapping a composite material at each cross point. With its simple design, terabit-scale memories excluding peripheral circuits can be realized with the width of the metal lines of the array less than 5 nm. The crossbar structure can be non-volatile based on the active material between the two sets of conducting nanowires. This non-volatile characteristic provides instant restart and longer standby operation resulting in protracted battery life.

However, it is not easy to implement general logic operations with the crossbar structure by itself since the structure has difficulty in achieving enough gain and inversion. This difficulty comes from the number of terminals of the crossbar structure. With two terminal devices, it is very difficult to make an inversion function. Even if inversion function can be achieved, the gain will not be sufficient for it to be used for general inversion logics. To overcome this difficulty, hybrid CMOS/crossbar structures have been studied [8, 9]. With these structures, some of the logic functions will be shifted to the underlying CMOS circuitry and the crossbar provides reconfigurable interconnects and wired-OR operation. The system as a whole is expected to present at least two orders of magnitude of higher function density at the same power per unit area and comparable logic delay compared with their CMOS counterparts fabricated with the same design rules. However, fabrication of the crossbar structures on top of the CMOS layers present significant challenges such as registration of the CMOS components with the crossbar array. Here, the constraint in attaching a CMOS circuit to the crossbar array is how to match the decoder length to the width/length of the array.

In this chapter, a Si-based crossbar memory device has been investigated in Professor Lu's group. Rigorous investigation on the feasibility of volatile and nonvolatile switching devices on Si substrate has been conducted [10–13] because such devices can be fully fabricated with CMOS processing technology. In the perspective of processing technology, these devices are not costly to fabricate and have a simple structure to operate.

2.1.2 Contrast with Competing Technologies

Among nonvolatile switching nanoelectronic devices, phase-change memory devices (PCRD), magnetic memory devices and molecular memory devices have been considered to be strong candidates to replace the conventional memory device.

Phase-change memory devices (PCRDs) have been rigorously investigated for commercialization since 2000. The basic function of the PCRDs is to convert the chalcogenide glass state between crystalline and amorphous with the application of heat. PCRDs show high performance arising from fast switching speed than conventional non-volatile memory devices. However, PCRDs have a notable drawback in temperature sensitivity. Since the temperature varies in fabrication processing time and even in memory operation time by users, the temperature sensitivity is a challenging characteristic to PCRDs. In addition, the device requires high programming power and the density of the PCRD is not compatible to a conventional non-volatile memory device, for instance, FLASH [14].

Magnetic memory devices (MMDs) have been studied for replacing the conventional memory devices since the 1990s and have made a continuous increase in density. MMDs use magnetic storage elements for storing data, while the conventional devices use electric charge for data memorization. MMD employs electric resistance change of the memory device cell. This change is triggered by the magnetic tunnel effect that changes the orientation of the fields in the two electrodes. Exploiting this characteristic, MMDs achieve fast writing and reading time and low power dissipation. The low power dissipation is achieved because they do not require refreshing all the cells as DRAM, and the applied voltage for writing does not need to be much higher than for reading which is common for FLASH memory. However, MMDs retain a notable drawback in fabrication cost. Since MMDs employ the magnetic tunnel effect, the fabrication process should be absolutely changed from the conventional manufacture. This increases the price of the MMDs and thus becomes the main factor of hesitation to investors in industry [14–17].

Molecular crossbar memory devices (MCMDs) are one of the strong candidates for a nanoelectronic memory device. The primary advantage of MCMDs is the potential to achieve high density; an ultimate device may be fabricated with just one molecule in principle. MCMDs can achieve a density of 10^{11} bit/cm^2 which is projected technology beyond 2015 for conventional DRAMs. However, MCMDs suffer from a number of problems such as low percentage of yield, low on/off ratio, poor thermal stability, and slow switching speed [18]. There is no known process to reliably fabricate high-density memory arrays based on molecules with high yield. The poor thermal stability further results in reliability problems [19].

Crossbar memory devices using amorphous Silicon is a striking device that overcomes many of the drawbacks of the other candidates described above. Amorphous Si crossbar memory device can achieve similar cell density but with much better yield percentage compared with MCMDs. Also, it retains temperature stability in processing time and operating time. The most attractive aspect of the amorphous Si crossbar memory device is the compatibility to conventional CMOS process, while other candidates require new processing technology which cannot be realized with the conventional CMOS process fabrication facilities. This makes amorphous Si crossbar memory device more attractive than other the crossbar memory devices.

2.1.3 Amorphous Si Crossbar Memory Cell

An M/a-Si/p-Si resistive switching device structure with two terminals was developed by the students in Professor Lu's group including Sung-Hyun Jo and others and has been shown to be a strong candidate for the crossbar architecture [20, 21]. These devices promise ultra-high density and an intrinsic defect-tolerable capability. Also, the M/a-Si/p-Si devices depict comparable scalability and significantly better performance compared with the crossbar structure devices based on molecules [19].

A-Si devices have been studied since the 1960s and 1970s to observe their feasibility for memory application based on resistance switching behavior of amorphous Si [22–30]. However, these conventional metal/a-Si/metal-based devices suffer from low yield during a forming process which requires a long and high voltage applied to the device and lead to reliability issues [26, 29, 30]. Micro-scale filaments have also been observed in these metal/a-Si/metal devices [31, 32] limiting their scaling potential. For these reasons, few studies on a-Si-resistance switching devices have been conducted for memory devices since the 1980s.

   In the following sections, the new structure of the M/a-Si/p-Si cross-bar devices and the feasibility to apply this structure to an ultra-high density memory architecture are studied. For application in the memory system, electrical modeling of the devices is conducted, and the hybrid CMOS/crossbar structure is used to implement the peripheral circuitry in the memory system.

   Figure 2.1 shows a basic structure of the amorphous Silicon crossbar memory cell and *I-V* curve of the device. The principle characteristic of the amorphous Silicon crossbar memory cell is programmable resistance of the amorphous Silicon in the crossbar point between Ag and Si, as shown in Figure 2.1(b). The resistance of the amorphous Silicon can be changed based on the voltages applied to the crossbar point. Positive high voltage (3.5 to 5V) on the top electrode and ground on the Si bottom electrode lowers the resistance of the amorphous Silicon. The device will remain in the low-resistance "1" state until a negative high voltage reverses

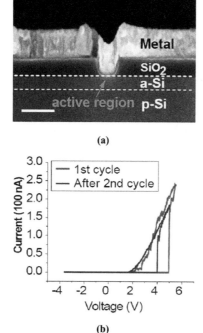

(a)

(b)

Figure 2.1   (a) A crosscut view of an amorphous Silicon crossbar memory cell and (b) *I-V* curve.

the resistance to a high resistance state, and this process can be repeated, as shown in Figure 2.1(b) [21]. Therefore, the high resistance of amorphous can be bit 0 and low resistance bit 1, and this bit assignment can be changed depending on the read circuitry. Compared with the earlier studies on metal/a-Si/metal devices, the M/a-Si/p-Si structure offers a well-controlled forming process, high yield, and excellent scalability down to < 20 nm [33]. The different behaviors between the M/a-Si/M devices and M/a-Si/p-Si devices are likely caused by the reduced density of defects at the a-Si/bottom-electrode interface in the M/a-Si/p-Si structure [33–35].

2.2 Structure

A floorplan of the architecture using the crossbar memory cells is shown in Figure 2.2. The crossbar memory units studied here consist of memory cells, row and column decoders and read and write circuitry. The difference between the amorphous crossbar memory architecture and conventional memory architectures are the basic memory cells, decoders and row and column voltage controllers.

In conventional CMOS memories, the cell consists of a capacitor in DRAM and a latch that is implemented with two inverters in SRAM. The bi-state values are determined by the charges on a capacitor in DRAM and charges on the latch in SRAM. In the decoding strategy, the conventional CMOS memories uses transistors that operate to select one word line out of n word lines in n addresses memories. The number of transistors used to decode increases as the number of word lines augment sharply. This results in incrementing the decoding time for conventional CMOS memories.

In the amorphous crossbar memory, the basic memory cells are implemented with two terminal resistive switching devices. The row and column decoders studied here are realized using the diode characteristics of the devices inside another crossbar to minimize the numbers of transistors used. Read circuitry using a trans-impedance amplifier is suggested for the crossbar memory design.

Two terminal resistive switching devices are the main part of the basic memory cells. Two terminal devices are fabricated with metal, amorphous Silicon and crystal Silicon materials. The switching devices are implemented by stacking materials serially in the order of crystal Silicon, amorphous Silicon and metal from the bottom to top. Device fabrication is conducted by CMOS processing with the exception of the active device area that is defined

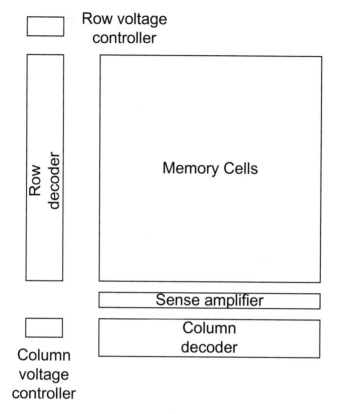

Figure 2.2   Floorplan of the amorphous crossbar memory architecture.

by electron-beam lithography. As-fabricated devices have characteristics of high resistance between the two electrodes, but when a voltage is applied on the top metal electrode, repeatable resistance switching can be observed, as shown in Figure 2.1(b). As a hysteretic resistance switching devices, the M/ a-Si/p-Si device demonstrates high yield, fast programming speed, high on/off ratio, long endurance, retention time and multi-bit/cell capability [20].

In memory design, decoders occupy more area as the memory size increases. To reduce the area overhead for large arrays, a new type of decoders, which is compatible with the amorphous Si crossbar array, is studied. In addition, to reduce the access time to the target memory as address bits are increased, a new decoder that uses the same device structure as the crossbar memory cell is studied. This decoder design takes advantage of the

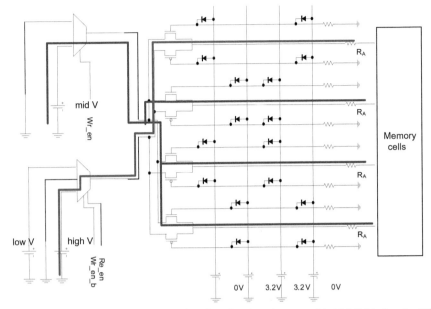

Figure 2.3   A schematic of the row decoder of a 4 × 4 array with highlighted write "1" operation.

diode characteristics from amorphous silicon devices that provide the wired-OR function and reduce the overall size of the decoder and the number of transistors used compared with a conventional design.

The row and column decoders consist of the diode characteristic devices, NMOS, PMOS and control signals. Figure 2.3 shows a schematic diagram of the row decoder of a simple 4 × 4 memory array. The decoder requires 4 types of voltage source to apply appropriate voltages across the crossbar points since 3 types for writing and 1 type for reading are used. For the write operation, high voltage, mid voltage and ground are needed. As shown in Figure 2.3, high voltage across the crossbar points needs to be applied to write "1". Negative high voltage across the crossbar points is required to write "0". A mid-voltage source is needed to prevent unnecessary cells from being written as "0" or "1". Not having mid voltage, the unselected cells may be accidentally written "0" when writing "1" to the selected cell. In the case of reading, low voltage should be applied to the row of the target cell. The diode model is used for the diode characteristic devices, as shown in Figure 2.4. The devices can be obtained using the same structure as the crossbar memory cell.

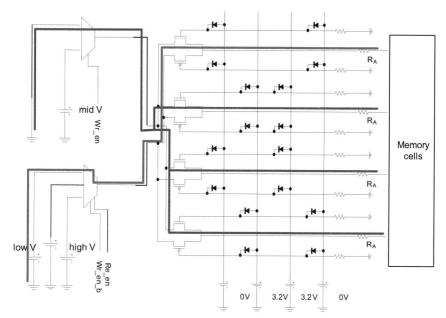

Figure 2.4 A schematic of the row decoder of a 4 × 4 array with highlighted read operation.

The row that is connected with PMOS is only enabled to apply voltages based on the function across the target crossbar point, and the other rows that are linked with NMOS are enabled so that different outside voltage sources are applied to the rows. This is because when the decoder inputs are assigned whatever inputs it may have, only one row is not enabled (00) since $V_{gs} = 0$ < Vth of the NMOS, while all other rows (11), (10) and (01) will be enabled. In Figure 2.3, the NMOS in the top first row is not enabled, while the other NMOS FETs are enabled by the decoder inputs. The selection of row for writing or reading is conducted by controlling the gate voltage to the NMOS or PMOS.

The main purpose of the decoder is to select the target row and apply appropriate voltages to the target crossbar point, while other rows are not selected for programming. In case of writing "1", the target crossbar point is applied with high voltage and the other points are applied with low voltage. To achieve this function, a high voltage is applied to the target row and mid voltage to the other rows through the selected NMOS FETs and the power supply, as shown in Figure 2.3. For the columns, low voltage is applied to the target row and mid voltage to the other rows.

When reading the value, the read circuitry consists of sense amplifiers, voltage controllers and output buffers. A read circuit based on a trans-impedance amplifier is suggested. The difference of resistances of crossbar points causes current difference so that the sense amplifier works based on the charge difference from the current flow of the crossbar points. The trans-impedance amplifier magnifies the current difference and converts it into the voltage outputs. When implementing the peripheral circuitry, the position of sense amplifiers affects the performance of the memory chip in terms of the area, power dissipation and speed. Also, the number of sense amplifiers can be adjusted using an MUX logic. The number of sense amplifiers also affects the performance of the memory. The performance comparison according to the location and number of sense amplifiers have been discussed later in this chapter.

2.2.1 Crossbar Modeling

Modeling the crossbar device as a memory device is required for simulation performance analysis. RC (resistance capacitance) modeling of the device is used for our simulation. To model the device with resistance and capacitance elements, a cross-cut view of the crossbar memory device, as shown in Figure 2.5, is convenient. For electrical modeling, it is needed to calculate intrinsic capacitance and resistance of the top metal layer and bottom Silicon electrode and then estimate the coupled capacitance between top metal and bottom Silicon electrode.

In Figure 2.5, the top electrode is fabricated with silver metal and the bottom electrode is made of Silicon. Amorphous Silicon (a-Si) is inserted in the crossing area of the two electrodes. The two electrodes and a-Si are surrounded with SiO_2 that is mounted on a Silicon wafer. The RC model consists of two constant resistors, one variable resistor and three capacitors as shown in Figure 2.6. The two constant resistors represent the resistive element of top electrode. The variable resistor demonstrates the electrical characteristic of the amorphous Silicon of the crossbar memory cell. Since the amorphous Silicon changes its resistance according to the applied voltages across it, it is represented as the variable resistor. The resistance values of the amorphous Silicon of the crossbar memory cell ranges from a few $K\Omega$ to a few $M\Omega$. The intrinsic resistance and capacitance of the top metal electrode is obtained as follows:

In Equation (2.2), C_1 represents the capacitance between the top electrodes to the ground. The first term in Equation (2.2) demonstrates the top

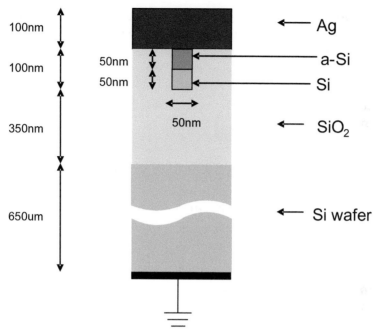

Figure 2.5 A cross-cut view of the crossbar resistive memory device.

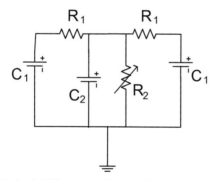

Figure 2.6 A simple RC model of the crossbar resistive memory device.

capacitance and bottom capacitance and the second term represents fringe capacitance between the sides of metal electrode and ground. Similarly, C_2 in Equation (2.3) shows the capacitance between the top electrode and the bottom electrode. The first term shows the top capacitance and bottom capacitance between the top electrode and the bottom electrode and the fringe capacitance between the sides of top electrode and the upper side of the bottom electrode and the fringe capacitance between the sides of the bottom electrode and the base side of the top electrode.

$$R_1 = \rho \frac{l}{S} = \frac{1.59 \times 10^{-8} \times 200 \times 10^{-9}}{(100 \times 10^{-9})^2} = 0.317 \ \Omega \tag{2.1}$$

$$\begin{aligned}
C_1 &= \varepsilon_1 L 1.15 \left(\frac{w}{h}\right) + \varepsilon_2 L 2.80 \left(\frac{t}{h}\right)^{0.222} \\
&= \frac{11.9 + 3.9}{2} \times 8.854 \times 10^{-12} \times 100 \times 10^{-9} \times 1.15 \left(\frac{100 \times 10^{-9}}{650.3 \times 10^{-6}}\right) \\
&\quad + 3.9 \times 8.854 \times 10^{-12} \times 100 \times 10^{-9} \times 2.80 \left(\frac{100 \times 10^{-9}}{650.3 \times 10^{-6}}\right)^{0.222} \\
&= 1.377 \ aF
\end{aligned} \tag{2.2}$$

$$\begin{aligned}
C_2 &= \varepsilon_1 L 1.15 \left(\frac{w}{h}\right) + \varepsilon_2 L 2.80 \left(\frac{t}{h}\right)^{0.222} \\
&= \frac{11.9 + 3.9}{2} \times 8.854 \times 10^{-12} \times 100 \times 10^{-9} \times 1.15 \left(\frac{100 \times 10^{-9}}{50 \times 10^{-9}}\right) \\
&\quad + 3.9 \times 8.854 \times 10^{-12} \times 100 \times 10^{-9} \\
&\quad \times 2.80 \left[\left(\frac{50 \times 10^{-9}}{50 \times 10^{-9}}\right)^{0.222} + \left(\frac{100 \times 10^{-9}}{50 \times 10^{-9}}\right)^{0.222} \right] \\
&= 37.02 \ aF
\end{aligned} \tag{2.3}$$

Modeling the active amorphous Silicon layer requires careful examination of the switching mechanism. Several mechanisms have been suggested in the past. One leading theory is that the position of the metallic particle inside the amorphous Silicon is shifted according to the applied voltages. The shift of the position affects the resistance of amorphous Silicon at the crossbar memory cell. This effect is demonstrated in Figure 2.7. When the crossbar memory device has a low resistance value, the metallic particles inside the amorphous Silicon are close to the bottom electrode; otherwise, the metallic particles have enough distance from the bottom electrode to lead to a high tunneling resistance. Since the switching element requires two distinct states, the resistance difference is the key. This device characteristic is electrically modeled as shown in Figure 2.8. The variable resistance from the shift of

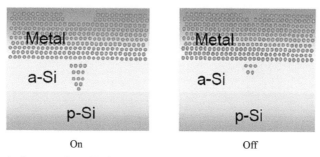

On   Off

Figure 2.7   A diagram of a cell of the crossbar resistive memory device with on state (left) and off state (right).

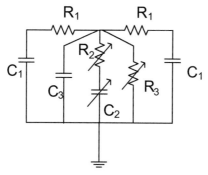

Figure 2.8   A simple RC model of the crossbar resistive memory device.

metallic particles is represented with R_2 and the variable capacitance induced from the distance between the metallic particles and the bottom electrode is demonstrated with C_2.

However, the resistance of the bottom p-Si electrode in the original memory cell is in the range between 200 Ω and 2 KΩ such that it is not suitable as an electrode because of the propagation delay, voltage drop and power dissipation caused by the high resistance of the bottom electrode, as shown in Figure 2.9(a) and (b). Hence, an improved design is studied by attaching a thin metal line to the Silicon bottom electrode in parallel [36]. Using this design, the resistance of the bottom electrode can be reduced in a memory cell to around 2 Ω. Figure 2.9(c) shows a two-dimensional view of the crossbar memory with the improved bottom electrode. Four cells of the crossbar memory are shown in Figure 2.9(c). Also, a circuit schematic of the

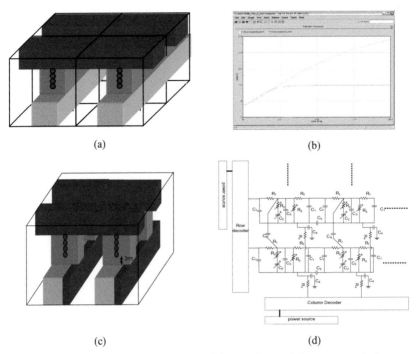

Figure 2.9   A simple 2-D RC model of the crossbar resistive memory device.

four crossbar memory cells is shown in Figure 2.9(d). In Figure 2.9(d), R_4 is calculated based on the bottom electrode which consists of a metal line and a semiconductor line connected in parallel.

For the 1-Kb crossbar memory architecture shown in this chapter using this model, the circuit elements for the simulation were fully implemented. All the elements described for one cell were included, and the coupling capacitance between the adjacent metal lines were included in the whole array.

2.3 Write Strategy and Circuit Implementation

Writing is the main function of a memory. By way of writing, the memory has data stored in a memory cell. Each memory has its own write strategy. For crossbar memory architecture, power assignment is used on both sides of the crossbar point for writing. The power assignment is attached with the decoder circuits. Hence, an appropriate power supply is applied to the crossbar point through the decoder circuit.

For writing "1", it is needed to apply a high voltage to the crossbar point. To implement this function, a high voltage is applied to one row and the other rows are assigned with intermediate voltage in the row decoder as shown in Figure 2.10. Therefore, it is needed to select a high voltage in the lower multiplexer (MUX) and an intermediate voltage in the upper MUX in the row decoder. Voltages to the column decoder oppositely is applied as shown in Figure 2.11. For example, to write "1" to the crossbar memory cell in the fourth row and the first column, high voltage is assigned to the sources of NMOS FETs that are connected to lower MUX, while middle voltage is assigned to the sources of PMOS FETs. For the decoder inputs, the fourth row is assigned to high voltage and the other three rows are assigned to middle voltage. Ground voltage is assigned to the lower MUX and an intermediate voltage to the upper MUX in the column decoder.

For writing "0", a negative high voltage is applied to the crossbar point. This function is implemented by assigning a ground voltage to one row and the other rows are assigned with intermediate voltage in the row decoder as shown in Figure 2.12. Therefore, the ground voltage is applied in the lower

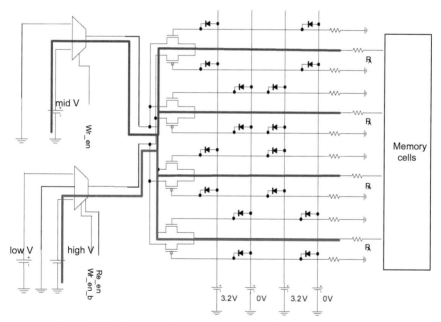

Figure 2.10 Row decoder and row power assignment for writing "1".

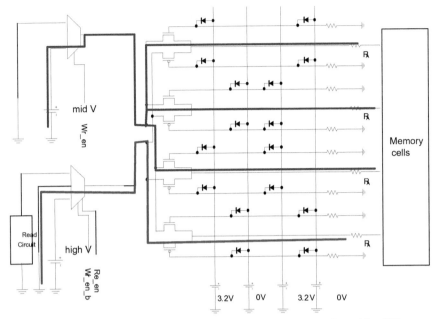

Figure 2.11 Column decoder and column power assignment for writing "1".

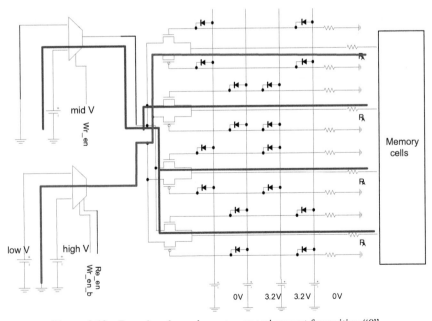

Figure 2.12 Row decoder and row power assignment for writing "0".

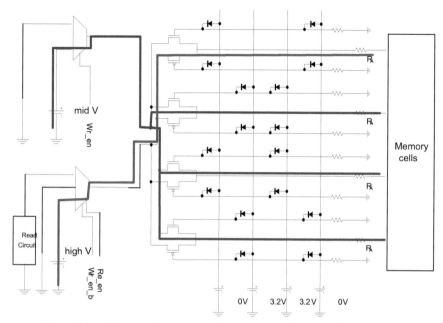

Figure 2.13 Column decoder and column power assignment for writing "0".

MUX and an intermediate voltage in the upper MUX in the row decoder. However, opposite voltages need to be assigned in the column decoder as shown in Figure 2.13. A high voltage in the lower MUX and an intermediate voltage in the upper MUX are applied in the column decoder.

2.4 Read Strategy and Circuit Implementation

Reading in the crossbar memory requires a different strategy. The main difference of the reading in the crossbar memory is that a low voltage is applied to measure the resistance value of the cell. In the new circuit design, zero voltage is assigned in the upper MUX and low voltage in the lower MUX in the row decoder in Figure 2.4. However, a ground voltage is applied in the upper MUX and a read circuit in the lower MUX in the column decoder in Figure 2.4.

As mentioned in Section 2.2, an amplifier can be implemented to detect the value of the resistances. Since a low voltage is applied in the row decoder, the different resistance values in the crossbar points result in differences of current through the crossbar points. Therefore, a differential trans-impedance

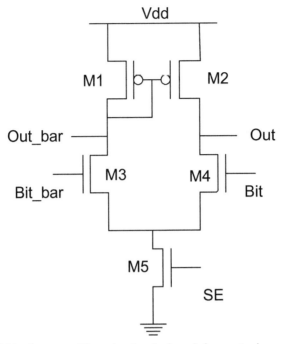

Figure 2.14   Sense amplifier using the single-ended current-mirror amplifier.

amplifier that detects the difference in the current and outputs in the form of voltages is investigated.

Figure 2.14 shows a sense amplifier for a conventional CMOS memory circuit. This sense amplifier consists of a current-mirror amplifier, and the bit lines are connected to the gate which controls the current flow from Vdd. When Bit_bar is higher than Bit, M1 and M3 turn on and increase the current through M1 and M3. This increases the current through M2 and decreases the current through M4. Therefore, the Out voltage increases and decreases Out_bar voltage. When Bit_bar is lower than Bit, M1 and M3 are turned off and the current through M1 and M3 decrease. This results in turning off M2 and makes M4 and M5 operate in the deep triode region. This results in turning off M2 and makes M4 and M5 operate in the deep triode region. Because there is no current flow from M2, M4 and M5, the voltage at Out is decreased to zero. This type of sense amplifier is used for a single-ended output.

Figure 2.15 illustrates a sense amplifier that is commonly used for conventional CMOS memory design. This sense amplifier employs a cross-coupled

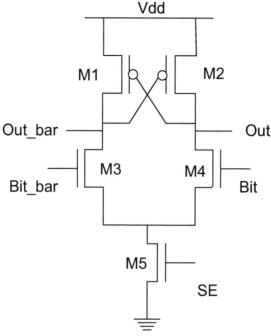

Figure 2.15 Sense amplifier using cross-coupled amplifier.

amplifier. The Bit line and the Bit_bar line are connected with gates of the pull-down differential pairs. This sense amplifier can be implemented with combining two single-ended current mirror amplifiers. When Bit is higher than Bit_bar, Bit, M1 and M3 are turned off and the current through M1 and M3 decrease. This will turn off M2 and makes M4 and M5 operate in the deep triode region and decrease the voltage at Out node to zero. However, when Bit is lower than Bit_bar, Bit, M2 and M4 are turned off and the current through M2 and M4 decrease. This will turn off M1 and makes M3 and M5 operate in the deep triode region and decrease the voltage at Out_bar node to zero.

Figure 2.16 demonstrates a sense amplifier that is commonly used for SRAM. This sense amplifier manipulates a latch that consists of two inverters. To operate this sense amplifier, equalization between Bit node and Bit_bar node should be conducted. Providing a PMOS that connects source to Bit and drain to Bit_bar, the equalization function can be operated. After the equalization, SE is enabled to make the sense amplifier function. When Bit is higher than Bit_bar, the latch makes Bit increase to Vdd and Bit_bar decrease

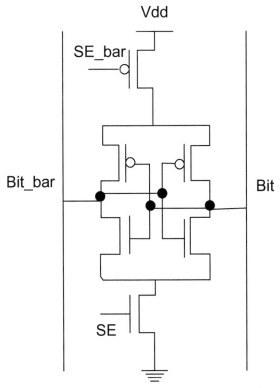

Figure 2.16   Sense amplifier using a cross-coupled CMOS inverter latch.

to Gnd. When Bit is lower than Bit_bar, the latch makes Bit increase to Gnd and Bit_bar decrease to Vdd.

However, these sense amplifiers cannot be used for the amorphous crossbar memory architecture. The cross-coupled amplifier is not suitable for the crossbar memory read circuitry since it operates based on the difference in input voltages. To detect the change of resistance, it is needed to differentiate the current change through the crossbar resistance with a supplied voltage. The cross-coupled CMOS inverter latch cannot be used by itself for the crossbar memory circuitry since the crossbar memory device does not have a latch-based cell as SRAMs. The latch-based cell can flip the Bit or Bit_bar depending on the stored value in the cell during the read mode.

Hence, a reference resistor and a differential amplifier that can produce the current difference are considered. Then, this current difference is detected through a latch that is connected to the output node.

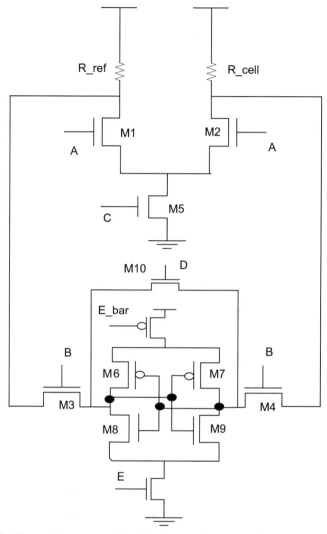

Figure 2.17   Suggested sense amplifier for the amorphous crossbar memory using a transimpedance differential amplifier that senses current difference.

Figure 2.17 shows the read circuit that senses the current difference through the resistance elements. This sense amplifier uses the principles of a cross-coupled inverter latch in Figure 2.16 and differential amplifiers in Figures 2.14 and 2.15. The suggested amplifier magnifies the current flow using the differential amplifier, and the magnified current affects the values inside a latch. Based on experimental results of the resistances of the crossbar points, the resistances of the crossbar points are applied. They have low resistance values in the range from 1 KΩ to 5 KΩ or high resistance values in the range from 1 MΩ to 5 MΩ. Hence, a reference resistance with a value of 100 KΩ between the two ranges are chosen. The currents through the resistance of the crossbar point and reference resistance are different and produce the signal. This difference is amplified through a sensing circuit, which consists of two inverters connected in series. The current flows from the resistances to the sensing circuit are controlled by external signals. This control is implemented with two pass gates in Figure 2.17. Finally, since the sensing circuit requires equalization before sensing, a pass gate that connects the two inverters is implemented.

Figure 2.18 demonstrates the HSPICE simulation results of the read circuit. The left figure depicts the simulation results with resistance of the crossbar point of 1 KΩ. As shown in Figure 2.18, the output values of the two inverters are equalized until 1 ns and then differentiated with the two current flows when the two pass gates are enabled. The output is stabilized after 1.6 ns. The right figure demonstrates the simulation results with resistance of the crossbar point of 1 MΩ.

Generally, sense amplifiers are sensitive to transistor sizing. The sizing influences the slew rate of the differential amplifier, the power dissipation and the speed of operation. At first, the slew rate depends on the ratio between PMOS and NMOS of the latch. The PMOS and NMOS ratios can be controlled by the width manipulation since the length is generally fixed based

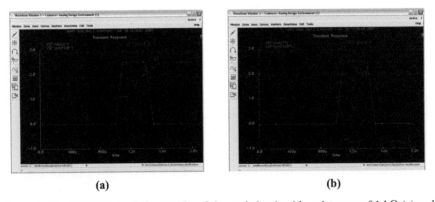

(a)   (b)

Figure 2.18   HSPICE simulation results of the read circuit with resistances of 1 kΩ (a) and 1 MΩ (b).

on the process. To control the slew rate, an equal width of the PMOS and NMOS was implemented. Then, the width of PMOS was gradually increased until the rising and falling time are equalized. In this simulation, the PMOS and NMOS width ratio was 1.6:1. Also, the equalization voltage is the major factor for the slew rate. A mid-voltage between Vdd and Gnd is applied for the equalization voltage.

After generating an optimal slew rate, gate sizing should be conducted for the optimal performance. In this simulation, sizing the gates M1, M2 and M5 are the main factors in the performance perspective. To be operated as a differential amplifier, the gate M5 should operate as a current source. The current source can be implemented with sizing gate M5 to be operated in the saturation region. In the saturation region, the current is not affected by Vds. In the sense amplifier circuit, it is needed to bias M5 so that the voltage fluctuation in the drain of M5 does not affect the current flow through M5. Then, when the M1 and M2 are enabled by the input A, there is current difference between the two paths that are attached to registers. Obviously, the bias voltage should be above the threshold voltage Vth. However, increasing the voltage near Vdd, the power dissipation will be worse. By contrast, biasing M5 near threshold voltage, there might be a chance for M5 operate in the triode region. Also, it will degrade the speed of the sense amplifier. In this sense, it is needed to find the optimal bias voltage of M5 for the power-delay product minimization. Based on the simulation results, M5 with 1.1V is biased. For the optimal power-delay product, the widths of the gates, M1, M2 and M5 should be chosen optimally. Increasing the widths of the gates, M1, M2 and M5, then the speed will be improved but this degrades the power dissipation and vice versa. In this simulation, the minimum size of width is selected for the savings because the writing time and reading time are more affected by the decoders and long array lines.

2.5 Memory Architecture

In this section, a 1-Kb (32 × 32) crossbar memory design using writing and reading circuits is presented. The 1-Kb crossbar memory design requires 5 bits for column decoder inputs and 5 bits for row decoder inputs. The column and row decoder inputs are decoded in 32 bits, respectively. The decoded bits are connected to 1-Kb arrays in the crossbar memory. By controlling the decoder inputs, one and only one of the 1-Kb array cells is selected.

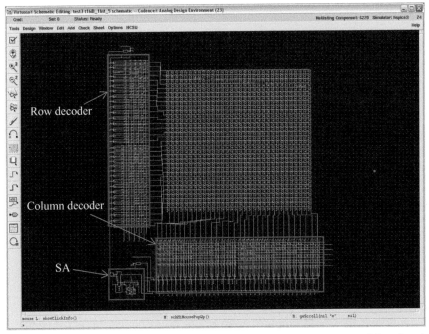

Figure 2.19   Schematic of a 1 Kb memory design with a single sense amplifier after the column decoder.

Two different approaches to design the 1-Kb crossbar memory design will be discussed. The first approach is to implement the memory with a single sense amplifier which is attached after the column decoder as shown in Figure 2.19. In the second approach, an array of sense amplifiers can be attached with each bit line, as shown in Figure 2.20.

The 1-Kb memory design in Figure 2.19 uses a single sense amplifier, and different bit lines are read after selection by the column decoder. This memory architecture uses fewer numbers of sense amplifiers than the memory in Figure 2.20 by the number of the bit line times. Hence, it has less area overhead and static power consumption than the 1-Kb memory design in Figure 2.20.

However, the memory as shown in Figure 2.20 requires less switching time for bit line changes. This is because of the sense amplifiers latch values that are ready for the decoder. Reading multiple bits sequentially, the design in Figure 2.20 only requires additional decoding time for additional read after initial sense amplification time. Therefore, when the bit line selection

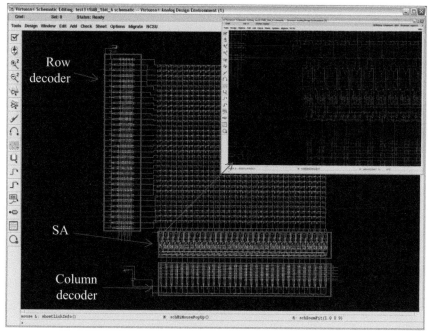

Figure 2.20 Schematic of a 1-Kb memory design with an array of sense amplifiers before the column decoder.

is changed, the output only requires an additional decoding time. However, in the memory design in Figure 2.19, when reading multiple bits in a word, extra time is needed for the sense amplifier to initialize when reading every single bit. Therefore, the memory design in Figure 2.19 is much slower than the design in Figure 2.20 when reading multiple bits sequentially.

Figure 2.21 describes a HSPICE simulation result of the memory, as shown in Figure 2.20. In Figure 2.21, curve (a) represents a signal of the sense amplifier connected with the selected decoder. The inverse signal of the sense amplifier is indicated as curve (b). The input signal that connects between sense amplifier and the column of the memory array is shown as curve (c). The output signal "high" is marked as curve (d).

The HSPICE simulation result of the memory in Figure 2.22 shows a case when the output signal is "low". In Figure 2.22, (a)–(c) can be referred to signals. However, (d) describes the output signal of the decoder. This is attributed to the threshold voltage of PMOS of the column decoder. Since the low output signal passes through the PMOS, the output signal of the decoder

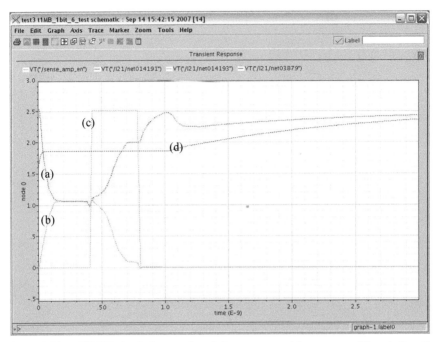

Figure 2.21   HSPICE simulation results of the memory with sense amplifiers before the column decoder for reading "high".

is not the ground voltage. To address this issue, a buffer is added after the decoder. The output signal after the output buffer is described as (e).

Table 2.1 shows area estimation for the 1-Kb crossbar memory design with sense amplifiers attached to all the bit lines. In this design, the sense amplifiers occupy $32 \times 20 \times 1.2$ um^2 of the area. The sense amplifiers occupy most area of the design as it uses 10 transistors to implement. The fact that horizontal width is much longer than vertical width is because the sense amplifiers are attached with all bit lines. Hence, the sense amplifiers determine the pitch width of the crossbar memory design.

Table 2.2 shows area estimation for the 1-Kb crossbar memory design with a single-sense amplifier in the final stage. In this design, the crossbar arrays occupy the most area of the design. Horizontal width and vertical width are similar in the design because the sense amplifier does not affect pitch width. Therefore, the 1-Kb crossbar memory design with a single-sense amplifier in the final stage occupies less area than the 1-Kb crossbar memory design with sense amplifiers before decoders, as discussed earlier.

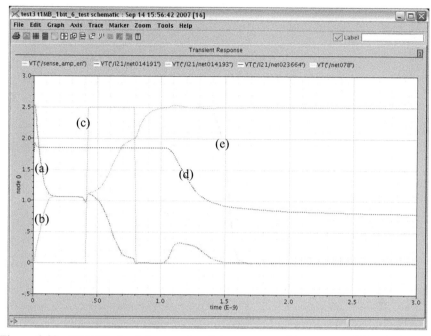

Figure 2.22 HSPICE simulation results of the memory with sense amplifiers before the column decoder for reading "low".

Table 2.1 Area composition of the 1 Kb crossbar memory design with a SA attached to all the bit lines

	Horizontal Width	Vertical Width
Array	$32 \times 1.2\ \mu$m	$6.4\ \mu$m
Decoder	$2\ \mu$m	$2\ \mu$m
SA	$32 \times 1.2\ \mu$m	$20\ \mu$m
Total	$68.8\ \mu$m	$28.4\ \mu$m

Table 2.2 Area composition of the 1 Kb crossbar memory design with a single SA at the final stage

	Horizontal Width	Vertical Width
Array	$6.4\ \mu$m	$6.4\ \mu$m
Decoder	$2\ \mu$m	$2\ \mu$m
SA	$6.6\ \mu$m	$2\ \mu$m
Total	$8.4\ \mu$m	$10.4\ \mu$m

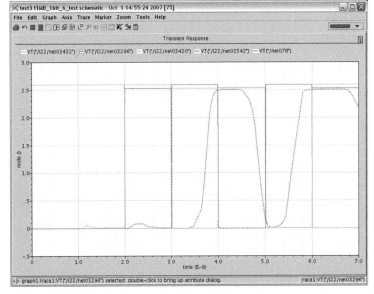

Figure 2.23   HSPICE simulation results of the 1 Kb crossbar memory design with sense amplifiers before the column decoder for reading.

The HSPICE simulation result on reading operation of the 1-Kb crossbar memory design with sense amplifiers before decoders, as shown in Figure 2.23. In Figure 2.23, the net078 with red line represents the output signal of the memory. The output signal changes its value according to the resistance of the crossbar memory cell. In this simulation, the crossbar memory cell is selected by the column address. In other words, the column address selects a crossbar memory cell, whose resistance affects the output signal. In Figure 2.23, the other nets represent column address bits. The column address bits change by 1 ns and the corresponding output also changes by 1 ns. In this simulation, the output can be read as "0101" and this output implies that the resistances of the target row are "low high low high".

Table 2.3 shows the comparison of reading times on both two crossbar memory architectures. As shown in Table 2.3, the crossbar memory architectures with sense amplifiers before decoders are approximately three times faster than the other architecture in reading 32 bits of 1-Kb crossbar memory. This originates from the difference in initial time for the sense amplifiers as discussed earlier. Since the crossbar memory design with a sense

Table 2.3 Comparison of data read time between the 1 Kb memories with an array of SAs before decoders and the 1-Kb memory design with a single SA after decoders

	SA after Decoders	SA before Decoders
Reading	32 ns + 4 ns	3.5 × 32 ns
Reading/bit	1.1 ns/bit	3.5 ns/bit

amplifier at the last stage requires initialization time for the sense amplifiers when reading a bit, this architecture demands initialization 32 times when reading 32 bits. On the other hand, the crossbar memory with sense amplifiers before decoders needs initialization only once when reading 32 bits. Hence, while the memory with sense amplifiers before decoders requires 3.5 ns per bit for reading, the other memory needs 1.1 ns per bit for reading.

The simulation results shown in Section 2.5 demonstrate the pros and cons between the two design approaches, SAs before decoder design and a SA after decoder design. Pros of one design become the cons for the other design and vice versa. Pros of SAs before decoder design are the fast processing time and better noise immunity and cons are the area overhead and increased power consumption due to larger number of SAs. The processing time for SAs before decoder design arises from early and multiple latches of the bit information. In this design, the latched information is decoded by decoders and as the decoder changes the address, the output only requires one time of SA processing for one row of information. However, the SA after decoder design requires the decoding time and SA processing time for every bit leading to much more time for reading and writing. In area overhead, the SA after decoder design requires one SA while the SAs before decoder design requires as many SAs as the number of bit lines which results in large area overhead. In the noise point of view, the SAs before decoder design are better than the other counterpart because the SAs before the decoder design hold the latched data before decoding that is connected to the power source passing through one closed transistor. However, the SA after decoder is vulnerable to noise because the decoded data are not connected to the power source directly through the transistor. In the power dissipation, the SAs before the decoder design induce a larger number of switching activities of SA for reading or writing within a period of reading a word because of the number of SA. In addition, the design consumes more static power due to leakage from gate or drain to source which will be discussed in the next section.

2.6 Power Dissipation

2.6.1 Power Estimation

Power dissipation of the memory system is one of the critical factors in performance measurement. The reason is that power dissipation affects how the circuit is designed, e.g. the number of circuits on a single chip, power-supply capacity, etc., and it also affects feasibility, cost and reliability.

In the crossbar memory system design, power dissipations of both the designs are compared. Since both designs have different circuit configuration, it is investigated how it affects the power dissipation. First, the power dissipation related with the sense amplifiers is measured. Subsequently, decoders and other circuits are examined for the power dissipation.

The power dissipation of the sense amplifiers can be categorized into three parts. One is the power dissipation from the reference voltage source that works for the differential amplification. This dissipation is continuous during the reading period because the reference voltage source should be enabled through the whole reading operation. Also, the reference voltage source is related with direct path power dissipation from the power supply to the ground. This direct path is constructed because the gate to source voltage is above the threshold voltage of the NMOS and there is no device that blocks the current path from power supply to the ground.

Another power dissipation source of the sense amplifiers is control signals on the equalization and the latch inside sense amplifiers. This power dissipation is correlated with dynamic power consumption since the control signal repeats on and off during the reading operation. The dynamic power consumption can be defined as

$$P_{dyn} = C \cdot V_{dd}^2 \cdot f$$

where C is load capacitance and f is switching frequency. The last power dissipation source of sense amplifier is static power consumption. Since the CMOS devices are not ideal, the NMOS and PMOS devices do not operate in steady state simultaneously.

Therefore, there is a leakage current from the drain to source or the substrate to source resulting from reverse-bias diode junctions of the transistors. Another source of leakage current is sub-threshold current of the CMOS devices. Even if the voltage across the gate and source is below the threshold voltage, there is a leakage current from drain to source. This leakage current depends on the threshold voltage. If the threshold voltage is close to zero volts, the leakage current is larger resulting in larger static power dissipation. For both sources of leakage current, the static power dissipation can be expressed as follows:

$$P_{stat} = I_{leakage} \cdot V_{dd}.$$

Figure 2.24 shows the static power dissipation of the 1-Kb crossbar memory design with a sense amplifier after the column decoders. The glitches in Figure 2.24 originate from the short period time when the input signal changes. The power dissipation of the sense amplifier part is shown in Figure 2.25.

Generally, width of PMOS and NMOS in a sense amplifier should be increased as size of an array increases because of the increased capacitance of bit line in an array. The current change during a read operation is decreased due to the increased capacitance of the bit line. It becomes more difficult for the sense amplifier to detect the current change, which affects the output voltage of the sense amplifier. However, the width of a sense amplifier is increased less than 1.5 times for the 1 Kb crossbar array compared with that of the smallest sense amplifier satisfying the design rules. It is observed that the amount of increment is enough to sense the current difference and does not increase time required for sense amplifier operation through the bit line

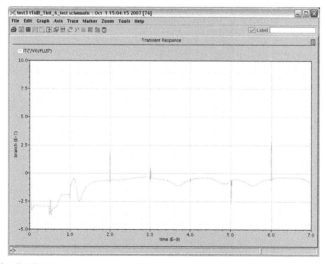

Figure 2.24 Static power dissipation of the SA part in the 1-Kb crossbar memory design with a SA after decoders from 0 ns to 7 ns.

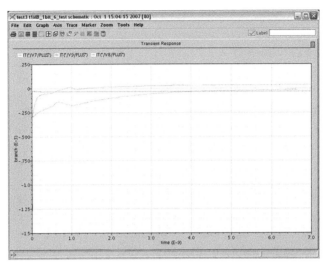

Figure 2.25   Power dissipation of the SA part in 1-Kb crossbar memories with a SA after decoders from 0 ns to 7 ns.

in the 1-Kb crossbar memory design. Therefore, it is needed to find the width of the sense amplifier in the 1-Kb crossbar memory designs that optimize the power-delay product for the crossbar memory design.

As shown in Table 2.4, total power dissipation of a sense amplifier after decoders is three times less than the other circuit configuration. The main reason is that the crossbar memory with sense amplifiers before decoders requires as many sense amplifiers as the bit lines. Since the number of sense amplifiers of crossbar memory design with sense amplifiers before decoders is 32 times more, the power dissipation of the memory is 32 times larger than the power dissipation of the counter design assuming that the changes of the input signal is equal during the operation period.

However, the variations of input signals can also be different during a reading operation. Input signals of the crossbar memory with sense amplifiers before decoders requires only one time equalization operation and one time enabling the latch operation for reading 32 bits. On the other hand, the crossbar memory with a sense amplifier at the final stage demands both equalization operation and enabling the latch operation every time reading a bit. Therefore, the time for reading 32 bits in the crossbar memory with a sense amplifier at the final stage is approximately 3 times longer than the counterpart. Data regarding the sense amplifiers in Table 2.4 shows what is mentioned above.

Table 2.4 Comparison of power dissipation between the two types of 1-Kb crossbar memory cell designs

	SA after Decoders	SA before Decoders
SA	1.5 mW	4.5 mW
Decoder	110 mW	310 mW
Others	4 mW	4 mW
Total	115.5 mW	318.5 mW

In the case of the decoder, power dissipation is dynamic power consumption. Variation of the decoder input affects the dynamic power dissipation. Variations of the decoder input in both the designs are different. The crossbar memory with a sense amplifier at the final stage alters its decoder input approximately three times less than the counterpart. Therefore, the power dissipation of the crossbar memory with a sense amplifier at the final stage is approximately three times less than the counterpart, as shown in Table 2.4. In total power dissipation, the crossbar memory with a sense amplifier at the final stage is approximately three times less than the counterpart design.

Considering the size of memory, the power dissipation illustrated in Table 2.4 is high. To build a larger memory array, it is needed to provide ways to reduce power dissipation at the expense of other performance. Since decoder consumes most of the power, reduction of power dissipation from the decoder will be most effective. One way is to replace the crossbar decoder with conventional CMOS decoders. Then, the area overhead advantage is lost but will save power more than 90% since the power dissipation in crossbar decoder comes from direct path current flow from Vdd to Gnd. Another way is to reduce the voltage source for reading. Then, saving the power, but the reading speed is reduced. Finally, power gating will be effective to the power saving for a large memory cells.

However, the power dissipation is measured for the worst case which means continuous writing to the crossbar memory cells. Also, when the crossbar memory is large (e.g. contains more than 50 billion memory cells), the power dissipation is not proportional to the size of memory directly since when reading or writing, it is needed to read or write a word each time. Hence, the issue is how effective minimization can be achieved in the static power dissipation while maintaining high performance. Since the array does not consume any power when the applied voltage is Gnd, the static power from decoder, SA and control logics are the only concerns. Therefore, the static power dissipation of the crossbar memory design is much less than the conventional memory designs.

As shown in Tables 2.3 and 2.4, there is a tradeoff between the reading time and the power dissipation. Superiority in the reading time results in inferiority in the power dissipation. The crossbar memory design with sense amplifiers before decoders shows better performance in the reading time, while it demonstrates the poorer performance in the power dissipation. Hence, the counterpart design shows the opposite performance in the reading time and the power dissipation.

2.6.2 Analytical Modeling on Static Power

In the crossbar memory design, power dissipation is one of important parameter for the memory. As discussed in Section 2.6.1, dynamic power dissipation and static power dissipation are the power dissipation types. In long channel devices, dynamic power is important since the leakage current is negligible in this case. However, as the device size scales down, static power dissipation becomes important. The absolute and the relative contribution of leakage power to the total power become more important since the leakage current increases exponentially with the technology scaling. Based on International Technology Roadmap for Semiconductors, the leakage current will contribute 50% of the total power dissipation in the next generation.

In the ideal transistor, current only flows when *Vgs* is greater than *Vt*. However, the current still flows in the real transistor even below the threshold voltage. This conduction can be expressed as given in Equations (2.4) and (2.5), where V_T is thermal voltage [37].

$$I_{ds} = I_{ds0} e^{\frac{Vgs-Vt}{nV_T}} \left(1 - e^{\frac{-Vds}{V_T}} \right) \tag{2.4}$$

$$I_{ds0} = \beta V_T^2 e^{1.8} \tag{2.5}$$

I_{ds0} is the current at threshold voltage and this value varies with process and device geometry. Also, *n* is a process-dependent term depending on the depletion region characteristics. For the equations, it is thought that the leakage current is a function of temperature and threshold voltage when turning off the device because the gate voltage is zero and V_{ds} is V_{dd} for the CMOS operations.

In the beginning, row and column decoders can be analytically modeled. To investigate the decoders, it is needed to look into the circuit at the gate level. As shown in Figure 2.26, when a selecting bit is enabled, the leakage power dissipation occurs in the upper NMOS and PMOS which are circled.

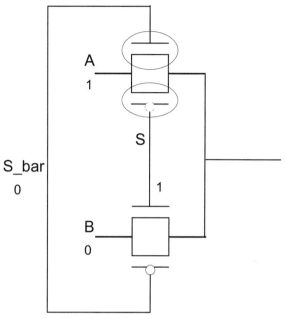

Figure 2.26 A 2 × 1 MUX for power selection which is connected to row decoders.

This power dissipation can be expressed as given in Equation (2.6).

$$I_{LMUX} = I_{LNMOS} + I_{LPMOS} \tag{2.6}$$

Considering the width and length of the device, the equation can be transformed as given in Equation (2.7).

$$I_{LMUX} = \frac{W_1}{L_1} I_{LNMOS} + \frac{W_2}{L_2} I_{LPMOS} \tag{2.7}$$

Since the middle voltage source is applied for the MUX, the leakage power dissipation of the MUX can be expressed as given in Equation (2.8).

$$P_{LMUX} = \frac{W_1}{L_1} V_M I_{LNMOS} + \frac{W_2}{L_2} V_M I_{LPMOS} \tag{2.8}$$

In the case of 3x1 mux shown in Figure 2.3, the leakage current can be calculated in the same way and can be expressed as given in Equation (2.9).

$$I_{LMUX3} = \frac{W_1}{L_1} I_{LNMOS} + \frac{W_2}{L_2} I_{LPMOS} \tag{2.9}$$

Since three power sources are used, the power dissipation for writing "1" is expressed as given in Equation (2.10).

$$P_{LMUX3} = \frac{W_1}{L_1}V_H I_{LNMOS} + \frac{W_2}{L_2}V_H I_{LPMOS} \tag{2.10}$$

For reading, the power dissipation can be expressed as given in Equation (2.11).

$$P_{LMUX3} = \frac{W_1}{L_1}V_L I_{LNMOS} + \frac{W_2}{L_2}V_L I_{LPMOS} \tag{2.11}$$

However, when writing "0", the leakage current is ignored since the power source is connected to Gnd in row decoders.

In the case of decoders, static power dissipation occurs since the power source is directly connected through diode. Assuming the resistance of the diode is R, the power dissipation through one diode can be expressed as given in Equation (2.12).

$$P_{DIODE} = \frac{V_H^2}{R} \tag{2.12}$$

The number of addresses, N_{addr}, decides the number of diodes in the decoder and the power dissipation from the diodes with N address decoders can be represented as given in Equation (2.13).

$$P_{DNaddr} = N \log_2 N \frac{V_H^2}{R} \tag{2.13}$$

For the column decoders, since the MUX uses high voltage when writing "0", the leakage power dissipation can be expressed as given in Equation (2.14).

$$P_{LCMUX} = \frac{W_1}{L_1}V_H I_{LNMOS} + \frac{W_2}{L_2}V_H I_{LPMOS} \tag{2.14}$$

Assuming the number of columns is the same as the number of rows, the power dissipation from the diodes in the column decoders can be represented as given in Equation (2.15).

$$P_{DCaddr} = N \log_2 N \frac{V_H^2}{R} \tag{2.15}$$

Secondly, the sense amplifier is analyzed for the leakage current. As shown in Figure 2.17, the sense amplifier should be analyzed based on the operation mode. At first, an idle mode is defined when A = 0, B = 0, C = 1, D = 0 and E = 0. In the idle mode, two NMOS FETs in the differential amplifier connected with A suffer from the leakage current and the NMOS connected with E also suffers from the leakage current. The leakage power dissipation in the idle mode can be expressed as given in Equation (2.16).

$$P_{SEIDLE} = 2\frac{W_1}{L_1}V_H I_{LNMOS} + \frac{W_2}{L_2}V_H I_{LNMOS} \qquad (2.16)$$

Next, the precharging mode should be analyzed and this mode is defined when A = 0, B = 0, C = 1, D = 1 and E = 0. In this mode, two NMOS FETs in the differential amplifier connected with A suffer from the leakage current and the NMOS connected with E also suffers from the leakage current. The leakage power dissipation in the idle mode can be expressed as same as the idle mode leakage power given in Equation (2.17).

$$P_{SEPRE} = 2\frac{W_1}{L_1}V_H I_{LNMOS} + \frac{W_2}{L_2}V_H I_{LNMOS} \qquad (2.17)$$

In the reading mode, the transistors that are connected with A, B, C, D and E are enabled. In this mode, the transistor M5 suffers from the leakage current since C is in between V_{th} and V_{dd}. The latch suffers from the leakage current since either of one node of the latch should be zero. Two transistors in the off state in the latch will suffer from the leakage current. M6 and M9 or M7 and M8 depending on the output value suffer from the leakage current in the latch. Therefore, the leakage power dissipation in the read mode can be expressed as given in Equation (2.18).

$$P_{SEREAD} = \frac{W_1}{L_1}V_H I_{2NMOS} + \frac{W_2}{L_2}V_H I_{LNMOS} + \frac{W_3}{L_3}V_H I_{LPMOS} \qquad (2.18)$$

In Equation (2.18), I_{L2NMOS} is much higher than I_{LNMOS} or I_{LPMOS} as V_{gs} is not zero in this device. Therefore, the sizing M5 is very significant to minimize the leakage current of the sense amplifier.

Considering the number of sense amplifiers connected to all the bit lines, the power dissipation will be increased as the number of bit lines are increased. Assuming that one bit in each read operation is read without any MUX and the sense amplifiers are connected to the bit lines, the leakage power dissipation can be represented as given in Equation (2.19).

$$P_{SEREAD} = N\left(\frac{W_1}{L_1}V_H I_{L2NMOS} + \frac{W_2}{L_2}V_H I_{LNMOS} + \frac{W_3}{L_3}V_H I_{LPMOS}\right)$$
$$(2.19)$$

Considering the frequencies in the idle, precharging and read mode, those frequencies are combined to generate the overall leakage dissipation in the sense amplifiers. Assuming the idle time, Ti, precharging time, Tp, and reading time, Tr, then the overall leakage power dissipation can be written as given in Equation (2.20).

$$P_{SE} = \frac{N \cdot Tr}{Ti + Tp + Tr} \left(\frac{W_1}{L_1} V_H I_{L2NMOS} + \frac{W_2}{L_2} V_H I_{LNMOS} + \frac{W_3}{L_3} V_H I_{LPMOS} \right)$$
$$\frac{N \cdot (Tp + Ti)}{Ti + Tp + Tr} \left(2 \frac{W_1}{L_1} V_H I_{LNMOS} + \frac{W_2}{L_2} V_H I_{LNMOS} \right) \tag{2.20}$$

It is intriguing to research the effect of the usage of MUX logics to reduce the number sense amplifier on the leakage current. A 2 by 1 MUX is used, then the number of sense amplifiers becomes half while additional leakage current may be added from the 2 by 1 MUX. Therefore, the leakage current of both sense amplifiers and the 2 by 1 MUX can be represented as given in Equation (2.21).

$$P_{SE\_MUX} = \frac{N \cdot Tr}{2(Ti + Tp + Tr)}$$
$$\cdot \left(\frac{W_1}{L_1} V_H I_{L2NMOS} + \frac{W_2}{L_2} V_H I_{LNMOS} + \frac{W_3}{L_3} V_H I_{LPMOS} \right)$$
$$+ \frac{N \cdot (Tp + Ti)}{2(Ti + Tp + Tr)} \left(2 \frac{W_1}{L_1} V_H I_{LNMOS} + \frac{W_2}{L_2} V_H I_{LNMOS} \right)$$
$$+ \frac{W_1}{L_1} V_M I_{LNMOS} + \frac{W_2}{L_2} V_M I_{LPMOS} \tag{2.21}$$

In a similar manner, extending to an N by 1 MUX will leads to the leakage power dissipation as given in Equation (2.22).

$$P_{SE\_MUX} = \frac{Tr}{(Ti + Tp + Tr)}$$
$$\cdot \left(\frac{W_1}{L_1} V_H I_{L2NMOS} + \frac{W_2}{L_2} V_H I_{LNMOS} + \frac{W_3}{L_3} V_H I_{LPMOS} \right)$$
$$+ \frac{(Tp + Ti)}{(Ti + Tp + Tr)} \left(2 \frac{W_1}{L_1} V_H I_{LNMOS} + \frac{W_2}{L_2} V_H I_{LNMOS} \right)$$
$$+ \log_2 N \cdot \left(\frac{W_1}{L_1} V_M I_{LNMOS} + \frac{W_2}{L_2} V_M I_{LPMOS} \right) \tag{2.22}$$

Based on the analytical modeling, it is imperative to properly size the transistor and the diode characteristic device to minimize the leakage power dissipation. Especially, the NMOS and PMOS of the latch can be sized to reduce leakage current, and the bias transistor for constant current source should be carefully sized to meet the performance criteria.

2.7 Noise Analysis

Noise has been an issue in the design of digital circuits of comparable importance to timing and power. Increased noise has been attributed to increasing interconnect densities, faster clock rates and scaling threshold voltages. Increasing interconnect densities represent a significant increment in coupling capacitance. Faster clock rates mean faster slew rate, and threshold voltages scale lower with scaling supply voltages. These effects merge to create more sources of on-chip noise.

In the crossbar memory architecture, the array of the memory cells contains dense metal lines. Interconnect density is increased due to the dense metal lines. The increased interconnect density becomes a source of on-chip noise which is called interconnect coupling noise.

The coupling noise is mainly due to capacitive coupling between metal lines. This noise is more relevant to the crossbar memory since it contains more cells than conventional memories resulting in denser interconnect capacitance.

To examine the interconnect capacitance in the crossbar memory design ignoring C_2, which simplifies the RC model of the amorphous Silicon crossbar memory as shown in Figure 2.27.

Assuming that the upper metal line is an aggressor while the bottom one is a victim, a circuit can be modeled as a capacitive voltage divider to compute the victim noise as shown in Figure 2.28. The aggressor's voltage becomes V_{high} as high voltage is applied to the selected row. Assuming a victim is a floating node, ΔV_{victim} without a victim's driver is represented in Equation (2.23).

$$\Delta V_{VICTIM} = \frac{C_6}{\frac{(2C_1+C_3)\cdot C_4}{(2C_1+C_3)+C_4} + C_6} \Delta(V_{AGGRESSOR}) \qquad (2.23)$$

Assuming an $N \times N$ array and V_{mid} is one half of V_{high}, Equation (2.23) can be modified as Equation (2.24).

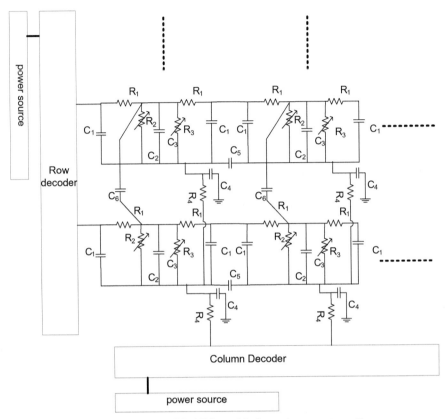

Figure 2.27   Simplified RC model of crossbar memory cells.

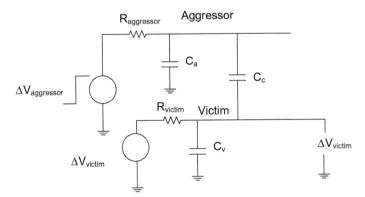

Figure 2.28   Interconnect coupling model.

Since V_{mid} to the victim is biased, V_{mid} is one half of V_{high}, Equation (2.23) can be modified as Equation (2.24).

$$\Delta V_{VICTIM} = \frac{C_6}{\frac{(2C_1+C3)\cdot C_4}{2C_1+C_3+C_4} + C_6} \Delta(V_{AGGRESSOR}/2) \frac{1}{\alpha+1} \qquad (2.24)$$

where

$$\alpha = \frac{\tau_{aggressor}}{\tau_{victim}} = \frac{R_{aggressor}}{R_{victim}}$$

Assuming $R_{aggressor}$ and R_{victim} are the same, Equation (2.24) is modified to Equation (2.25).

$$\Delta V_{VICTIM} = \frac{C_6}{2\left(\frac{(2C_1 + C_3) \cdot C_4}{(2C_1 + C_3) + C_4} + C_6\right)} \Delta(V_{AGGRESSOR}/2) \qquad (2.25)$$

Based on Equation (2.25), the victim's voltage deviation by the aggressor depends on the capacitance C_1, C_3 and C_6. Since C_3 is at least three times bigger than the other capacitances, the worst case victim's voltage deviation by the aggressor will be around $0.064 V_{aggressor}$. Hence, the crossbar memory is quite immune to the interconnect noise even if it has higher cell density than the competing memories.

2.8 Area Overhead

Since the crossbar memory with sense amplifiers attached to all the bit lines uses large area, MUX logics is added to reduce the number of sense amplifiers. Proper choice of the MUX size can lead to memory with high performance in terms of reading time, power dissipation and area overhead. Before simulating the whole 1-Kb crossbar memory design with this modification, a simplified memory circuit is discussed for analytical understanding.

Figure 2.29 represents a simplified sense amplifier part of a 1 Kb crossbar memory design with sense amplifiers before decoders. In Figure 2.29, the crossbar memory cell is modeled with a resistor for this simulation. The MUX is used to select one of the two cells which are represented as resistors. Figure 2.29 uses 2 by 1 MUX so that the number of the sense amplifiers can be reduced by half.

Simulation results in Figure 2.30 are transient results of the circuit in Figure 2.29. The first graph of Figure 2.30 is the output signal of Figure 2.29.

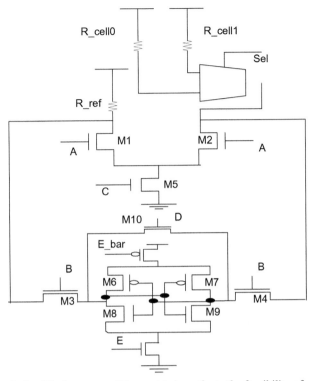

Figure 2.29   A simplified sense amplifier part to investigate the feasibility of usage of MUX logics to reduce the numbers of the sense amplifiers attached to all the bit lines.

As shown in Figure 2.30, the output is changed based on selection of the MUX signal in the third graph of Figure 2.30. The second graph of Figure 2.30 is a control signal for the equalization.

Figure 2.31 is a schematic of a 1-Kb crossbar memory design using 2 by 1 MUX logics for reducing the number of sense amplifiers by half. Principles of the circuit in Figure 2.29 are applied to a 1 Kb crossbar memory circuit. In Figure 2.31, the former 16 bits are read and change the selection bit of the MUX logics, and then the latter 16 bits are read sequentially. Also, the column decoder is reduced by one half and the decoder inputs are reduced from 10 to 8. There should be a routing overhead to connect the MUX logics, the array, and the column decoder. The reading time is increased compared with the memory in Figure 2.20 because it requires the additional time to select the MUX logics and initialization time for the sense amplifiers.

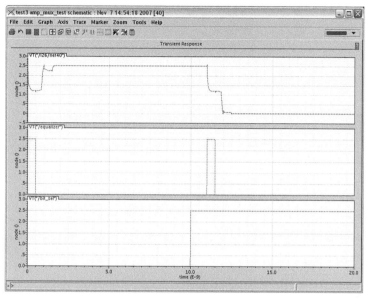

Figure 2.30 Simulation results of the simplified sense amplifier with MUX logics.

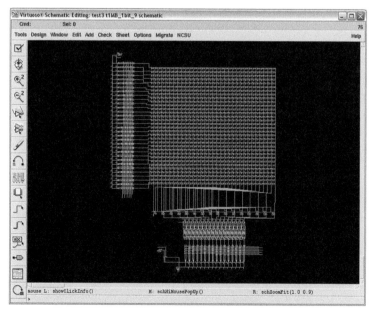

Figure 2.31 Schematic of the 1-Kb crossbar memory design with using 2 by 1 MUX logics to reduce the numbers of sense amplifiers by one half.

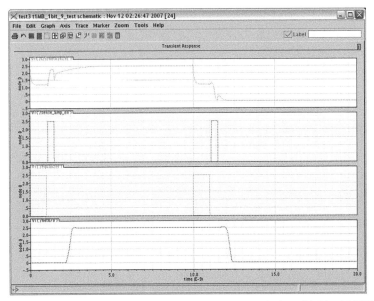

Figure 2.32   Simulation results on the 1-Kb crossbar memory design with 2 by 1 MUX logics to reduce the numbers of sense amplifiers.

Figure 2.32 represents the simulation results on the crossbar memory shown in Figure 2.31. The simulation results in Figure 2.32 are the transient results of the circuit in Figure 2.31. The first graph of Figure 2.32 is the output signal of Figure 2.31. As shown in Figure 2.32, the output is changed based on the selection of the MUX signal in the fourth graph of Figure 2.32. The third graph of Figure 2.32 is a control signal for the equalization, and the second graph of Figure 2.32 is a control signal for the latch inside of the sense amplifier.

Figure 2.33 is a schematic of the 1-Kb crossbar memory design using 4 by 1 MUX logics for reducing the number of sense amplifiers to one quarter. In Figure 2.33, the former 8 bits are read and change the selection bit of the MUX logics. The latter 8 bits are read sequentially until 32 bits are read. Also, the column decoder is reduced to one quarter and the decoder inputs are reduced from 10 to 6. The routing overhead to connect the MUX logics, the array and the column decoder is increased more than that of the memory design in Figure 2.31. The reading time is increased compared with the memory in Figure 2.31 because it requires a longer time to select the MUX logics and a longer initialization time for the sense amplifiers.

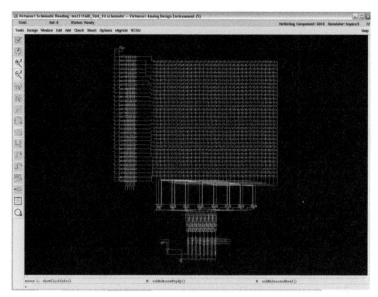

Figure 2.33 Schematic of the 1-Kb crossbar memory design with using 4 by 1 MUX logics to reduce the numbers of sense amplifiers by one quarter.

Figure 2.34 represents the simulation results on the crossbar memory shown in Figure 2.33. The simulation results in Figure 2.34 are the transient results of the circuit in Figure 2.33. The first graph of Figure 2.34 is the output signal of Figure 2.33. As shown in Figure 2.34, the output is changed based on the selection of the MUX signal in the fourth graph of Figure 2.34. The third graph of Figure 2.34 is a control signal for the equalization, and the second graph of Figure 2.34 is a control signal for the latch inside the sense amplifier. As shown in Figure 2.34, the crossbar memory using 4 by 1 MUX logics requires longer time to initialize and enable the latch inside the sense amplifier.

Figure 2.35 is a schematic of the 1-Kb crossbar memory design using 8 by 1 MUX logics for reducing the number of sense amplifiers to one eighth. In Figure 2.35, the former 4 bits are read and change the selection bit of the MUX logics, and then the latter 4 bits are read sequentially until 32 bits are read. Also, the column decoder is reduced to one eighth, and the decoder inputs are reduced from 10 to 4. The routing overhead to connect the MUX logics, the array and the column decoder is increased more than that of the memory in Figure 2.33. The reading time is increased compared to the

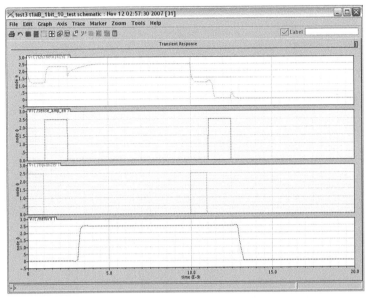

Figure 2.34   Simulation results on the 1-Kb crossbar memory design with 4 by 1 MUX logics to reduce the numbers of sense amplifiers.

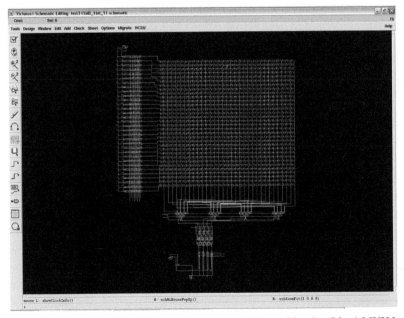

Figure 2.35   Schematic of the 1-Kb crossbar memory design with using 8 by 1 MUX logics to reduce the numbers of sense amplifiers by one quarter.

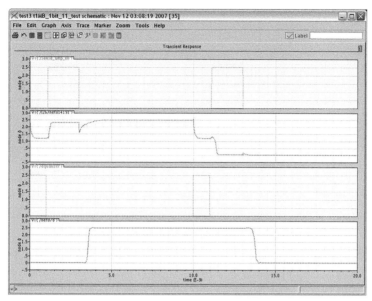

Figure 2.36 Simulation results on the 1-Kb crossbar memory design with 8 by 1 MUX logics to reduce the numbers of sense amplifiers.

memory in Figure 2.33 because it requires a longer time to select the MUX logics and a longer initialization time for the sense amplifiers.

Figure 2.36 represents the simulation results on the crossbar memory shown in Figure 2.35. The simulation results in Figure 2.36 are the transient results of the circuit in Figure 2.35. The first graph of Figure 2.36 is the output signal of Figure 2.35. As shown in Figure 2.36, the output is changed based on the selection of the MUX signal in the fourth graph of Figure 2.36. The third graph of Figure 2.36 is a control signal for the equalization, and the second graph of Figure 2.36 is a control signal for the latch inside the sense amplifier. As shown in Figure 2.36, the crossbar memory using 8 by 1 MUX logics requires a longer time to initialize and enable the latch inside the sense amplifier than does the crossbar memory using 4 by 1 MUX logics.

Table 2.5 shows the area comparison among the 1-Kb crossbar memory design using various sizes of MUX logics. As shown in Table 2.5, the best memory design for area overhead is the crossbar memory using 8 by 1 MUX logics. The larger size of MUX will reduce the pitch width of the 1-Kb crossbar memory design. However, it will also increase the area overhead of routing. If the size of MUX is above 8 by 1, the pitch width of the 1 Kb crossbar memory design will not be further decreased since the width of sense

Table 2.5 Area comparison among the 1-Kb crossbar memory designs with various sizes of MUX

Unit: μm^2	No Mux	2 by 1 Mux	4 by 1 Mux	8 by 1 Mux	16 by 1 Mux
Array	$32^2 \times 0.2 \times 4.8$	$32^2 \times 0.2 \times 2.4$	$32^2 \times 0.2 \times 1.2$	$32^2 \times 0.2 \times 0.6$	$32^2 \times 0.2 \times 0.3$
Decoders	$(32 \times 1 \times 0.2)$ $+(32 \times 1 \times 4.8)$	$(32 \times 1 \times 0.2)$ $+(32 \times 0.8 \times 2.4)$	$(32 \times 1 \times 0.2)$ $+(32 \times 0.6 \times 1.2)$	$(32 \times 1 \times 0.2)$ $+(32 \times 0.4 \times 0.6)$	$(32 \times 1 \times 0.2)$ $+(32 \times 0.2 \times 0.3)$
SA	$32 \times 4.8 \times 10$	$32 \times 10 \times 2.4$	$32 \times 10 \times 1.2$	$32 \times 10 \times 0.6$	$32 \times 10 \times 0.3$
Mux	0	$32 \times 10 \times 2.4$	$32 \times 25 \times 1.2$	$32 \times 61 \times 0.6$	$32 \times 153 \times 0.3$
Global Routing	0	$32 \times 1 \times 2.4$	$32 \times 2 \times 1.2$	$32 \times 4 \times 0.6$	$32 \times 8 \times 0.3$
Total	2679.04	2172.16	1696	1557.76	1682.56
Cell/ peripheral	0.58	0.29	0.17	0.08	0.03

amplifiers will become less than the width of the array. Therefore, the 1 Kb crossbar memory design using 8 by 1 MUX logics occupies the smallest area with 67 percent of the area being reduced. However, this ratio between cell area to peripheral circuit area is proper to be used as memory chips.

Figure 2.37 represents the simulation results with the changes of the MUX selection bits and addresses on the 1Kb crossbar memory design using 8 by 1 MUX logics. The first graph of Figure 2.37 depicts the first selection bit of the MUX logics and the second graph of Figure 2.37 shows the second selection bit of the 8 by 1 MUX logics. The third graph of Figure 2.37 represents output signal of the 1Kb crossbar memory design. Also, the fourth graph and last graph of Figure 2.37 demonstrate the first decoder bit and the control signal for enabling the latch inside the sense amplifiers, respectively.

Finally, Table 2.6 shows the area comparison among the 1-Mb crossbar memory designs using various sizes of MUX logic. As shown in Table 2.5, the area overhead is decreased as the size of MUX is increased. Reduction in the area overhead is achieved since the area portion of the array in the memory is increased as the size of the memory is enlarged. In other words, the area portion of the peripheral circuitry is decreased so that the area overhead of MUX logic is not critical in the overall area overhead. Also, the ratio between cell area and peripheral circuit area is appropriate for a 1-Mb crossbar memory circuit with MUX logic.

Table 2.7 shows the estimated read time of 1-Mb crossbar memory designs by extrapolating the read time of 1 Kb crossbar memory designs. The read time difference between the worst case and best case is around 1.6 ns. Compared with area deviation with various sizes of MUX, the read

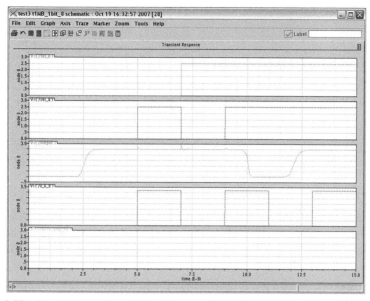

Figure 2.37 Simulation results with the changes of the MUX selection bits and addresses on the 1Kb crossbar memory design with 8 by 1 MUX logics.

Table 2.6 Area comparison among the 1-Mb crossbar memory designs with various sizes of MUX

Unit: μm^2	No Mux	2 by 1 Mux	4 by 1 Mux	8 by 1 Mux	16 by 1 Mux
Array	$1K^2 \times 0.2 \times 4.8$	$1K^2 \times 0.2 \times 2.4$	$1K^2 \times 0.2 \times 1.2$	$1K^2 \times 0.2 \times 0.6$	$1K^2 \times 0.2 \times 0.4$
Decoders	$(1K \times 2 \times 0.2)$ $+(1K \times 2 \times 4.8)$	$(1K \times 2 \times 0.2)$ $+(1K \times 1.6 \times 2.4)$	$(1K \times 2 \times 0.2)$ $+(1K \times 1.2 \times 1.2)$	$(1K \times 2 \times 0.2)$ $+(1K \times 0.8 \times 0.6)$	$(1K \times 2 \times 0.2)$ $+(1K \times 0.4 \times 0.4)$
SA	$1K \times 4.8 \times 10$	$1K \times 10 \times 2.4$	$1K \times 10 \times 1.2$	$1K \times 10 \times 0.6$	$1K \times 10 \times 0.4$
Mux	0	$1K \times 10 \times 2.4$	$1K \times 25 \times 1.2$	$1K \times 63 \times 0.6$	$1K \times 200 \times 0.4$
Global Routing	0	$1K \times 1 \times 2.4$	$1K \times 2 \times 1.2$	$1K \times 4 \times 0.6$	$1K \times 8 \times 0.6$
Total	1017600	534640	286240	167080	169360
Cell/ peripheral	16.7	8.78	5.2	2.6	0.9

Table 2.7 Read time comparison among the 1-Mb crossbar memory designs with various sizes of MUX

	No Mux	2 by 1 Mux	4 by 1 Mux	8 by 1 Mux	16 by 1 Mux
Read time	∼6.4 ns	∼6.6 ns	∼6.8 ns	∼7.2 ns	∼8.0 ns

Table 2.8 Comparison of total area of 1-Mb crossbar memory designs according to CMOS process technology

Unit: μm^2	No Mux	2 by 1 Mux	4 by 1 Mux	8 by 1 Mux	16 by 1 Mux
250 nm	1060905	534640	286240	165880	113800
130 nm	1020201	516761.6	263260.2	147404.8	90134.8
45 nm	1008442	505109.2	253205.2	128705.9	69526.19

time deviation is not so large. Therefore the 1 Mb crossbar memory design with 8 by 1 MUX is the best choice compromising the area overhead and fast read time.

Since a 0.25-μm process library is used for the peripheral circuit design, it is worthwhile noting how the total area is changed as process libraries change. Extrapolating the peripheral circuit area based on the gate length ratio between the process libraries, it is observed that the total area is minimized using 16 by 1 MUX logics in Table 2.8. Based on this result, the total area can be changed depending on the process libraries and the optimal MUX logic is also changed as the process library is changed.

2.8.1 Bank-based System Design

From the completed 1-Kb crossbar memory design demonstrated in the previous sections, how much larger systems can be designed is shown. A large system design is studied using bank based system design. A complete 4-Kb crossbar memory system has been simulated due to the working memory space limitation for simulation; however, this technique can also be applied to make a large array such as 1 Gb or even more with the same design technique. Figure 2.38 shows the 4-Kb crossbar memory implemented with 4 banks. Each bank is selected by the 4 by 1 MUX on the top. Since bank-based design removes the huge delay to access a large array, this technique is very essential for a large memory design.

The reading time comparison is demonstrated between the 1-Kb crossbar memory design and the 4-Kb crossbar memory design with four banks, where each bank consists of a 1-Kb crossbar memory design. The 4-Kb crossbar memory can be implemented with four 1-Kb memory designs using a 4 by 1 MUX to select each 1-Kb memory design which is called a bank. As shown in Table 2.9, the reading time per bit is not changed significantly with the addition of the 4 by 1 MUX. This is because the reading time of 32 × 4 bits is much longer than the propagation delay time of the 4 by 1 MUX.

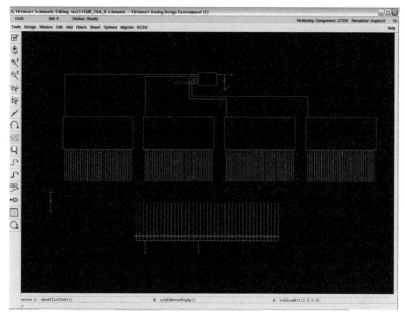

Figure 2.38 Schematic view of 4 Kb crossbar memories with 4 banks.

Table 2.9 Read time comparison between 1 Kb and 4 Kb with 4 banks

	1 Kb	4 Kb (4 banks)
Read	32 ns + 4 ns	$4 + 32 + 3(0.5 + 4 + 32)$ ns
Read per bit	1.1 ns/bit	1.14 ns/bit

Using the bank-based system design technique for a large memory system design, it is needed to have basic memory cells, decoders, bank selection address, column and row addresses, and global data bus and global amplifier and driver for input and output and control circuit which is usuallyimplemented in CPU. For a large memory system design with crossbar memory technique, all the core components which consist of the complete memory system are provided in this chapter.

Assuming using a bank design system with 1 ns of global routing delay and three stage hierarchies to make 1 Gb crossbar memories with a cell size of 200 nm by 200 nm, the read time and area overhead with an unit array size are described in Table 2.10. The read time is increased as the size of the unit array is increased due to the increased time to access the target crossbar cell for read or write through decoders. In contrast, the delay due to the MUX

Table 2.10   Read time and area overhead comparison for the optimal size of an unit array to make 1 Gb crossbar memories with a cell size of 200 nm × 200 nm

	256 Kb	1 Mb	4 Mb	16 Mb
Read time	∼ 7.4 ns	∼9 ns	∼ 2.5 ns	∼17 ns
Area	∼1.1 cm^2	∼0.9 cm^2	∼0.7 cm^2	0.6 cm^2
Read time × Area	8.14 ns · cm^2	8.1 ns · cm^2	8.75 ns · cm^2	10.2 ns · cm^2

is decreased as the size of the unit array is increased because of the reduced size of the MUX. The delay due to the MUX is not comparable to the delay due to the increased decoder size and the array size so that the overall delay is increased as the size of the unit array is increased. The area overhead is decreased as the unit array size is increased since the area overhead due to the peripheral circuitry is decreased. Therefore, the optimal unit array size can be obtained by finding the minimum product of the read time and the area overhead. The estimated optimal unit array size to make 1 Gb memories is 1 Mb as shown in Table 2.10.

Since the peripheral CMOS circuit takes around 800 ps to read or write in a 1-Kb crossbar memory, the read time can be extrapolated for a 1-Mb crossbar memory design. As the time for crossbar memory cell access excluding CMOS circuit will be around 200 ps, the read time can be around 7.2 ns. Assuming a bank design system is used with 1 ns of global routing delay and four stage hierarchies with 8 by 1 MUXes which depends on the process library, read time for a 4-Gb crossbar memory can be calculated as shown in Table 2.11. Also, read time for a 32-Gb crossbar memory can be calculated in the same manner, as shown in Table 2.11. From the area data from Table 2.8, it is estimated how many cells can be inserted in 1×1 cm^2 using 16 by 1 MUX logics as shown in Table 2.12.

For power dissipation, accessing one word each read or write operation, the total power dissipation is only increased by MUXes for 1 Gb compared with 1 Mb. Considering the power measurement for 1 Kb, the power dissipation for 1Gb is estimated as around 7 W. However, this power dissipation depends on the decoder designs and the design structure. The power dissipation can be lowered to be less than 1 W by sizing the SA and changing the decoder design at the expense of read or write time. This will be the future work of this chapter.

However, it is problematic to incorporate library cells to the crossbar memory design as crossbar memory metal line width is narrower than the feature size for the 250 nm process library. This problem can be solved as the

Table 2.11 Estimated read time for 4-Gb and 32-Gb crossbar memory designs

	4 Gb	32 Gb
250 nm	~9 ns	~10 ns
130 nm	~6.6 ns	~7.6 ns
45 nm	~5 ns	~6 ns

Table 2.12 Estimation of memory capacity to be realized in a 1 × 1-cm die

Cell size	200 nm × 200 nm	100 nm × 100 nm	10 nm × 10 nm
2-D technology	~1.3 G	~4 G	~350 G
3-D technology	~2.6 G	~8 G	~700 G

feature size is decreased. With cutting-edge technology, the 45-nm process, the crossbar memory cell with the library cell can be incorporated. This situation will become even better as the feature size is decreased in the next technology generation.

2.9 Technology Comparison

Finally, the technology parameters of the amorphous crossbar memory are compared with the conventional memories and other types of crossbar memory. In the cell area perspective, the amorphous crossbar memory demonstrates better size than other memories. Even if the conventional memories are scaled down to 45 nm in 2008, the cell area of the amorphous crossbar memory is still less than the conventional memories. MRAM has been scaled down to 90 nm with toggle mode, but it has half-select problem which is caused by the induced field overlaps between adjacent cells in the small area. This problem is the barrier for MRAM scaling even if an advanced technique such as spin-torque-transfer is researched. MRAM has another problem in the cost as described in the introduction. It requires fundamental changes in the fabrication facilities resulting in a barrier to the investors. In terms of writing and reading times, the amorphous silicon represents similar performance compared with CMOS-based or other emerging memory technologies. The power dissipation for data retention of the amorphous crossbar memory is zero as the other nonvolatile memories.

PRAM (not shown in Table 2.13) may provide comparable cell area and writing and reading performance. However, the most difficult problem of PRAM is that it requires high programming density. Additionally, it is vulnerable to temperature in the fabrication process and the user operation. This affects the performance of the device severely [14–17].

Table 2.13   Technology comparison between the amorphous crossbar memories and the conventional memory or other candidate memory for the future [14–17]

Category	Parameter	SRAM (130 nm)	DRAM (130 nm)	NOR Flash (130 nm)	MRAM (180 nm)	Crossbar
Cost	Cell area	$0.16\,\mu m^2$	$0.14\,\mu m^2$	$0.19\,\mu m^2$	$0.7–1.4\,\mu m^2$	$0.005–0.01\mu m^2$
	Cost/Mb	~80 cent	~20 cent	~3 cent	50 \$	~1 cent
	Process cost adder	0%	25%	25%	25%	<5%
Performance	Read access	5–10 ns	10–20 ns	80 ns	5–20 ns	5–10 ns
	Write cycle	3.4 ns	20 ns	1 ns	5–20 ns	5–10 ns
Power	Data retention	0.6 nA per bit at 85 °C	0.2 nA per bit at 85 °C	0	0	0
Miscellaneous	Data retention	Volatile	Volatile	Nonvolatile	Nonvolatile	Nonvolatile

References

[1] P. J. Kuekes and R. S. Williams and J. R. Heath, "Molecular Wire Crossbar Memory", US Patent 6,128,214, 2000.

[2] M. R. Stan, F. D. Franzon, S. C. Goldstein and J. C. Lach and M. M. Ziegler, "Molecular Electronics: From Devices and Interconnect to Circuits and Architecture," *Proceedings of the IEEE*, vol. 91, 2003.

[3] P. J. Kuekes and R. S. Williams and J. R. Heath, "Demultiplexer for a Molecular Wire Crossbar Network", US Patent 6,256,767, 2001.

[4] P. J. Kuekes and R. S. Williams and J. R. Heath, "Molecular-wire Crossbar Interconnect for Signal Routing and Communications", US Patent 6,314,019, 2001.

[5] T. Hogg and G. Snider, "Defect-tolerant Logic with Nanoscale Crossbar Circuits", *Journal of Electronic Testing*, 2007.

[6] R. J. Luyken and F. Hofmann, "Concepts for hybrid CMOS-molecular non-volatile memories," *Nanotechnology*, 14, 2003, 273–276.

[7] A. DeHon, P. Lincoln and J. E. Savage, "Stochastic assembly of sublithographic nanoscale interfaces," *IEEE Transactions on Nanotechnology*, vol. 2, 2003, 165–174.

[8] D. B. Strukov and K. K. Likharev, "CMOL FPGA: a reconfigurable architecture for hybrid digital circuits with two-terminal nanodevices," *Nanotechnology*, vol. 16, 2005, 888–900.

[9] D. B. Strukov and K. K. Likharev, "Prospects for terabit-scale nanoelectronic memories," *Nanotechnology*, vol. 16, 2005, 137–148.

[10] A. E. Owen, P. G. Le Comber, W. E. Spear and J. Hajto, "Memory switching in amorphous silicon devices," *J. Non-Cryst. Solids* 59 & 60, 1983, 1273–1280.

[11] M. Jafar and D. Haneman, "Switching in amorphous-silicon devices," *Phys. Rev. B*, 49, 1994, 13611–13615.

[12] P. G. Lecomber et al., "The switching mechanism in amorphous silicon junctions," *J. of Non-Cry. Solids 77 & 78*, 1985, 1373–1382.

[13] Jerzy Kanicki Ed., "Amorphous & microcrystalline semiconductor devices," Artech House, Boston, 1992.

[14] H. Horn et al., 2003 Symposium on VLSI Technology, 177–178.

[15] M. Durlam et al., "A low power 1Mbit MRAM based on 1T1MTJ bit cell integrated with copper interconnect" Motorolar presentation 2002.

[16] Saied Tehrani, "Toggle MRAM performance, reliability, and scalability" Freescale presentation 2005.

[17] http://www.research.ibm.com/journal/rd/501/gallagher.html.

[18] Y. Chen, G. Y. Jung, D. A. A. Ohlberg, X. M. Li, D. R. Stewart, J.O. Jeppesen, K. A. Nielsen, J. F. Stoddart, and R. S. Williams, "Nanoscale molecular-switch crossbar circuits," *Nanotechnology*. 14, 462, 2003.

[19] Stephen Y. Chou, Peter R. Krauss and Preston J. Renstrom, "Imprint Lithography with 25-Nanometer Resolution," *Science 5*, vol. 272, 1996, 85–87.

[20] Sung Hyun Jo and Wei Lu, "Nanovolatile Resistive Switching Devices based on Nanoscale Metal/Amorphous Silicon/Crystalline Silicon Juctions," *Material Research Society*, 2007.

[21] Sung Hyun Jo and Wei Lu, "CMOS Compatible Nanoscale Nonvolatile Resistive Switching Memory", *Nano letters*, Vol. 8, No. 2, 2008, 392–397.

[22] H. Frizsche, "Physics of instabilities in amorphous semiconductors", *IBM J. Res. Develop.*, vol. 13, Sept. 1969.

[23] R. G. Neale and John A. Aseltine, "The Application of Amorphous Materials to Computer Memories", *IEEE Transactions on Electron Devices*, vol. ED-20, 1973, 195–205.

[24] P. G. Le Comber, W. E. Spear and A. Ghaith, "Amorphous-Silicon Field-Effect Device and Possible Application", *Electronics Letters*, vol. 15, 1979, 179–181.

[25] W. Den Boer, *Applied Physics Letters*, 1982, 40, 812.

[26] P. G. Le Comber, A. E. Owen, W. E. Spear, J. Hajto, A. J. Snell, W. K. Choi, M. J. Rose, S. Reynolds, *Journal of Non-Crystal Solids*, 1985, 1373, 77–78.

[27] A. E. Owen, P. G. Le Comber, J. Hajto, A. J. Snell, *International Journal of Electron*, 1992, 73, 897.

[28] M. Jafar, D. Hanemann, *Physical Reiview B*, 1994, 49, 13611.

[29] A. Avila, R. Asomoza, *Solid-State Electron*, 2000, 44, 17.

[30] J. Hu, H. M. Branz, R. S. Crandall, S. Ward, Q. Wang, *Thin Solid Films*, 2003, 430, 249.

[31] Y. Shacham, "Filament Formation and the Final Resistance Modeling in Amorphous-Silicon Vertical Programmable Element," *IEEE Transactions on Electron Devices*, vol. 40, 1993, 1780–1788.

[32] R. D. Gould, "The effects of filamentary conduction through uniform and flawed filaments and insulators and semiconductors," *J. Non-Crystalline Solids*, vol. 55, 1983, 363.

[33] W. Lu and C. Lieber, "Nanoelectronics from the bottom up," *Nature Materials*, 6, 2007, 841–850.

[34] J. Hu, H. M. Branz, R. S. Crandall, S. Ward, Q. Wang, "Switching and filament formation in hot-wire CVD p-type a-Si:H devices," *Thin Solid Films*, 430, 2003, 249–252.

[35] Jafar, M. and Haneman, D., "Switching in amorphous-silicon devices," *Phys. Rev. B*, 49, 1994, 13611–13615.

[36] Sung Hyun Jo and Wei Lu, unpublished.

[37] Y. Cheng and C. Hu, MOSFET Modeling and BSIM3 User's Guide, Boston: Kluwer Academic Publishers, 1999.

3

Memristor Digital Memory

Idongesit Ebong and Pinaki Mazumder

In this chapter, a procedure to program and erase a memristor memory is presented. The procedure is proven to have an adaptive scheme that stems from the device properties and makes accessing the memristor memory more reliable.

3.1 Introduction

In the process of developing highly dense computing systems, the problem of dense, low power, non-volatile memory still remains. This section discusses the difficulties currently present in memristor memories and provides an adaptive method to tackle those difficulties.

The memristor memory is a viable candidate for future memory due to the difficulties encountered with CMOS scaling. However, memristors have their own complications to realizing this memory system. The patent database provides a myriad of methods to deal with difficulties (resistance drift, nonuniform resistance profile across the crossbar array, leaky crossbar devices, etc.) that arise from working with these resistive memory elements. These difficulties (problems) are addressed within the database using the correcting pulses to mitigate the effect of resistance drift due to normal usage [1], using a temperature-compensating circuit to counter resistance drift due to temperature variation [2], using an adaptive method to read and write to an array with nonuniform resistance profile [3] and introducing diodes [4] or metal-insulator-metal (MIM) diodes to reduce leaky paths within the crossbar memory array [5].

With every proposed solution to solve a problem, there are drawbacks that need to be considered. The work in this chapter exposes a view that may lead to the realization of memristor-based memory in the face of low device yield

95

and the aforementioned problems that plague memristor memory. Section 3.3 describes the reading, writing, and erasing methodology; Section 3.4 shows the simulation results; Section 3.5 explains the results; and Section 3.6 provides concluding remarks.

Figure 3.1 shows the top-level block diagram of the envisioned memory architecture and the connections between the crossbar array and the periphery circuitry. The Row Address and Column Address signals allow a selected row or column to be transparent to either the RC (Read Circuitry) or the Data sections.

The nature of the muxes may prove to make design more difficult due to the stringent requirements of their functionality. These requirements do not affect the muxes controlled by the RP (Reverse Polarity) signal; these muxes are simpler, as they are essentially transmission gate muxes that switch between two paths. For the Row and Column Address muxes, the mux requirements extend beyond switching paths for unselected and selected

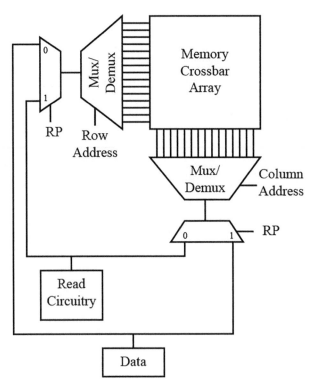

Figure 3.1   Memory system top-level block diagram.

lines. Our preliminary simulations agree with the results from [6], proposing active bias for unselected lines (columns and/or rows). When a line is unselected, a reference bias must be set on all the unselected lines, thereby limiting the leakage paths that may affect read and write integrity. For more details on this problem, refer to [7]; this reference discusses in detail the effect on noise margin of floating the unselected lines in a resistive memory.

In this implementation, the selected and unselected lines have two different references corresponding to when the memory is in use and when the memory is not in use. When in use, the unselected lines are held at VREF voltage, and when not in use, the lines are grounded. The selected lines pulsate between VREF and VDD when memory in use but is held to ground when memory is not in use. The signal flow is unidirectional from Data, through an RP mux, through a Mux/Demux, through the MCA (Memristor Crossbar Array), through another Mux/Demux, then another RP mux, and finally to the RC. The signal flow direction is controlled by which RP mux is connected to the RC and which is connected to Data.

Data is a small driver that asserts VDD. The length of time VDD is asserted is controlled by timing circuits that determine when to open the signal path from Data to RC. The RC block is essentially a generic block that implements the flow diagram represented in Figure 3.2, specifically, the "Calculate δ" and the steps that lead to determining the logic state of the selected memristor device. This signal flow is used to avoid negative-pulse generation signals as seen in [8] and [9].

3.2 Adaptive Reading and Writing in Memristor Memory

This section delves into the operations of the RC division with respect to the flow diagram in Figure 3.2. Figure 3.2a shows the decision process for a read, while Figure 3.2b shows the decision process for a write or erase operation. The write and erase operations are extensions of a single-cycle read operation. The double cycle read is given in the flow diagram, and this is a dubbed double cycle because the memristor is read in one direction and then read in the other direction to restore state if necessary. The read process is designed this way in order to prevent read disturbance in the memory device. Since each memory device in the crossbar array is different, the chosen pulses utilized for the read may cause destructive reads, thereby requiring a data refresh after read. The refresh process is essentially built into the read just in case it is necessary.

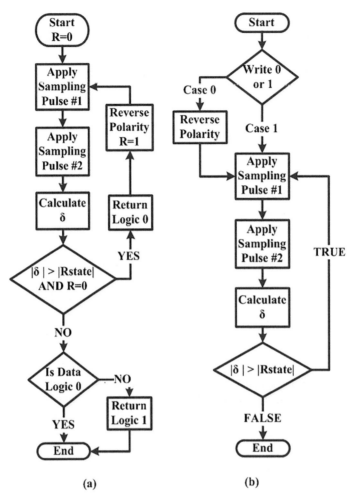

Figure 3.2 (a) Read flow diagram. (b)Write/erase flow diagram.

Referring back to Figure 3.2a, bias is applied to the memristor to sample its current value (Apply Sampling Pulse #1), then another bias is applied to the memristor to sample its value again (Apply Sampling Pulse #2), then calculate δ. δ signifies the amount of change that has occurred within the memristor between the two sampled pulses. The pulses are chosen in a manner that will change the conductance of the memristor. Depending on the magnitude of δ, the read circuitry will return either a "Logic 0" or a "Logic 1". The definitions of both the "Logic 0" and "Logic 1" states depend on the designer. In one state, the sampling pulses push against an upper (lower) limit, while in the other state, the sampling pulses move the memristor in a direction opposite to its current state. In the latter case, a correction is necessary if considering that each memristor is different within the crossbar array. The pulses used will disturb memory state based on the location of memristor within the crossbar array and also the low/high resistance boundaries of specific memristors. The unknown memristor resistance response to the applied pulses puts a requirement of a loop back requiring a polarity reversal.

The read process described in Figure 3.2a is extended to create Figure 3.2b. The goal of the latter figure is to reuse circuitry for the erase and write operations. The erase operation is defined as taking the memristor from a "Logic 0" to a "Logic 1", while the write operation changes the memristor from a "Logic 1" to a "Logic 0".

These states can be interchanged depending on definition, as long as the definition is consistent across the read, write, and erase operations.

The adaptive write process is similar to that in [10]. While the presented process is a discrete process that requires multiple steps, the process in [10] continuously changes the memristor until a latch stops the write process. This method is appropriate for single devices, but using a control to stop an applied bias may be tricky to implement in a crossbar because signal delays will come into play. The delays may cause an over-programming or over-erasing of a device, or even over-disturbing unselected devices.

The advantages of reading, writing and erasing using this scheme include tolerance to crossbar variation resistance, adaptive method to write and erase a crossbar memory and circuitry reuse for read, write and erase. Figure 3.3 shows the different tasks (equalize, charge v1, charge v2, no op and sense enable) that compose a read. The circuit that produces these signals is shown in Figure 3.4 to make sense of the different tasks. Two sampling signals, ϕ_1 and ϕ_2, control the conversion of current to voltage samples on capacitors C_1 and C_2. But before any sampling, an equalize operation is performed to balance the charges on both capacitors by asserting EQ signal high. Once the signals are sampled, then the sense enable operation is performed by first asserting NS high then later PS high. The sense amplifier in Figure 3.4b is modified from the sense amplifiers found in literature. The amplifier is purposefully made unbalanced to produce a default output of LOW resistance.

The unbalanced attribute of the sense amplifier can be achieved in multiple ways, but the chosen method in this implementation is to make

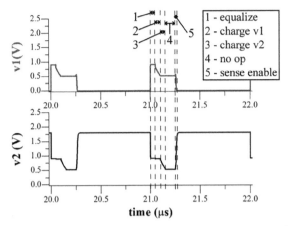

Figure 3.3 Memory cell read operation showing the different phases of read: equalize, charge v1, charge v2, no op, and sense enable.

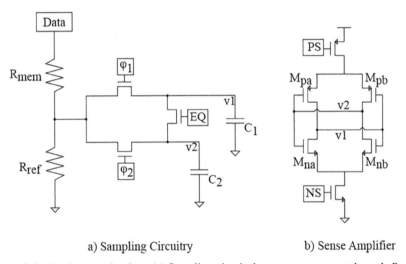

a) Sampling Circuitry   b) Sense Amplifier

Figure 3.4 Read sense circuitry: (a) Sampling circuit that converts current through R_{mem} to voltage (b) Sense amplifier that determines HIGH or LOW resistive state.

the W/L ratio of both M_{pa} and M_{pb} 320 nm/180 nm, the W/L ratio of M_{na} 1 μm/500 nm and W/L ratio of M_{nb} 1.2 μm/500 nm. The NMOS devices are unbalanced while the PMOS devices are balanced. The transistor controlled by NS has a ratio of 280 nm/180 nm, while the one controlled by PS has 400 nm/180 nm. R_{ref} is an 80 kΩ resistor while R_{mem}'s default value is expected to vary from 20 kΩ to 20 MΩ.

3.3 Simulation Results

The simulation approach consists of considering different memory conditions on a 16×16 array. The device of interest is situated in the center of the array, but all verifications were done with a worst case device at the corner with minor changes in the results. The crossbar array, unless otherwise specified, contains all memristors with the ability to change states.

3.3.1 High State Simulation (HSS)

In HSS, the memristor crossbar array has all devices initialized to a high conductive state (worst case scenario). The device of interest to be written to has a resistive range between 20 kΩ and 20 MΩ, and its initial resistance is $\sim$18 MΩ. The device accessed for the write operation is located at the center of the array (8th row, 8th column). Figure 3.5 provides a sample number for read cycles necessary to perform the write operation.

Figure 3.5a shows the number of cycles required for a write, while Figure 3.5b shows the change in memristance of the accessed device each read cycle. Each read operation provides device state feedback, and the device only changes from high resistance to low resistance when the device is written to its lowest resistance level, i.e., 20 kΩ. The number of read cycles necessary to write in this case is $\sim$21. The signals v1 and v2 presented in Figures 3.3 and 3.4 are appropriately renamed to help facilitate the understanding of the simulation results. vHighRes and vLowRes are the logically renamed signals

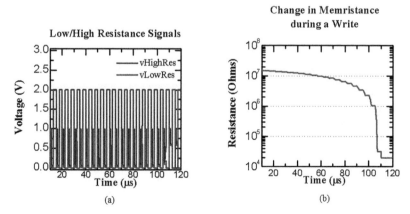

(a)　　　　　　　　　　　　(b)

Figure 3.5　Simulation Results Writing to an RRAM cell (a) Low/High Resistance Signals; (b) Memristance High Resistance to Low Resistance switch.

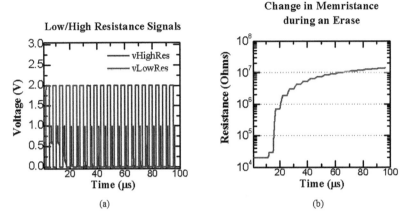

Figure 3.6　Simulation Results Erasing an RRAM Cell (a) Low/High Resistance Signals; (b) Memristance Low Resistance to High Resistance switch.

to denote when the device of interest is in a high resistance state and a low resistance state. When the signal vHighRes is high, the memristor is in a high resistance state, but when vLowRes is high, the memristor is in a low resistance state. Both vHighRes and vLowRes are always opposite to each other in the sense enable phase.

Figure 3.6a shows the number of cycles required for an erase, while Figure 3.6b shows the change in memristance of the accessed device. Just like the write cycle, the erase cycle is performed through read operations. The erase cycle takes six read cycles to go from a low resistive state to a high resistive state. The sense amplifier recognizes the switch to high resistive state when the resistance is about 4.21 MΩ.

This implies that during memory operation, the number of read operations necessary for a write after an erase may be different. And this adaptive method will prevent any over-erasing or over-writing (over-programming).

3.3.2 Background Resistance Sweep (BRS)

In the BRS-simulated state, the background resistance for all devices are swept from 20 kΩ to 20 MΩ. The device of interest is kept the same as the HSS case: its resistance range is from 20 kΩ to 20 MΩ. The goal of the simulation is to show the effect of current memory state on reading, erasing, and writing to a selected memristor. Figures 3.7 and 3.8 show the simulation result for a broad spectrum (20 kΩ, 200 kΩ, 2 MΩ and 20 MΩ),

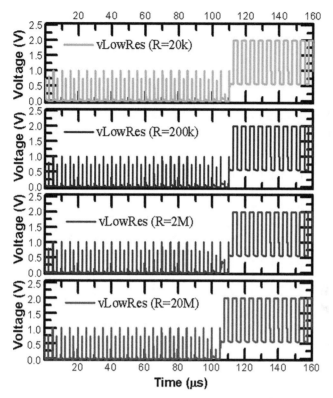

Figure 3.7 Writing in the BRS Case showing that resistance background has minimal effect on the number of read cycles required for a write.

from top to bottom. Since tuning memristors to specific resistances is a time consuming process, the background resistance for all devices are achieved with static resistors. Figure 3.7 shows the simulation results for the write case, while Figure 3.8 shows the simulation results for the erase case.

From Figure 3.7, the starting resistance is about 16 MΩ, and ~21 read operations are necessary for a write. In the 20 MΩ case, one less read is required. The simulation results show only vLowRes signal for clarity (vHighRes is its opposite as shown earlier in Figures 3.5 and 3.6. The BRS experiment is performed for the erase case to show that using the memristor, with proper diode isolation, a similar result is obtained. The same number of read cycles is necessary to erase the memristor in all four background resistance sweeps.

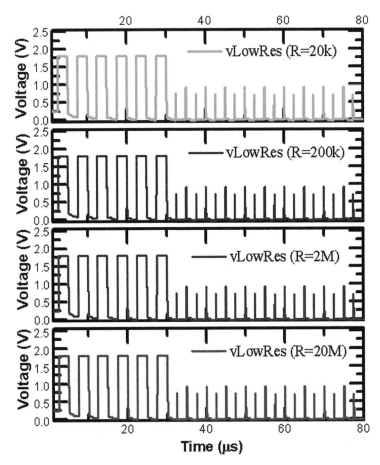

Figure 3.8   Erasing in the BRS Case showing that resistance background has minimal effect on the number of read cycles required for an erase.

Another concern aside from the background resistance is the effect of reading, writing, and erasing on unselected devices. A BRS experiment was performed but instead of using static devices around a memristor, the memory array was composed of all memristors with background resistances around 20 kΩ, 40 kΩ and 200 kΩ. The maximum resistance for all devices still remained at 20 MΩ. Figure 3.9 provides the results for the change in unselected devices during an erase operation.

In Figure 3.9, the larger the minimum resistance, the larger the percentage of change undergone by the unselected memristors. This simulation hints that

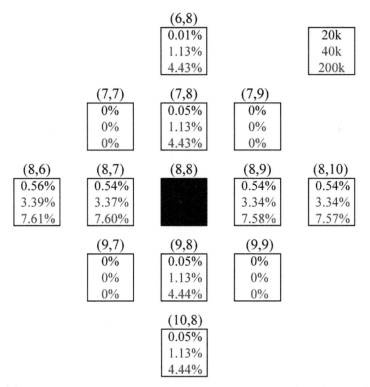

Figure 3.9   Percent change in unselected devices during an erase for different minimum resistances.

the larger the spread between minimum and maximum resistance, the less likely unselected memristors will change. Another factor that may contribute to the results of Figure 3.9 is that the lower the minimum resistance is compared to the resistance of an OFF diode, the less likely the memristor will change. This is because of the voltage divider set up by the memristors in series with the diode, whereby most of the voltage drop is on the diode, thereby causing very little voltage drop on the unselected memristor.

3.3.3 Minimum Resistance Sweep (MRS)

For the MRS case, the resistance range for the memristor of interest is modified. Since the BRS case has shown that the background resistance is really no factor with proper diode isolation, the HSS simulation conditions are used, whereby unselected devices are initialized to low resistance and

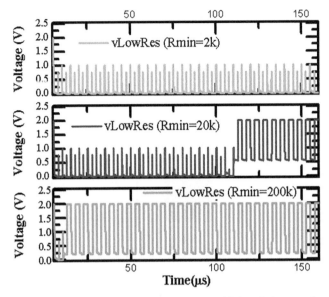

Figure 3.10   Writing to memristor devices with the same high resistive state but varying low resistive states (coarse spread). Minimum resistance affects number of read cycles necessary before a write occurs.

may change during writing operation. Figure 3.10 shows a coarse spread of low resistances and the number of read cycles necessary to complete a write. This result suggests that with set pulse duration for sampling, there exists a continuum on the number of read cycles necessary before a write occurs. The farther the lowest resistance is from 20 MΩ, the more number of read cycles necessary for a write to occur. In the 2 kΩ case, the switch to the low resistive state does not occur. In the 20 kΩ case, the switch to the low resistive state occurs after $\sim$21 read cycles, and in the 200 kΩ case, the switch to low resistive state occurs after 1 read cycle. This trend implies that the current parameters chosen for sensing may be limited to the range currently provided. For cases where the low resistive state is greater than 200 kΩ, the sensing circuit might only give vLowRes as high. The sensing resolution takes a hit here, but this can be adjusted by using a shorter pulse width.

The implication of an upper end only means that for devices with low resistance states closer to their high resistance states, shorter sampling pulses will need to be used in order to detect the memory state. Shorter pulses will provide the resolution necessary to avoid over-writing. Figure 3.10 might show a coarse sweep, but Figure 3.14 shows a finer sweep of the minimum resistance. The trend already mentioned holds true when the low resistance state is varied from 28 kΩ to 100 kΩ. As the low resistance state value increases, the number of pulses required to reach this value decreases.

3.3.4 Diode Leakage Current (DLC)

The goal of this simulation is to determine how much diode leakage the 16 by 16 network's sensing scheme can handle. The graphs shown in Figure 3.11 depict multiple read cycles under different diode saturation currents, I_S. The saturation currents going from left to right are: 2.2 fA, 4.34 fA, 8.57 fA, 16.9 fA, 33.4 fA, 65.9 fA, 130 fA, 257 fA, and 507 fA. For the first seven I_S values, the sensing scheme works as expected. For the lowest saturation current, 2.2 fA, it takes about three more read cycles for a write to occur as opposed to the highest saturation current, 130 fA. The sensing scheme fails for the 257 fA and 507 fA case.

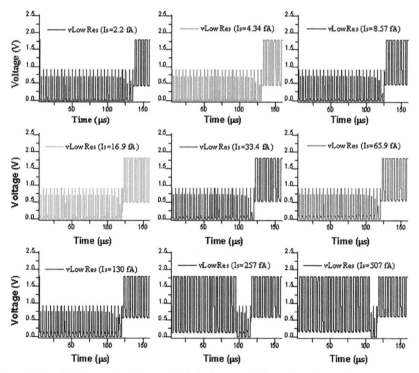

Figure 3.11 Writing under different diode leakage conditions (from left to right: 2.2 fA, 4.34 fA, 8.57 fA, 16.9 fA, 33.4 fA, 65.9 fA, 130 fA, 257 fA, and 507 fA) to show that under heavy leakage, the Read/Write circuitry fails to correctly determine logic state of memristor.

In Figure 3.11, the higher leakage cases actually switch the memristor device state more quickly than the lower leakage case. The failed cases (257 fA and 507 fA) do not signify a change in memristor characteristic behavior, but they signify a drawback in the sensing mechanism. This view is supported in the simulation results of Figure 3.12. The memristor responses to the pulses provide the same general shape, therefore, the sensing method should be able to determine the resistive state. The high leakage cases take the memristor to low resistive state quicker than the low leakage cases, and this is verified also in the memristance profiles.

A redesign of the sensing circuit can overcome this drawback and only suggests that the circuit only responds to certain limits. By resizing the sense amplifiers, a better leakage range can be accommodated at the cost of lower precision.

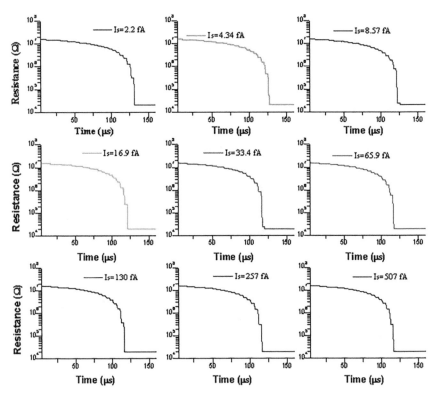

Figure 3.12   Memristor changes under the different leakage conditions showing that the Read/Write failure in Figure 3.11 is not because of characteristic deviation but because of sensing methodology drawback.

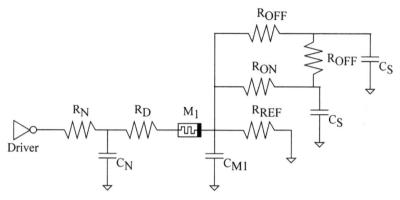

Figure 3.13 Equivalent circuit schematic showing the components considered in power analysis (note that series diode RD $<<$ M$_1$).

3.3.5 Power Modeling

For hand analysis, a lumped wire model is used for the nanowire as shown in Figure 3.13, but for simulation, a distributed pi model is used. The capacitance C_N is in the femtofarad (fF) range while C_{M1} is in the attofarad (aF) range. The capacitors of interest that contribute most to the transient behavior of the chosen method are the C_S transistors that have capacitance in the hundreds of fF range.

Using a Delta-Wye conversion and ignoring some capacitors, the time constants related to the OFF and ON resistance paths are derived. The small capacitors, C_N and C_{M1}, are ignored in this analysis for the sake of simplicity, since they are much smaller than C_S. The ON and OFF paths relate to the switches in Figure 3.4 that are controlled by the sampling signals, ϕ_1 and ϕ_2, and the EQ signal.

There are four noticeable sources of power consumption. The first comes in the form of power dissipated by the resistive nature of the nanowires, transistors and memristors. The second comes in the form of dynamic power needed each cycle due to capacitances that charge and discharge. The third source comes from non-ideal isolations and leakage, i.e., diode leakage in the nano-crossbar array or OFF transistor leakage. The last source of power comes from the static and dynamic nature of the driving circuitry used to drive the crossbar array. The third and fourth sources of power severely depend on implementation and will not be considered in the following analysis; note, though, that with CMOS scaling, these may dominate future power consumption. The power analysis is done for one complete read cycle, and depending on the amount of read cycles necessary for a write or an erase, the equations can be iterated through N cycles to estimate the power for the necessary number of cycles.

Worst case C_S charging and discharging energy: $C_S \times (V_{REF})^2$ Worst case energy dissipated in resistor reference: $(I_N(M_1))^2 \cdot (R_N + M_1) \cdot l_s + (V_{REF})^2/R_{REF} \cdot t_s$ where t_s is average time the resistor combination is under bias.

Programming and Erasing Sequence: During programming and erasing, the value of M_1 changes with applied bias. For hand analysis and verification of programming and erasing sequence, a model is necessary that will account for memristance change from high to low and from low to high depending on the sample voltage pulses. The change in memristance is discretized in (3.1) through N read cycles necessary for programming or erasing.

$$M_T = R_0 \sqrt{1 - \frac{2 \cdot \eta \cdot \Delta R \cdot \phi(t)}{Q_0 \cdot R_0^2}} \cong R_0 \sqrt{1 - \frac{2 \cdot \eta \cdot \Delta R \cdot \sum_{n=1}^{N} v_n \cdot t_s}{Q_0 \cdot R_0^2}}$$

(3.1)

The memristance values over time follow the definition of M_T in (3.1), where M_T is the total memristance, R_0 is the initial resistance of the memristor, η is related to applied bias (+1 for positive and −1 for negative), ΔR is the memristor's resistive range (difference between maximum resistance and minimum resistance), $\varphi(t)$ is the total flux through the device, Q_0 is the charge required to pass through the memristor for dopant boundary to move a distance comparable to the device width and v_n is the voltage across the memristor.

For programming, the adaptive method registers a change from high resistance to low resistance when the memristor hits 20 kΩ. For erasing, the change from low resistance to high resistance occurs around 4.21 MΩ. Iteratively, the power and energy is determined using constant time steps of t_s.

For simulation/hand analysis, the values used are $R_{REF} = 80$ kΩ, $R_N = 26$ kΩ, $C_S = 320$ fF, $t_s = 2$ μs and $M_1 = 18$ MΩ for high resistive state and 20 kΩ for low resistive state. The V_{DD} value for this simulation was chosen as 1.8 V and adjusted down to 1.1 V to account for drops on the MIM diode. With these parameters, the power consumed for each read cycle in the low resistive state is 9.68 μW, while the power consumed in the high resistive state is 0.07 μW. For the SPICE-simulated case, the power consumed for each read cycle in the low resistive state is 10.5 μW, while the power consumed in the low resistive state was 0.67 μW. The values for the low resistive state for both SPICE and calculated are similar, but the calculated value for the high resistive state is a great underestimation (89.6 % error)!

Leakage inclusion: The high resistive state is definitely a victim of the leakage power. The simulation in the work is done in a low resistive memory state to account for the worst case condition. In this memory state, the measured leakage value for devices in the selected rows and selected columns is around 20 nA each. In our 16 × 16 array, this accounts for 30 devices biased to around 0.9 V (lower than the MIM diode threshold), therefore the leakage increases due to the applied bias. The diodes are modeled with two P-N diodes in series for worst case performance while actual MIM characteristics will be better.

In order to estimate the energy more efficiently, this leakage power must be accounted for. This was done by using the P-N diode equation in (3.12), with $I_0 = 2.2$ fA, $kT/q = 25.85$ mV, $V_D = 0.45$ V (0.9 V divided equally by two identical P-N diodes) and $n = 1.08$, $I_{Diode} = 22$ nA. Assuming each path on the selected rows and columns takes a diode current of this magnitude, then the total power consumed by leakage in the 16 × 16 array is 30×22 nA $\times 0.9$ V $= 0.59$ μW. Adding this value to the hand calculated values in the previous section gives better agreement with simulation in both resistive states: 10.27 μW and 0.66 μW.

To summarize, the energy per bit for the memristor memory compared to flash memory looks very promising. The numbers from flash include the periphery circuitry and driving circuitry. Most energy consumption in flash is usually attributed to the charge pumps which are unnecessary in the resistive memory case. In flash memory product comparison, the lowest read energy for single level cells is 5.6 pJ/bit, program energy 410 pJ/bit, and erase energy 25 pJ/bit [11]. These values are from different single level cells (one product could not boast to be the lowest in all categories). The read and erase energy per bit for the resistive memory is given in Table 3.1. There is potential to reduce the program energy significantly by shifting to resistive memory technology. The erase energy between this technology and flash are similar, and the read energy depends on the state of the memristor being read.

3.4 Adaptive Methods Results and Discussion

The RRAM (Resistive Random Access Memory) is a structure that strives on the isolation provided from one cell to the next cell. The ability to selectively access one device without disturbing the other is the most vital trait of the technology. The results from the diode leakage current (DLC) simulation show the vulnerability of sensing in the resistive memory when the leakage current is too high. One way to combat this effect is to allow for an adjustable reference resistor and design for specific leakage tolerance. The background resistance sweep (BRS) results show that as long as the diode isolation is intact, the memory state does not dominate device state sensing. In essence, the proposition of more tolerable sensing methods does not eliminate the need for tighter device processes with respect to isolation.

The power results are given in Table 3.1; the energy per bit for the memristor memory compared to flash looks very promising. In flash memory product comparison, the lowest read energy for single-level cells is 5.6 pJ/bit, program energy 410 pJ/bit and erase energy 25 pJ/bit [11]. The flash numbers are from a study of different flash memories optimized for different applications. Usually, when optimized for read energy, the other two values suffer. Hence, the quoted values are from different single level cells (one product could not boast to be the lowest in all categories). There is potential of reducing the program energy significantly by shifting to resistive memory technology. The drawback to this move is the inability to perform block erasures which allow flash to have low erase energy per bit.

The adaptive method proposed provides a sensible way to deal with errors (defects) in the crossbar structure. Errors can be classified in three ways: firstly, the memristor is in a stuck open state; secondly, the memristor is in a stuck closed state; and thirdly, the lower-bound or upper-bound resistance

Table 3.1 Power and Energy results

	Power (μW)		
	Calculated	Simulated	% Error
Read high resistance	0.66 μW	0.67 μW	−1.49
Read low resistance	10.27 μW	10.5 μW	−2.19
Program*	23.83 μW	35.9 μW	−33.62
Erase**	13.21 μW	15.3 μW	−13.7
	Energy per bit (pJ/bit)***		
Read high resistance	1.32	1.34	−1.49
Read low resistance	20.55	21	−2.14
Program*	47.67	71.8	−33.62
Erase**	26.41	30.6	−13.7

*26 read cycles necessary for a write in simulation while this number is less in hand calculation.
**Calculated changed to match number of cycles necessary to exceed 4.21 MΩ and not number of cycles necessary to erase device to ~20 MΩ.
***2 μs total pulse width used for each read cycle.

targets are not met. In the first two errors (stuck open or stuck closed), an attempt to write the opposite data to the memristor will fail. In either case, as long as the memristor is static, the write methodology will only attempt the write process once. The read process will always produce a "Logic 1" as defined in the flow diagram in Figure 3.2b. The stuck open or stuck closed case will not take multiple write cycles in order to determine if the memristor is functional. To determine if the device works or not, a read in one direction is performed, an opposite data write is tried (again lasting only one read cycle due to the static nature of the failed device) and a read verify is performed. If both reads yield the same result, then the device is non-operational. This method removes the guesswork from setting hard thresholds and setting maximum write tries before a memory storage cell is deemed defective.

The defective nature of a stuck open or stuck closed cell is different from a device that misses the target high and low resistances for memristor devices. These devices behave in a way that exhibit hysteresis, but they may have larger or smaller ratios of the resistance in the high state to the resistance in the low state compared to the design target. Since the proposed method does not deal directly with absolute resistance values, the exact extremes of the resistance of a certain device is not of interest. Resistance extremes are dealt with in ratio (Figure 3.14). The larger the range between the high and low resistive states, the more number of read cycles necessary to perform a write or erase operation. Also, depending on the resistance range, the pulse widths used for the design may not be enough to distinguish high and low states. For example, in Figure 3.10, any low level greater than 200 kΩ does not provide enough separation between the high and low resistive states. The chosen 1 μs pulse widths would already change the device state from one extreme

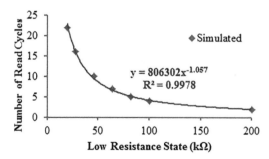

Figure 3.14   Writing to memristor devices with the same high resistive state but varying low resistive states. The larger the minimum resistance, the lower number of read cycles necessary to reach the low resistive state.

to another during a read operation. The analyses done in this work examines the memory limitations for a chosen pulse width, but the values presented can be improved upon with shorter pulses (<1 μs) based on improved memristor switching performance.

The advantage of using this method for read/write is to combat the effects of process variation within the crossbar structure. The exact low level does not matter except that the level is within operational limits imposed by the 1 μs pulse. The nature of the low level and high levels of memristive devices to change during operation requires that the sensing method take this into account. During operation, as long as the pulses do not change the memristor device to an extreme, then a device that may have been deemed a failed device under another sensing scheme is salvaged for further use. This method provides an insightful scheme to combat the effects of resistance drift as memristors' absolute extreme resistances change over their lifetime.

The power and energy numbers in Table 3.1 show some disagreement between calculated and simulated values. Eliminating the assumptions made due to the low time constant values for the different capacitor paths may lead to an agreement. Essentially, the storage capacitors, although their access transistors are in the OFF state, are leaking and charging depending on the cycle presented by data. Also, the peripheral circuitry consumes power not included in the calculated values. Considering that the same driving circuitry is used to drive the memristor in both its high and low resistive states, the low current achieved in the high resistive state suggests that the time constants of the OFF and the ON paths have similar power characteristics, which account for 0.01 μW. However, in the low resistance state, the OFF and the ON paths have differing power profiles leading to 0.23 μW disagreement between simulation and calculation.

The program and erase numbers have a larger error differential because two different models are used to determine the weight change in the memristor. In the calculated case, the weight change is determined through an approximated linear diffusion model whereby boundary effects are not taken into consideration. In the simulated model, boundary effects are modeled with a window function, which is why when the device is in a low resistive state at a boundary, albeit high current, the memristance does not change as drastically as predicted by the linear model.

The current method proposed takes into account problems that may be more pronounced in a higher-dimension grid, i.e., 4-Kb block size as used in many commercial flash devices. The resistive nature of the nanowire will be more pronounced for devices not very close to the driver. This method of determining memory state adjusts to the resistive drops that may be made when the nanowires are more resistive than expected. The problem that may affect a larger memory size is excessive voltage drops which would require tuning the voltage level to accommodate all devices in the crossbar array. Devices far from the drivers will essentially take longer to write or erase compared to devices closer to the driver. Essentially, an adaptive read, write and erase method allows for a more flexible process technology and will enable the adoption of the memristor memory sooner since devices that do not meet high and low resistance criteria may still be used with confidence.

3.5 Chapter Summary

The memristor memory showcased extols the advantages of using the new technology in memory applications. The method of achieving the read, write and erase relate adaptively to each memristor device thereby allowing for increased yield when it comes to using devices that have differing high to low resistance range. The memristor memory also exhibits lower power and energy consumption when compared to flash memory. Unfortunately, the proposed method cannot be directly applied to the multi-bit memory, since this method depends on writing the memristor to an extreme.

New methods will need to be devised that will allow for reliably writing to the device in the multi-bit case, as well as perform flash-like operations, such as block erasures. The latter is not necessary, but it would improve the operation per bit statistics when it comes to power and energy consumption.

References

[1] Moore, J. T., and K. A. Campbell (2005), Memory device and methods of controlling resistance variation and resistance profile drift, patent Number: US 6,930,909.

[2] Hsu, S. T. (2005), Temperature compensated rram circuit, patent number: US 6 868 025.

[3] Straznicky, J. (2008), Method and system for reading the resistance state of junctions in crossbar memory, patent Number: US 7 340 356.

[4] Myoung-Jae, L., et al. (2007), 2-stack 1d-1r cross-point structure with oxide diodes as switch elements for high density resistance ram applications, in *Electron Devices Meeting, 2007. IEDM 2007. IEEE International*, 771–774.

[5] Rinerson, D., C. J. Chevallier, S. W. Longcor, W. Kinney, E. R. Ward, and S. K. Hsia (2005), Re-writable memory with non-linear memory element, patent Number: US 6 870 755.

[6] Rinerson, D., C. J. Chevallier, S. W. Longcor, E. R. Ward, W. Kinney, and S. K. Hsia (2004), Cross point memory array using multiple modes of operation, patent Number: US 6 834 008.

[7] Csaba, G., and P. Lugli (2009), Read-out design rules for molecular crossbar architectures, *Nanotechnology, IEEE Transactions on*, 8(3), 369–374.

[8] Ho, Y., G. M. Huang, and P. Li (2009), Nonvolatile memristor memory: device characteristics and design implications, in *IEEE/ACM International Conference on Computer-Aided Design-Digest of Technical Papers*, ICCAD 2009, 485–490, IEEE.

[9] Niu, D., Y. Chen, and Y. Xie (2010), Low-power dual-element memristor based memory design, in *Proceedings of the 16th ACM/IEEE international symposium on Low power electronics and design*, ISLPED '10, 25–30, ACM, New York, NY, USA, doi:10.1145/1840845.1840851.

[10] Yi, W., F. Perner, M. Qureshi, H. Abdalla, M. Pickett, J. Yang, M.-X. Zhang, G. Medeiros-Ribeiro, and R. Williams (2011), Feedback write scheme for memristive switching devices, *Applied Physics A: Materials Science and Processing*, 102(4), 973–982.

[11] Grupp, L. M., A. M. Caulfield, J. Coburn, E. Yaakobi, S. Swanson, P. Siegel, and J. Wolf (2009), Characterizing flash memory: Anomalies, observations, and applications, in *Proceedings of the 42nd International Symposium on Microarchitecture*, 24–33.

4

Multi-Level Memory Architecture

Yalcin Yilmaz and Pinaki Mazumder

In this chapter, a crossbar memory architecture that utilizes a reduced constraint read-monitored-write scheme is presented. The proposed scheme supports multi-bit storage per cell and utilizes reduced hardware, aiming to decrease the feedback complexity and latency while still operating with CMOS compatible voltages. Additionally, a read technique that can successfully distinguish resistive states under the existence of resistance drift due to read/write disturbances in the array is presented. Derivations of analytical relations are provided to set forth a design methodology in selecting peripheral device parameters.

4.1 Introduction

Nonvolatile memory technologies led by NAND Flash have been generating increased market revenues due to the increased usage of these devices, especially in portable consumer electronics and solid state drives (SSDs) [1]. The trend toward the cloud storage and computing is continually demanding enterprises to invest especially in SSD-based storages, as these provide higher performance compared to hard disk drives (HDDs) [2].

Flash memories have been providing solutions to the ever-increasing high-performance storage demands with continued feature scaling. However, flash scaling is reaching its limits due to the increased reliability problems such as aging of the oxide, charge leakage, retention problems and the increased capacitive coupling between the floating gates of the neighboring cells [3].

The approaching end of the flash scaling has led researchers to look for alternative nonvolatile memory technologies that can sustain the scaling trend [4]. Many promising emerging technologies have been proposed, each with its own advantages and challenges. Magnetoresistive random-access memory (MRAM) [5], spin-transfer torque random-access memory (STT-RAM) [6], phase-change memory (PCRAM) [7] and resistive random-access memory (RRAM) [8] that is also commonly referred to as memristive crossbar memory, have been the major candidates to supersede the flash technology.

The successor technology has to be dense, has to be scalable, has to have high write endurance and has to support multi-level cell structure as this has been the trend in flash. Variable resistance devices (memristors) as predicted by Chua in his 1971 paper [9] and realized by Hewlett-Packard Labs in [10] meet all these requirements with their CMOS compatibility, write endurance, data retention, multilevel storage capability and scalability down to molecular dimensions [11].

Ever since the discovery of the missing variable resistance devices [10], they have attracted great interest not only due to their nonvolatile nature but also due to their hysteretic variable resistance characteristics which allowed for the realization of unconventional circuits and systems. They have found their applications in logic circuits [12, 13], neural computing [14–16], image processing [17], analog circuits [18, 19], field-programmable gate arrays (FPGAs) [20, 21] and nonvolatile memory [22–24]. However, among these, the most commercially promising application is the nonvolatile crossbar memory due to existing consumer market.

The crossbar memory has attracted more attention, due to its increased cell density compared to the other architectures that have been proposed, such as the unfolded architecture presented in [25], which requires more metal connections to be routed. In order to further increase the storage density, there has been much research effort in terms of achieving multi-bit storage per cell [24, 26] rather than single-bit. Multi-bit provides increased storage per unit area, reducing fabrication costs.

To achieve single or multi-bit storage per cell, various write schemes have been proposed. These schemes are split into two main categories: Pulse-based schemes, where a predetermined duration and amplitude pulse is applied to the cell vs feedback-based schemes, where the pulse duration depends on the feedback circuit, indicating if the cell has reached the desired state [27].

Feedback schemes are shown to have advantages over the pulse-based schemes, as they limit the resistive distributions of the programmed cells. In [28], it has been shown that a feedback-based scheme shows narrowing of the resistance distributions compared to a pulse-based scheme. However, the use of DAC, ADC [23, 24], or multi-stage comparisons [27] in feedback circuitry can introduce significant peripheral circuitry overhead and can introduce latency in response time that can be significant when the memory device is highly non-linear. Thus, a more simplistic approach is required to reduce the circuit overhead, and to reduce latency to avoid over-programming.

Aside from the read/write techniques, another important factor that plays a role in the design of the resistive crossbar memory is the cell structure that is used in the memory array. There are three major types of cell structures that have been proposed: 1T1R, where a selection transistor is integrated in series with a resistive device [29], 1D1R structure, where a series diode is integrated with a resistive device [30] or the device itself shows diode-like behavior [31] and 1R structure [32], where a resistive device does not have any series selection device or diode-like behavior. Other device types are also proposed such as the 3-terminal resistive devices as in [33].

1T1R structure has density problem. With this structure, the memory density is dictated by the scaling of the series transistor, which has bigger feature size than the memristor cell itself. 1R structure has so called "sneak paths" problem which limit the array size due to the deterioration of sensing margins. In fact, relatively smaller-size arrays have been shown to provide enough margin for sensing [34]. In these structures, so called "half-selected cells" still see VDD/2 voltage levels across them [29] and their resistances can drift over time due to the read/write disturbances [35]. Although some methods claim the disturbance is not significant [27], it is more pronounced in 1R architectures as there is no selection device to reduce or eliminate the leakage through these cells. Some proposed methods such as grounding of unselected rows and columns can even cause the half-selected cells to see a higher voltage across them than the selected cell as shown in [36], significantly disturbing the half-selected cells.

Thus, it is believed that 1D1R structure is the most promising solution as it does not suffer from density issues like 1T1R and it does not suffer from the sneak paths problem as much as the 1R structure. Indeed, Crossbar Inc. recently presented a RRAM structure that utilizes series selector devices that provide programming thresholds [37], strengthening the belief that 1D1R structures will be widely adopted in RRAM designs.

To realize these structures, there have been various approaches. The introduction of series diodes [30] or metal-insulator-metal (MIM) diodes [38, 39] and the engineering of the devices to integrate diode-like behavior [31, 40] in the device itself is presented in literature. Even with the series diode, cell-to-cell isolation is not perfect. Most works in literature fail to consider what happens to unselected or half-selected cells as the other cells are being programmed. In this work, how the programmed resistance distributions change as all the cells in the array are programmed is observed, and a read method that compensates for the expanded and shifted distributions is presented.

A read/write scheme where the voltage references are derived from the intermediate node via means of a combination of active and passive devices, such as diodes, diode-connected transistors and resistors that generate distinct thresholds is put forward. The contributions presented in this chapter include a reduced-constraint read-monitored-write scheme which supports multi-bit storage per cell and utilizes reduced hardware, aiming to reduce the feedback complexity and latency while still operating with CMOS-compatible voltages, a read technique that provides enough margins for state detection while allowing certain amount of resistance drift in the array cells (thus reducing the need for frequent refresh operations), a relaxed array biasing scheme that aims to facilitate read/write operations while reducing cell disturbances and derivations of analytical relations to pave the path for a design methodology in selecting peripheral device parameters. The outlined read/write methodology applies generally to 1D1R structures but can be generalized to other structures with minor modifications.

In Section 4.2, the approach adopted for modeling of the memory cells was presented. Herein, Section 4.3 details the memory architecture, our read/write methodology, and analytical expressions that guide the peripheral circuitry design. Section 4.4 and 4.5 present our simulation results for read/write operations as well as the effects of variations on the programming voltages and the series resistance.

4.2 Multi-State Memory Architecture

4.2.1 Architecture

The proposed multi-level memory architecture is presented in Figure 4.1. The crossbar memory array is the main storage area that is composed of metal crossbars and resistive cells located at every intersection of these crossbars.

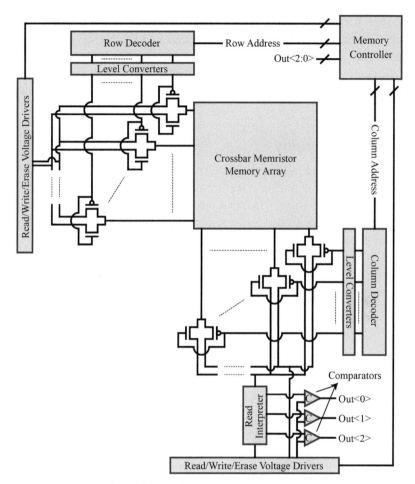

Figure 4.1 Multi-level RRAM architecture.

N-type and p-type access transistors enable the driving of the cross-bars with adequate voltage levels. Row and column decoders activate the relevant access transistors depending on the location of the selected cell in the array. The voltage drivers provide various voltage levels to adequately bias the selected or unselected cells. The read interpreter circuitry is serially connected to the selected column through the access transistors and is capable of actively monitoring the resistance of the selected cell. The interpretation results encoded in voltage levels are then fed into the comparators to detect whether the desired resistive state is reached. The memory controller is responsible for coordinating which memory operation to perform and generating relevant control signals to activate peripheral blocks.

In this work, we present the multiplexed read and write circuitries as parallel reads can introduce circuit overhead [41]; however, our methodology can be modified to support parallel reading and writing of the cells on the selected row.

4.2.2 Read/Write Circuitry

Figure 4.2 shows the read interpreter circuitry that actively monitors the resistance change in real time during a write operation and detects the encoded state during a read operation. The read circuitry is composed of a voltage division stage employing diodes and resistors to interpret the voltage change across the series resistor, and a comparison state employing fast comparators to detect if a desired state is reached. R_{series} is the series resistor, V_{int} is the voltage level on the intermediate node that is the node connecting the read interpreter and the selected column through the selection circuitry,

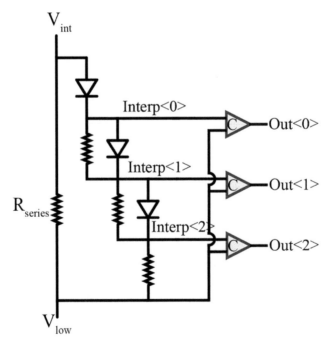

Figure 4.2   Read interpreter circuitry block.

Interp<2:0>signals are the outputs generated by the voltage division stage and Out<2:0>signals are the corresponding comparator outputs generated in the comparison stage.

The series diode in the voltage division stage provides a close-to-constant voltage reduction in the voltage to be interpreted by the circuitry, providing compaction of the resistive states. The interpreter circuitry can be expanded depending on the number of bits to be stored in the memory cell. For n-bit storage, 2^n-1 diodes and comparators are needed. The resistors connected to the diode outputs have high resistances (1M Ω) in order to minimize the effect of the interpreter circuitry on the intermediate node voltage.

The Interp<2:0>signals are unique analog outputs, and their values decrease as the resistance of the cell increases. The read interpreter circuity ensures that the Interp<2>signal falls below the comparator threshold before the Interp<1>signal does, and the Interp<1>signal falls below the threshold before the Interp<0>signal. Each Interp signal falling below the comparator threshold indicates that a particular resistive state is reached. We have adopted the convention such that the resistive states are '00', '01', '10' and '11'; where the states are listed in the order of increasing resistance.

However, the adoption of the reverse convention where the states are in the order of decreasing resistance is also possible.

4.2.3 Array Voltage Bias Scheme

When performing a write operation, the voltage bias across the selected memory cell should exceed the cell threshold to achieve a fast write operation, whereas in order to minimize the resistance change of the unselected or half-selected cells, the voltage bias across these cells should be kept lower than the threshold of the memory cells.

To achieve this goal, we bias the array with four different voltage levels as shown in Figure 4.3. The selected row is applied $V_{select-row}$, the unselected rows are applied $V_{unselect-row}$, the unselected columns are applied $V_{unselect-col}$, and the selected column is applied $V_{select-col}$ through the interpreter circuitry which yields an applied voltage value of V_{int} on the intermediate node. These conditions are summarized as follows:

$$|V_{select-row} - V_{int}| > V_{mth} \tag{4.1a}$$

$$|V_{select-row} - V_{unselect-col}| < V_{mth} \tag{4.1b}$$

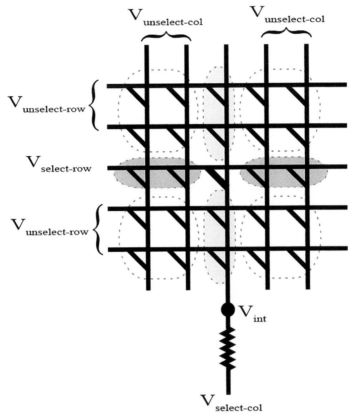

Figure 4.3 Memory array biasing scheme. The cells that lie in white regions see.

$$|V_{unselect-row} - V_{int}| < V_{mth} \qquad (4.1c)$$
$$|V_{unselect-row} - V_{unselect-col}| < V_{mth} \qquad (4.1d)$$

where V_{mth} is the memory cell threshold of the 1D1R cell. The above-stated conditions can be satisfied by choosing voltage values that follow the following inequality:

$$V_{select-row} > V_{unselect-col}$$
$$> V_{unselect-row} > V_{int} > V_{select-col} \qquad (4.2)$$

Uneven biasing of the unselected rows and columns are also proposed in [42], where the array is biased with voltage levels that are VDD/3 apart. However, the scheme we propose does not have strict rules on the voltage levels, as long as the difference between the two consecutive voltage values is selected so that it is smaller than the magnitude of the memory cell threshold.

This biasing scheme yields four groups of cells that observe different voltage levels at their terminals as shown in Figure 4.3. Since V_{int} value changes during programming, the voltage difference at the terminals of the unselected cells connected to the same column as the selected cell is not constant. Therefore, it is important to pick the voltage levels such that the worst-case voltage difference across these cells is below the cell threshold.

$V_{unselect-row} - V_{unselect-col}$, the cells that lie in the light grey regions see $V_{unselect-row} - V_{int}$, the cells that lie in dark grey regions see $V_{select-row} - V_{unselect-col}$ across them. The selected cell represented with a thick line sees $V_{select-row} - V_{int}$ across it.

4.2.4 Read/Write Operations Flow

At the beginning of the write operation, it is assumed that the selected cell in the array is at the erase state, which corresponds to the lowest resistive state ('00') in our convention. After the controller is prompted to perform a write, it signals the voltage drivers to apply relevant voltage levels on the array and enables the row and column decoders to facilitate the application of the voltages on the selected and unselected rows and columns. As the voltages are applied, the interpreter circuitry generates distinct analog voltage levels (Interp) that directly depend on the resistance of the selected memory cell. As the cell resistance increases, the Interp signal levels begin to decrease. As soon as one of these signals reaches the comparator threshold, the corresponding comparator output (Out) signal flips. The controller checks if the comparator outputs indicate that the desired state is reached. If the state is reached, the controller immediately terminates application of voltages; if not, the controller keeps enabling the application of the write voltages on the array. The flow chart visualizing these steps is shown in Figure 4.4(a). The chart also includes the possible write protection and failed cell detection mechanisms that can be adopted similar to flash memories.

Figure 4.4(b) shows the flow chart for the read operation. The read operation is similar to a write operation, except the voltage levels and the interpreter circuitry used can have different characteristics, as will be discussed in the following sections.

When the controller is prompted to perform a read, it signals the voltage drivers to apply read voltage levels on the array and enables the row and column decoders to facilitate the application of these voltages on the selected and unselected rows and columns. As the voltages are applied, the interpreter

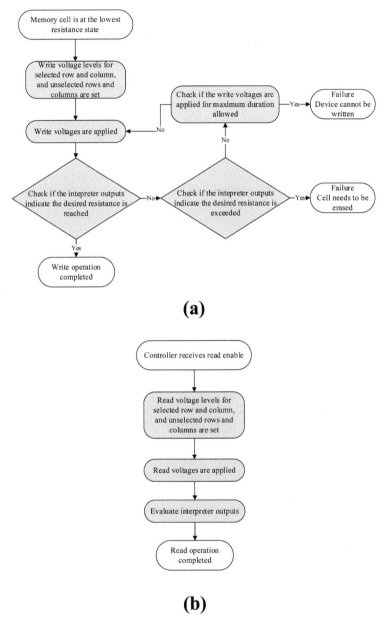

Figure 4.4 Flow chart explaining memory operations: a) Write Operation, b) Read Operation.

circuitry generates three distinct Interp signals that are dependent on the resistance of the memory cell. Interp signals are fed into the comparators. Out signals generated by the comparators indicate which resistive state the memory is in, and the read operation completes. Unlike the write operation, the read operation has a fixed duration, and this duration should be kept as short as possible to reduce the read disturbance which can cause the memory resistance to drift.

4.2.5 State Derivations

It is important to characterize the behavior of the read interpreter circuitry together with the array elements to understand how the component parameters should be selected.

First, we characterize the dependence of the resistive state of the memory cell on the intermediate node voltage. Then, we characterize the dependence of the intermediate node voltage on the series diode threshold based on the detection threshold requirements of the comparison stage and the current-voltage (IV) characteristics of the series diode in the interpreter circuitry.

The analog voltage levels in the intermediate node that correspond to the encoded resistive states can be characterized by the following equation:

$$V_{int} = V_{select-col} + \frac{R_{series}(V_{select-row} - V_{select-col} - V_{thm})}{R_{parasitics} + R_{cell(s)} + R_{series}} \qquad (4.3)$$

where $R_{cell(s)}$ is the resistance of the selected cell at a given state and V_{thm} is the cell threshold. Parasitic resistance sources represented by lumped $R_{parasitics}$ term consist of the effective resistances of the n-type and p-type access transistors and the crossbar resistance seen by the interpreter circuitry. The read interpreter voltage division stage outputs meet the following equality when a particular memory state is being programmed:

$$V_{Interp3-k} = V_{select-col} + \frac{(V_{int} - V_{diode1} - V_{select-col})}{3}k$$
$$= V_{c-res} + V_{select-col} \qquad (4.4)$$

where $V_{Interp3-k}$ represents the voltage level at the corresponding voltage division stage output (Interp), and k is the index of the corresponding voltage division stage output, V_{diode1} is the threshold of the series diode in the interpreter circuitry and V_{c-res} is the comparator threshold. For analysis simplicity, the resistors connected to the outputs of diodes are assumed to have same resistances. However, it is possible to alter these resistances separately to tune the separation between resistive states of the memory cell. The decision on where to set the resistive states is highly dependent on the non-linear behavior of the device. The design decisions can also be made based on the trade-off between the programming time and the resistive margins. Solving for V_{diode1} in (4.4) yields:

$$V_{diode1} = V_{int} - V_{select-col}\frac{3}{k}V_{c-res} \tag{4.5}$$

The above expression relates the intermediate node voltage with the series diode threshold. We need one more expression that relates the two to be able to numerically solve both to obtain V_{int} value, which we can plug in (4.3) to obtain the corresponding resistive state level. The additional expression can be obtained using the diode current equation and solving for the diode threshold. The final expression is:

$$V_{diode1} = 3I_0R_h + V_{int} - V_{select-col}$$
$$- \frac{kT}{nq}LambertW\left[\frac{3e^{\frac{3I_0R_h + V_{int} - V select - col}{kT/nq}I_0R_h}}{kT/nq}\right] \tag{4.6}$$

where LambertW is the lambert omega function, R_h is the magnitude of the resistances connected to the diode outputs, I_0 is the reverse bias saturation current, n is the ideality factor, k is the Boltzmann constant, T is the absolute temperature and q is the magnitude of the charge of an electron.

Since I_0R_h term is extremely small, it can be omitted where it is an additive factor. The new simplified expression yields:

$$V_{diode1} = V_{int} - V_{select-col}$$
$$- \frac{kT}{nq}LambertW\left[\frac{3e^{\frac{V_{int} - V select - col}{kT/nq}I_0R_h}}{kT/nq}\right] \tag{4.7}$$

The expressions (4.5) and (4.7) form a pair of equations with two unknown parameters. They can be evaluated together numerically to obtain a unique pair of V_{int} and V_{diode1} values which satisfy both. A unique pair

Table 4.1 Calculated vs. Simulated Resistance Levels

	Calculated (Ω)	Simulated (Ω)	% Error
'01'	10064	10166	1.003
'10'	20227	20590	1.763
'11'	25653	26067	1.588

is obtained for each resistive state since expression (4.5) depends on the state to be encoded. After obtaining unique pairs, the next step is to evaluate expression (4.3) to obtain corresponding resistive states.

Table 4.1 lists the resistances calculated using expressions (4.3), (4.5) and (4.7) vs the results obtained through SPECTRE simulations of a 16 × 16 array. The percent errors in calculations are also listed. The disagreement between the simulated and the calculated results are less than 2% for each programmed state. The leakage through the half-selected cells also contributes to the intermediate node voltage V_{int}; however, the agreement between the simulated and calculated results indicate that this contribution to voltage mode reading is minimized especially owing to the 1D1R structure and the array biasing scheme used.

Another point to note is that the analytical models derived in this section are not dependent on the resistive device model used. The proposed circuitry behaves the same as long as the model meets the minimum and maximum resistances required for state encoding.

4.3 Read/Write Operations

4.3.1 Read/Write Simulations

Our simulations are performed on a 16 × 16 array with the adoption of the distributed PI-model for the metal crossbars. The read interpreter circuitry is capable of performing a read operation during a write operation by actively monitoring the voltage change across the series resistor connected to the selected column via the selection circuitry.

The bias voltages applied to the array during a write operation is shown in Figure 4.5. As explained earlier, this biasing scheme ensures that the disturbances to the unselected cells are minimized, while the selected cell is exposed to a large voltage bias higher than the cell threshold. Example voltage levels shown in Figure 4.5 are:

$$V_{select-row} = -V_{select-col} = 1.6V, V_{unselect-row}$$
$$= 0.82V \; and \; V_{unselect-col} = 0.045V.$$

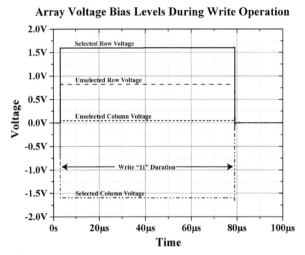

Figure 4.5   Array biasing levels during a write operation.

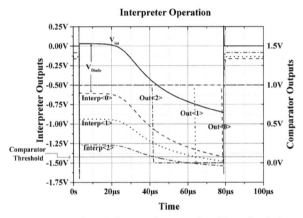

Figure 4.6   Interpreter operation during a write operation. The value being written is '11'.

The operation of the interpreter circuitry is shown in Figure 4.6. The voltage division stage generates three distinct Interp signal levels which are then compared to the selection voltage that is also applied through the interpreter circuitry. Each Interp signal is connected to a comparator, and the comparator generates the corresponding Out signal that indicates if a certain state is reached. Interp signal levels decrease as the resistance of the cell increases. Once the signal level reaches below the comparator threshold, the comparator output (Out) signal becomes low, signaling to the controller.

In our simulations Out<2>, Out<1> and Out<0> indicate that the states '01', '10' and '11' are reached, respectively. The series diode provides a close-to-constant voltage reduction of the intermediate voltage to be interpreted, allowing Interp signals to reach the comparator threshold more quickly, thus, reducing the write time and compacting the resistance levels.

Figure 4.7 shows writing of states '00', '01', '10' and '11'. Since state 00 is the erase state of the cell, the controller does not apply any voltage to the array to change the state, and the Out signals remain high.

In '01', '10' and '11' cases, as the resistance of the cell increases, the Out signals start to become low. Once the desired resistances are reached, the controller stops applying write voltages to the array, and the Out signals become high again.

Since the read interpreter circuitry monitors the state of the cell as it is being written, the read operation can be implemented using the same biasing scheme as the write operation while keeping the pulse duration short to prevent significantly altering the resistive state of the memory. Since the write operation programs each cell to the point of detection, there is a significant read margin when the cell drifts toward a higher resistance; however, there is no margin if the cell drifts toward a lower resistance and a wrong value can be read. In order to mitigate this problem, modifications on the read scheme by reducing the pulse magnitudes used, and/or increasing the series resistance value during a read operation are proposed. The optimizations to the read operation will be discussed further in the following sections.

4.3.2 Read Disturbances to the Neighboring Cells

Even though each memory cell has a built-in threshold, a voltage bias below this value still induces a trivial resistance change in the cell. This means that the cell resistance can drift over time due to the repeated reading of the cell or due to the read and write operations performed on other cells in the array.

In order to quantify the effects of read disturbances in the presented architecture, the resistance drift of the neighboring cells to a selected cell is simulated while the selected cell is read repeatedly. Since the amount of drift is state dependent, the simulations for different values stored in the memory cell and its neighbors are performed. We have assumed the same voltage levels and the series resistance as used during a write operation to simulate the worst-case drift. As any reduction in read voltage and increase in series resistance would yield better drift characteristics as shown in Figure 4.8.

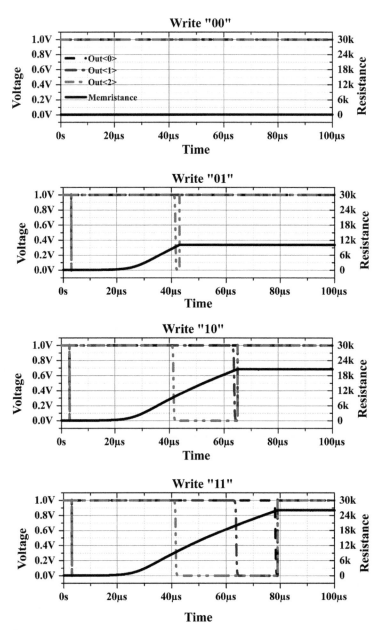

Figure 4.7 Writing of various values into the selected memory cell. Resistance change is overlapped with interpreter output signals.

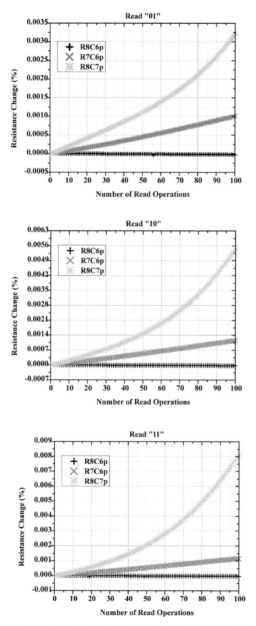

Figure 4.8 Percent change of resistance in neighboring cells vs. number of read operations. The cell located at the crossing of row 7 and column 7 (R7C7) is selected.

The results indicate that for the cases of '01', '10' and '11', the greatest resistance drift is observed in the cells that are connected to the same column as the selected cell. The cells connected to the same row as the selected cell observe a drift that is less-however, still non-zero. The remaining cells observe a close-to-zero drift within the simulated 100 consecutive read operations. Among these three resistive states, the greatest change is observed when the selected cell is storing '11' due to the fact that the voltage bias across the unselected cells connected to the same column as the selected cell is highest.

In the case when the memory cell is storing '00', no meaningful change in resistance was observed, hence the results are not listed.

4.4 Effects of Variations

4.4.1 Variations in Programming Voltage

As indicated by expressions (4.3), (4.5) and (4.7), the programmed states are dependent on the voltage levels used. If the voltage bias levels are scaled, the programmed states change accordingly as shown in Figure 4.9.

The selected row and column pulse magnitudes are kept equal and are listed in the x-axis while their signs are opposite. Although it is possible to scale the unselected row and column voltages separately, they were reduced

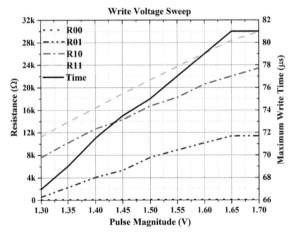

Figure 4.9   Write voltage sweep. How the resistive states change is shown with dashed and dotted lines. Maximum write time which corresponds to writing of state '11' is shown with straight line.

or increased the same amount as the selected row and column voltage magnitudes. It was found that this scaling scheme provided more even scaling of the applied voltages to the unselected and half selected cells.

The results show that as the voltage levels are increased, the programmed resistance levels increase. This also results in increase in the programming time of the cell. As the resistance levels increase, the separation between the states increases, which can allow for better margin of detection.

Since a read operation is a write operation with an extremely short duration, the results reported in Figure 4.9 have important consequences in terms of improving the tolerance for resistance drift in both increasing and decreasing directions. If a cell is written at a pulse magnitude of 1.6V, it can still be read at, for example, 1.55V while allowing drift in its resistance. This can be further clarified by comparing the programmed resistances for both voltage levels as shown in Table 4.2.

The '11' state programmed at 1.6 V has resistance of 26067 Ω. However, if this programmed resistance is read at 1.55 V, it would yield a value of '11' since 26067 Ω is greater than the '11' value written at 1.55 V which is 23715 Ω. In fact, reading at 1.55 V allows the programmed resistance of 26067 Ω to drift up to the device maximum, which is assumed to be 100 $K\Omega$, or down to the boundary of state '10', which is 23715 Ω.

Table 4.2 lists the high and low resistance margins that are the distances of the programmed resistances of the memory states at higher voltage level to the detection boundaries of the lower voltage reads. It is possible to optimize these margins by scaling the circuit parameters that contribute to the resistive states as characterized in Section 3.2.5.

4.4.2 Variations in Series Resistance

The programmed states are also dependent on the series resistance used, similar to the case of pulse magnitudes. The change in programmed resistances versus the series resistance is shown in Figure 4.10.

For fixed write pulse magnitude of 1.6 V, the increase in series resistance results in increase in programmed resistances. The same trend as in the case of increased voltage levels is observed. The separation between the states increases with increased series resistance, which can allow for better margin of detection.

Similar to the case of voltage reduction, the reduction of series resistance when reading allows for generation of high and low margins, which in turn allows for drift in both increasing and decreasing directions. An example is

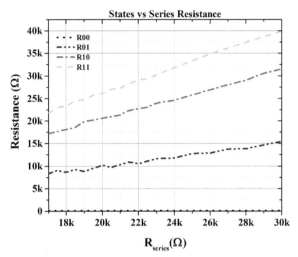

Figure 4.10   Series Resistance Sweep. The change in programmed states vs. the series resistance value is shown for fixed write voltage levels.

not provided for this case in order to avoid repetition; however, similar results as in Table 4.2 can be deduced from values plotted in Figure 4.10.

4.4.3 Reduced-Impact Read Scheme

Even more powerful benefits can be obtained when both the read voltages and the series resistance are scaled simultaneously.

Figure 4.9 indicates that it is not possible to scale the read pulse magnitudes to 1.5 V because the lower boundary for '11' state is 18891 Ω which is lower than the '10' boundary (20590 Ω) for 1.6 V writes. This means '10' written at 1.6 V would be evaluated as '11' at 1.5 V reads. However, the increase in series resistance shows the opposite trend as in the case of decrease in read voltage. Therefore, it is possible to perform reads even below 1.5 V pulse magnitude with the help of increased series resistance.

The simulations indicated that it is possible to write the values at 1.6 V with 20 $K\Omega$ resistance, and then read at 1.5 V pulse magnitude with 25 $K\Omega$ series resistance or at 1.25 V pulse magnitude with 45 $K\Omega$ series resistance.

Figure 4.11 aims to visualize how the reduced-impact read scheme works together with the write scheme. When a single cell is programmed in the array, its resistance can land anywhere in the distributions shown in Figure 4.11(a), depending on its location in the array. In this case, the

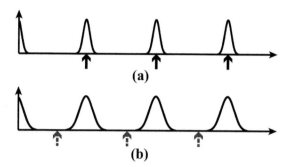

Figure 4.11 Resistance distributions when a single device is programmed in the whole array (a), after all the cells are programmed in the whole array (b). The black arrows indicate a sample cell resistances. The grey dashed arrows indicate where the lower boundaries of the resistive states shift when reduced-impact read scheme is used.

variation in resistances is due to the change in the lumped crossbar parasitic resistance seen by the interpreter circuitry, depending on the location of the memory cell in the array. As more cells are programmed in the array, the resistances of the previously-programmed cells start to drift due to write disturbances, causing the spreading of the resistance distributions as shown in Figure 4.11(b). Most approaches in literature fail to address this spreading; however, write simulations on the whole array are performed sequentially to show the spreading and shifting of the states, and the results are presented in the next part of this section.

The reduced-impact scheme aims to shift the lower detection margins such that the new distributions fall completely within their intended resistive states.

Figure 4.12 shows the reduced-impact read operations performed on cells storing the four possible values. A cell located at the middle of the array is selected. The scheme uses 1.25 V pulse magnitude with 45 $K\Omega$ series resistance. In each case, when the controller receives the read enable signal, it facilitates the application of the bias voltages on the array. How Interp signals settle depending on the value that the cell is storing is shown. Interp signals settle in decreasing order, hence Out signals flip in the decreasing order as well. Signals R11, R10, R01 and R00 are the controller signals indicating whether a particular state is detected. The sequential flipping of Out signals also cause these signals to flip; however, the correct result is obtained at the end of the read operation.

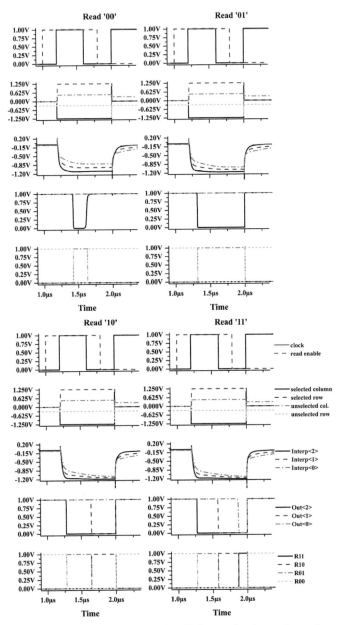

Figure 4.12   Reduced impact read operation. Relevant signals are shown for each state.

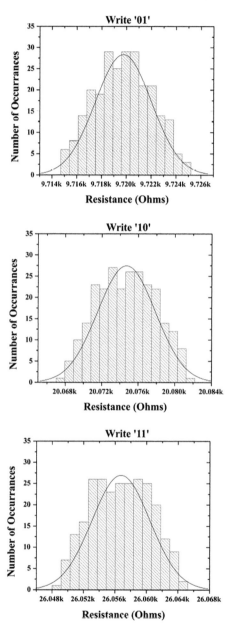

Figure 4.13   Resistance distributions after programming the entire array with the same value.

4.4.4 Resistance Distributions after Array Writes

The consecutive write operations on the neighboring cells cause the cell resistances to drift over time. Even though unselected and half selected cells are biased with low voltages, the leakages during the lengthy write durations add up and cause the array resistances to drift.

In order to quantify this drift, consecutive write operations on every cell of the array are simulated. The resulting resistance distributions are listed in Figure 4.13. The distributions include the effects of writes to the neighboring cells and the parasitic crossbar resistances as these values change depending on the location of the selected memory cell. The variations from the nominal value are within the margins that can be obtainable with the reduced-impact read scheme proposed in this section. Therefore, it is concluded that it is possible to write to the cells in the whole array, and then read the values using the presented read scheme while allowing margin for additional drift that can be caused by repeated read operations.

4.5 Conclusion

In this chapter, a multi-level memristor memory architecture that incorporates a novel read-monitored-write scheme was presented. The architecture allows for programming of the cells to the detection thresholds with very tight state distributions. In addition, an analytical model for state derivations which can be utilized to decide on the component parameters to be used in the design was presented. Various read schemes where voltage reduction or series resistance increase or a combination of both can be used to evaluate the resistive state of the memory cell were also presented. Finally, the resistive distributions after the entire array is written were shown. These distributions fall within the detection margins of the proposed read methods.

References

[1] Borghetti, J., et al. (2009), A hybrid nanomemristor/transistor logic circuit capable of self-programming, *Proceedings of the National Academy of Sciences of the United States of America*, *106*(6), 1699–1703.

[2] Burr, G. W., B. N. Kurdi, J. C. Scott, C. H. Lam, K. Gopalakrishnan, and R. S. Shenoy (2008), Overview of candidate device technologies for storageclass memory, *IBM Journal of Research and Development*, *52*(4.5), 449–464.

[3] Cassenaer, S., and G. Laurent (2007), Hebbian stdp in mushroom bodies facilitates the synchronous flow of olfactory information in locusts, *Nature*, *448*(7154), 709–713.

[4] Cauwenberghs, G. (1998), Neuromorphic learning vlsi systems: A survey, in *Neuromorphic Systems Engineering, The Kluwer International Series in Engineering and Computer Science*, vol. 447, edited by T. S. Lande, 381–408, Springer US.

[5] Chen, E., et al. (2010), Advances and future prospects of spin-transfer torque random access memory, *Magnetics, IEEE Transactions on*, *46*(6), 1873–1878, doi:10.1109/TMAG.2010.2042041.

[6] Choi, J., and B. Sheu (1993), A high-precision vlsi winner-take-all circuit for self-organizing neural networks, *Solid-State Circuits, IEEE Journal of*, *28*(5), 576–584, doi:10.1109/4.229397.

[7] Choi, S.-J., et al. (2011), Synaptic behaviors of a single metal-oxide-metal resistive device, *Applied Physics A: Materials Science and Processing*, *102*(4), 1019–1025.

[8] Chua, L., and S. M. Kang (1976), Memristive devices and systems, *Proceedings of the IEEE*, *64*(2), 209–223, doi:10.1109/PROC.1976.10092.

[9] Bao, B., Z. Liu, and J. Xu (2010), Steady periodic memristor oscillator with transient chaotic behaviours, *Electronics Letters*, *46*(3), 237–238, doi: 10.1049/el.2010.3114.

[10] Afifi, A., A. Ayatollahi, and F. Raissi (2009), Implementation of biologically plausible spiking neural network models on the memristor crossbar-based cmos/nano circuits, in *European Conference on Circuit Theory and Design*, ECCTD '09, IEEE, New York, doi:10.1109/ECCTD.2009.5275035.

[11] Chua, L. O. (1971), Memristor - missing circuit element, *IEEE Transactions on Circuit Theory*, *CT18*(5), 507–519.

[12] Cong, J., and B. Xiao (2011), mrfpga: A novel fpga architecture with memristor-based reconfiguration, in *Nanoscale Architectures (NANOARCH), 2011 IEEE/ACM International Symposium on*, 1–8, doi: 10.1109/NANOARCH.2011.5941476.

[13] Csaba, G., and P. Lugli (2009), Read-out design rules for molecular crossbar architectures, *Nanotechnology, IEEE Transactions on*, *8*(3), 369–374.

[14] Cutsuridis, V., S. Cobb, and B. P. Graham (2008), A ca2 + dynamics model of the stdp symmetry-to-asymmetry transition in the ca1 pyramidal cell of the hippocampus, in *Proceedings of the 18th international*

conference on Artificial Neural Networks, Part II, ICANN '08, 627–635, Springer-Verlag, Berlin, Heidelberg.

[15] Dan, Y., and M.-m. Poo (2004), Spike timing-dependent plasticity of neural circuits, *Neuron*, *44*(1), 23–30, doi: 10.1016/j.neuron.2004.09.007.

[16] Driscoll, T., Y. Pershin, D. Basov, and M. Di Ventra (2011), Chaotic memristor, *Applied Physics A: Materials Science and Processing*, *102*(4), 885–889, doi: 10.1007/s00339-011-6318-z.

[17] Ebong, I., and P. Mazumder (2010), Memristor based stdp learning network for position detection, in *Microelectronics (ICM), 2010 International Conference on*, 292–295, doi:10.1109/ICM.2010.5696142.

[18] Ebong, I., and P. Mazumder (2012), Cmos and memristor-based neural network design for position detection, *Proceedings of the IEEE*, *100*(6), 2050–2060, doi: 10.1109/JPROC.2011.2173089.

[19] Ebong, I., D. Deshpande, Y. Yilmaz, and P. Mazumder (2011), Multi-purpose neuro-architecture with memristors, in *Nanotechnology (IEEE-NANO), 2011 11th IEEE Conference on*, 431–435, doi:10.1109/NANO.2011.6144522.

[20] Ebong, I. E., and P. Mazumder (2011), Self-controlled writing and erasing in a memristor crossbar memory, *Nanotechnology, IEEE Transactions on*, *10*(6), 1454–1463.

[21] Eshraghian, K., K.-R. Cho, O. Kavehei, S.-K. Kang, D. Abbott, and S.-M. S. Kang (2011), Memristor mos content addressable memory (mcam): Hybrid architecture for future high performance search engines, *Very Large Scale Integration (VLSI) Systems, IEEE Transactions on*, *19*(8), 1407–1417, doi: 10.1109/TVLSI.2010.2049867.

[22] Ferrari, S., and R. F. Stengel (2004), Online adaptive critic flight control, *Journal of Guidance Control and Dynamics*, *27*(5), 777–786.

[23] Grupp, L. M., A. M. Caulfield, J. Coburn, E. Yaakobi, S. Swanson, P. Siegel, and J. Wolf (2009), Characterizing flash memory: Anomalies, observations, and applications, in *Proceedings of the 42nd International Symposium on Microarchitecture*, 24–33.

[24] Ho, Y., G. M. Huang, and P. Li (2009), Nonvolatile memristor memory: device characteristics and design implications, in *IEEE/ACM International Conference on Computer-Aided Design-Digest of Technical Papers*, ICCAD 2009, 485–490, IEEE.

[25] Hopfield, J. J., and D. W. Tank (1985), "Neural" computation of decisions in optimization problems, *Biological Cybernetics*, *52*(3), 141–152, doi: 10.1007/BF00339943.

[26] Hsu, S. T. (2005), Temperature compensated rram circuit, patent number: US 6 868 025.

[27] Indiveri, G. (2001), A current-mode hysteretic winner-take-all network, with excitatory and inhibitory coupling, *Analog Integrated Circuits and Signal Processing*, 28(3), 279–291.

[28] Ishikawa, M., et al. (2008), Analog cmos circuits implementing neural segmentation model based on symmetric stdp learning, in *Neural Information Processing, Lecture Notes in Computer Science*, vol. 4985, 117–126, Springer, Berlin, Heidelberg.

[29] Itoh, M., and L. O. Chua (2008), Memristor oscillators, *International Journal of Bifurcation and Chaos*, 18(11), 3183–3206, doi:10.1142/S0218127408022354.

[30] Itoh, M., and L. O. Chua (2009), Memristor cellular automata and memristor discrete-time cellular neural networks, *International Journal of Bifurcation and Chaos*, 19(11), 3605–3656, doi:10.1142/S0218127409025031.

[31] Jo, S. H., and W. Lu (2008), Cmos compatible nanoscale nonvolatile resistance, switching memory, *Nano Letters*, 8(2), 392–397.

[32] Jo, S. H., T. Chang, I. Ebong, B. B. Bhadviya, P. Mazumder, and W. Lu (2010), Nanoscale memristor device as synapse in neuromorphic systems, *Nano Letters*, 10(4), 1297–1301, doi: 10.1021/nl904092h.

[33] Joglekar, Y. N., and S. J. Wolf (2009), The elusive memristor: properties of basic electrical circuits, *European Journal of Physics*, 30(4), 661–675.

[34] Kaelbling, L. P., M. L. Littman, and A. W. Moore (1996), Reinforcement learning: A survey, *Journal of Artificial Intelligence Research*, 4, 237–285.

[35] Klein, R. M. (2000), Inhibition of return, *Trends in Cognitive Sciences*, 4(4), 138–147, doi: DOI: 10.1016/S1364-6613(00)01452-2.

[36] Koickal, T. J., A. Hamilton, S. L. Tan, J. A. Covington, J. W. Gardner, and T. C. Pearce (2007), Analog vlsi circuit implementation of an adaptive neuromorphic olfaction chip, *Circuits and Systems I: Regular Papers, IEEE Transactions on*, 54(1), 60–73.

[37] Kozicki, M., M. Park, and M. Mitkova (2005), Nanoscale memory elements based on solid-state electrolytes, *Nanotechnology, IEEE Transactions on*, 4(3), 331–338, doi:10.1109/TNANO.2005.846936.

[38] Lehtonen, E., and M. Laiho (2009), Stateful implication logic with memristors, in *Proceedings of the 2009 IEEE/ACM International Symposium on Nanoscale Architectures*, NANOARCH '09,

33–36, IEEE Computer Society, Washington, DC, USA, doi:10.1109/NANOARCH.2009.5226356.

[39] Lehtonen, E., and M. Laiho (2010), Cnn using memristors for neighborhood connections, in *Cellular Nanoscale Networks and Their Applications (CNNA), 2010 12th International Workshop on*, 1–4, doi:10.1109/CNNA.2010.5430304.

[40] Lin, Z.-H., and H.-X. Wang (2009), Image encryption based on chaos with pwl memristor in chua's circuit, in *International Conference on Communications, Circuits and Systems (ICCCAS)*, 964–968, doi:10.1109/ICCCAS.2009.5250354.

[41] Linares-Barranco, B., and T. Serrano-Gotarredona (2009), Memristance can explain spike-time-dependent-plasticity in neural synapses, *Nature precedings*, 1–4.

[42] Manem, H., G. S. Rose, X. He, and W. Wang (2010), Design considerations for variation tolerant multilevel cmos/nano memristor memory, in *Proceedings of the 20th symposium on Great lakes symposium on VLSI*, GLSVLSI '10, 287–292, ACM, New York, NY, USA, doi:10.1145/1785481.1785548.

5

Neuromorphic Building Blocks with Memristors

Idongesit Ebong and Pinaki Mazumder

In this chapter, a neuromorphic approach is presented, whereby spike-timing-dependent-plasticity (STDP) can be combined with memristors in order to withstand noise in circuits. It is shown that the analog approach to STDP implementation with memristors is superior to a digital-only approach.

5.1 Introduction

Neuromorphic engineering is not a new approach to information processing systems. It particularly gained momentum in the 1980s with the amalgamation of learning rules and VLSI technology [1]. The growing transistor integration density in CMOS enabled better simulation of the neural systems in order to verify models and nurture new bio-inspired ideas. Since then, the neuromorphic landscape has changed and neuromorphic chips and programs are now available that cater to specific applications and tasks.

Technological advancement has always been both friend and foe to neuromorphic networks. Neuromorphic networks are essentially more valuable in instances where parallel computing is necessary. In order to perform neuromorphic computing effectively, a large number of processing elements (PE) is needed [1]. In current CMOS technology, the density and connectivity required for more sophisticated neuromorphic systems does not exist. This has led many neuromorphic chips to implement various schemes that utilize virtual connectivity between processing elements.

The shortcomings of CMOS in terms of density and parallel computing encouraged more complex neuromorphic system techniques and designs. Although design complexity increased, the number of neurons, synapses and connections that can be simulated are orders of magnitude below the integration density of neurons in the human brain. Human beings, possessing neurons that operate in the millisecond range, can perform arbitrary image recognition tasks in tens to hundreds of milliseconds, while very powerful computers would take hours, if not days, to perform similar tasks. This lapse between digital computing and biology (specifically, the human brain) gives motivation for exploring technologies with connection densities that surpass anything CMOS can offer.

Low power and high device integration in nanotechnology have reignited a spark in the advancement of neuromorphic network in hardware as shown by Türel in [2] and Zhao in [3]. The "Crossnets" approach shown in [2] provides evidence of the design problems and methods of incorporation of resistive nanoscale devices in crossbar topology with CMOS circuitry to design neuromorphic circuitry. Nanotechnology, specifically the memristor as postulated by Chua, shows much promise in this area because it may overcome the inability to reach densities found in biological systems. This inability is reduced by two factors: the first is the small size of the memristors with respect to their functionality, and the second is the ability to connect the memristors with crossbars. Connecting these nano-devices (memristors) with nano-wires (crossbars) has been shown to increase device integration significantly [4]. Device integration in MMOST (Memristor-MOS Technology) is expected to improve in the age of memristors and crossbar scaling. A hypothetical study of a cortex-scale hardware, performed in [5], shows the use of nano-devices in a crossbar structure has the potential of implementing large-scale spiking neural systems. More complex algorithms like Bayesian inference [6] have also been studied for crossbar implementation, but these studies limit the crossbar array to digital storage. Analog use of the array would be ideal to reap its full benefits.

Neuromorphic networks derive their behavior from learning rules [7]. The networks have inherent governance that maintains relationships between neurons and synapses. Based on the myriad of combinations of synaptic weights and neuron behavior, the network at any given point in time is unique.

The goal of this chapter is to show that memristors are valuable in the development of biologically inspired adaptable circuitry. Three identified behaviors, well documented with biological neurons, will be introduced. These are lateral inhibition, spike timing-dependent plasticity (STDP), and inhibition of return (IOR). An approach to implementing these behaviors with memristors will be discussed. These behaviors are fundamental building blocks for neural hardware that have been well demonstrated in CMOS. This chapter will show a new, compact way of implementing STDP compared to pure CMOS. In addition, the chapter will also provide a method of realizing a reconfigurable XOR gate. The XOR gate is provided as an example, for in order to build more complex systems, both analog and digital methods will most likely be implemented. No specific recommendation is made for integration of analog/digital neuromorphic circuit blocks.

5.2 Implementing Neuromorphic Functions with Memristors

5.2.1 Lateral Inhibition

Lateral inhibition is seen prevalently in the biological world. This phenomenon has been credited with playing a part in amplifying variation in gradients [8], signaling orientation for vision processing and sensations [9] and providing form and structure during development and neurogenesis [10]. The inhibition process, a simple idea, seems to play a role in bio-logical processing to create complex schemes and structures such as leaf patterns on trees, branch formations and limbs on various organisms. The importance of inhibition for biological processing cannot be discounted. Although the inhibition process might seem a simple idea, its deconstruction from biological systems has not been so straightforward.

Inhibition plays a key role in neuron processing, so artificial neurons need to exhibit this behavior to closely approximate their biological counterparts. The lateral inhibition in artificial neural architectures exists as either total inhibition (as in the case of McCulloch-Pitts neurons [11]) or partial inhibition (as in the case of the perceptron [12]). Examples of these include the contrast enhancer using cross coupled transistors [13] and the winner-take-all (WTA) circuitry [14–18] that may be used for self-organizing maps [19]. These examples show that lateral inhibition has progressed and has been realized in neuromorphic hardware research. Adoption of memristor crossbar should further encourage and support the ease with which the inhibition pro-cess can be achieved since the lateral inhibition with memristors in crossbar simplify the circuitry and wire connections necessary with CMOS.

Lateral inhibition as well as recurrent network configurations can be achieved with memristors as shown in Figure 5.1. The memristor crossbar

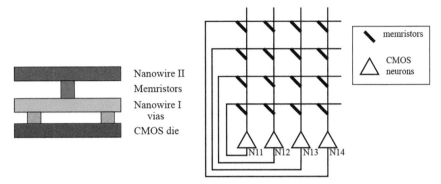

Figure 5.1   Recurrent network architecture showing an example of how Winner-TakeAll, lateral inhibition, or inhibition of return can be connected using crossbars.

allows massive connectivity from one neuron to another through modifiable weights. Neurons in the same functional vicinity can be made to inhibit one another through the crossbar configuration. For example, N11 is connected to N12 through some synapse M1112; the signal injected through this synapse M1112 from N11 will be an inhibitory signal that will disturb the internal state of neuron N12. This crossbar method can also be extended to excite neighboring neurons. In this neuromorphic approach, two memristor crossbars can be stacked upon one another: one for excitatory synapses and the other for inhibitory synapses.

In addition to lateral inhibition, self-enhancement seems to play a key role in neurogenesis [10]. An effect measured in biological neurons seems to be lateral inhibition of neighboring neurons but self-enhancement of oneself. This effect prevents a feature or neuron from inhibiting itself. For example, when a leaf forms on one part of a branch, an area around the leaf receives an inhibitory effect that suppresses the formation of other leaves too close. Since this inhibitory effect applies to an area that includes the inhibiting leaf itself, in order to combat its inhibitory effect, the leaf has a positive feedback loop that reinforces its continued development and existence. This self feedback loop can be made with the memristor crossbar as shown in Figure 5.1.

5.2.2 Inhibition of Return

IOR, in its hardware implementation, is a neuromorphic algorithm used to allow different neurons to spike [20]. From the previous section, lateral inhibition implementing WTA only allows for one neuron to be considered the winner when in competitive spiking with its neighbors. By combining

WTA with IOR, the behavior of the winner changes, for successive neurons will take the winner's place after a designed time period. By implementing this combination, the winner inhibits itself after an allowed spike duration and gives rise for another spiking neuron to win. This algorithm can be used to map network activity as well as compare different input pattern intensities. No surprise, it is mostly used in visual neuromorphic applications, such as attention shifts [21].

Memristor MOS Technology (MMOST) design of IOR can be accomplished in a similar way as the WTA. The self-feedback parameter (synapse) would be strengthened so the neuron will inhibit itself strongly as its spiking frequency increases. The neurons with the strongest synaptic inhibitions (lowest synaptic weights) can be compared with one another with respect to synaptic strength in order to determine the current relationship between them.

5.2.3 Coincidence Detection

Coincidence detection occurs when two spiking events are linked and coded for in a certain way. This algorithm is usually found in pattern recognition or classification systems, whereby the neuromorphic network codes differently an input train of pulses or spikes. Based on the level of coincidence between different inputs to the network, the neural network responds appropriately. This realization is not the only way to use coincidence detection.

Another way to use coincidence detection is to update synaptic weights based on coincidence. This relates to the plasticity of the synapse and governs the learning rule of the synapse locally. In this form, the coincidence detection is known as STDP [22].

There are two main forms of STDP: symmetric STDP and asymmetric STDP (as depicted in Figure 5.2). Symmetric STDP performs the same weight adjustments, independent of the spike order between the pre-neuron and the post-neuron, while asymmetric STDP reverses weight adjustment based on the spike time difference between the pre-neuron and the post-neuron.

STDP implementations utilizing the crossbar structure have been proposed [23–25].

In their current state, they do not provide much density gains when comparing MMOST to CMOS. The implementations require pulse/signal generations in both the positive and negative directions across the memristor. Snider [25] proposes a decaying pulse width while Linares-Barranco and

Serrano-Gotarredona [24] and Afifi et al. [23] propose decaying sig-nal amplitudes. All three suggested implementations rely on the additive effect of the signals across the memristor to control the synaptic weight changes. The STDP synaptic weight implementation in this thesis is realized with a different approach; pulses are used to make a linear approximation of the STDP curve in order to reduce the size of the neuron.

The proposed STDP implementations are usually of the form in Figure 5.2. These synaptic behaviors, both asymmetric and symmetric, have been implemented in CMOS [23, 26, 27]. In the asymmetric STDP case, if the pre-neuron spikes before the post-neuron, the synaptic weight is increased. If the order of spikes is reversed, the synaptic weight is decreased. In both cases, the larger the duration between the preneuron and the post-neuron spikes, the lesser the magnitude of the synaptic change. Most circuit implementations take advantage of the asymmetric implementation.

The STDP implementation in this work is asymmetric and is based on the equation in the form of (5.1):

$$\Delta W(t_2 - t_1) = \begin{cases} A_+ e^{(t_2 - t_1)/\tau_+}, & t_2 - t_1 > 0 \\ -A_- e^{(t_2 - t_1)/\tau_-}, & t_2 - t_1 < 0 \end{cases} \tag{5.1}$$

The change in synaptic weight, ΔW, is dependent on spike time dif-ference between the pre-neuron and the post-neuron, $t_2 - t_1$. A_+ is the maximum change in the positive direction, A_- is the maximum change in the negative direction, and both changes decay with time constants τ_+ and τ_-, respectively. Most implementations use capacitors and weak inversion transistors to adjust τ_+ and τ_- in order to obtain decay times in the hundreds of milliseconds [28]. An alternate way to realize STDP in CMOS when working under a lower area budget is to incorporate digital storage units that can help remember spike states instead of using huge analog capacitors to set time constants.

The total change in weight for a given synapse is the summation of all positive and negative weight changes. Over the learning period, the synapse will converge to a certain weight value and will remain stable at that value. The STDP concept was tested through Verilog simulations, whereby STDP was pitted against digital computation to do a comparison under noisy conditions.

The network of interest for simulation was that of a 1D position detec-tor, where the location of an object is determined by the two-layered neural network presented in Figure 5.3. The network consists of an input

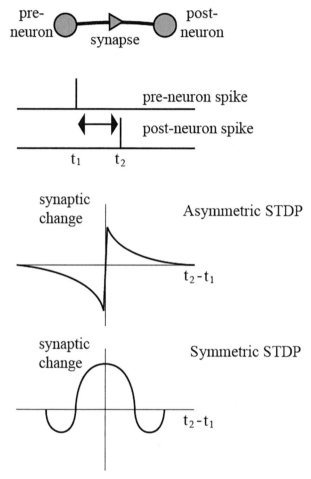

Figure 5.2 STDP curves showing relationship between synaptic weight change and the difference in spike times between the pre-neuron and the post-neuron. Symmetric STDP and Asymmetric STDP are both found in nature [29].

neuron layer (neurons labeled n_{11} through n_{15}) connected through feedforward excitatory synapses to an output neuron layer (neurons labeled n_{21} through n_{25}). At the output layer, each output neuron is connected to every other output neuron through inhibitory synapses.

The network shown in Figure 5.3 updates its synaptic weights through STDP. Both excitatory (gray triangles) and inhibitory (red triangles) synaptic weights are modified through STDP. The inherent competition resulting when

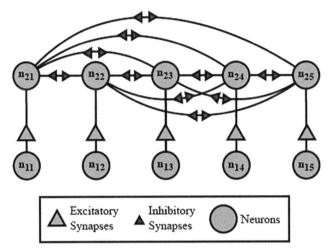

Figure 5.3   Neural network implemented in Verilog in order to determine noisy performance of STDP in comparison to digital logic.

the output neurons spike help establish the weights for all 20 inhibitory synapses. An object is presented to the line of input neurons shown in Figure 5.3. The object's presence generates signals that affect the closest neurons to its position. For example, if the object is directly in front of n_{13}, then only n_{13} receives the object's generated signals, but if the object lies between n_{13} and n_{14}, then both n_{13} and n_{14} receive the input signals. The object's position is deciphered from the output neuron based on the relative spiking frequency (or period) of the output neurons.

The 1D position detection was simulated for two noise conditions – noise-free condition and noisy condition – with different object locations. The noise-free case results are trivial. If there is no noise in the input of the system, then the output neuron results can be reduced to binary outputs – spike or no spike. For example, in the noise free case, an object placed next to n_{13} causes n_{23} to spike while the other input and output neurons do not spike. In this noise free case, the implementation of this position detection function could have been accomplished with digital logic where input signals exceeding some threshold would provide the desired output. In the noise-free case, when the object is placed between n_{12} and n_{13}, both n_{22} and n_{23} spike but the relationship between their spiking frequencies is proportional to the input object's exact location between both n_{12} and n_{13}. If the object is closer to n_{13}, then the spiking frequency of n_{23} is a little greater than that of n_{22}.

Table 5.1 Verilog STDP output neuron results for an object placed at different locations on the 1D position detection line is exactly midway between n_{12} and n_{13}, then both n_{22} and n_{23} spike with the same spiking period

Output Neurons	Object at n_{13}	Period (Time between Successive Spikes) Object between n_{12} & n_{13} but closer to n_{13}	Object Midway between n_{12} & n_{13}
n_{21}	1746	2046	1014
n_{22}	786	684	660
n_{23}	636	642	660
n_{24}	786	3030	1506
n_{25}	1746	7242	7266

The noise free condition provides direct mapping of either a spike or a no spike with neurons involved in receiving the object's input and those not receiving the object's input.

The noisy condition case is a bit more interesting, and the results are summarized in Table 5.1. Table 5.1 provides results for the noisy case whereby all neurons in the output layer spike due to the noise background effect fed in through the input layer. The units in the simulation are time units or simulation time steps. Period is determined after weight stabilization has occurred and the time between successive spikes becomes fairly regular. The object's position can be determined in all three cases presented in the table. When the object is at n_{13}, n_{23} spiking period is the lowest (n_{23} is spiking the most). When the object is between n_{12} and n_{23} but closer to n_{13}, n_{23} spikes the most but its spiking period is comparable to n_{22}. A second level processing can compare these two neurons' spiking period to determine the object's location relative to the two neurons that spike the most. Lastly, when the object is exactly midway between n_{12} and n_{13}, then both n_{22} and n_{23} spike with the same spiking period.

An extension of these results may be used for motion detection. Looking at the spiking response of n_{23}, we may conclude that the spiking period decreases as the object moves away from n_{13}. The advantages therefore seen in using STDP is that by determining the object's position using the spiking frequency, the neural network can withstand the effects in a noisy background while digital threshold logic fails.

5.3 CMOS-Memristor Neuromorphic Chips

The validity of memristors as processing elements is investigated using two neuromorphic architectures that exhibit lateral inhibition as well as STDP.

The first architecture is for a local "position detector" and the second architecture is a multifunction chip that can be trained to perform digital gate functions such as the XOR function. The XOR function is later extended to perform edge detection.

5.3.1 Analog Example: Position Detector

Procedure: Given a two dimensional area, split up the area into a 5 × 5 grid shown in Figure 5.4. Each square on the grid represents the resolution for the detector. A neuron resides at the center of each square on the grid. The detector has a two-dimensional layer of neurons. Each neuron is connected to its immediate neighbor through synapses. Each synaptic connection is unidirectional, so by having two connections, there is a bidirectional information flow between neighboring neurons. Each neuron is a leaky-integrate-and fire (LIF) neuron. Each has a leaky capacitor that stores integrated input information.

Two design methodologies were taken in order to achieve STDP. The first is the CMOS design which is based on previous work in literature in order to provide a basis for the state of the art, while the second is the MMOST design used to specifically provide a new way of achieving STDP with area-conscious neuron design. The CMOS design will be explained briefly because the implementation is not exactly new, and the MMOST design decisions will be expanded upon to show that STDP really can be implemented in a

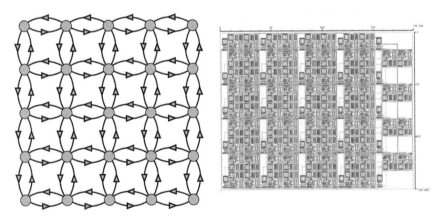

Figure 5.4   Neuron layer connectivity showing position detector architecture (circles are neurons and triangles are synapses). The left figure shows the connectivity matrix while the right shows the CMOS layout (190 μm × 152 μm).

Table 5.2 Design summary for both proposed WTA CMOS and MMOST 5X5 Position Detector Arrays

	CMOS	MMOST
Timing	Asynchronous	Clocked (1 kHz)
Power (Static, Dynamic max)	0.2 μW, 55 μW	5.28 μW, 15.6 μW
Chip Area	$2.89 \times 10^{-4} \text{cm}^2$	$6.1 \times 10^{-5} \text{cm}^2$
Input Noise(0.3V noise level)	> 3 dB SNR	> 4.8 dB SNR

way that does not consume too much area. Lastly, the comparison results will be explicated in context so that apples are not compared to oranges due to different design decisions. The design summary is given in Table 5.2.

5.3.1.1 CMOS Design Description

The CMOS design has an LIF neuron with multiple inputs depending on the location within the position detection fabric. The neuron is inspired by designs with complimentary inputs, which has PMOS (pull ups) for excitatory inputs and NMOS (pull downs) for inhibitory inputs. Each neuron has only one pull up and multiple pull downs depending on the location in the position detector fabric, e.g., four pull downs for neurons surrounded by four neighbors. The STDP synapse approach is similar to those already presented in literature [18, 28] and the synapse schematic is shown in Figure 5.5.

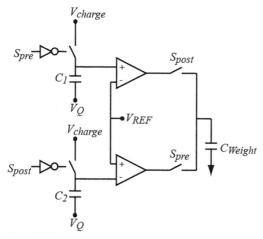

Figure 5.5 STDP synapse circuit diagram implemented in CMOS.

When the pre-neuron spikes, S_{pre} activates a switch that charges C_1. When S_{pre} deactivates, C_1 discharges exponentially, but the capacitor C_{Weight} is not updated until there is a post-neuron spike event. A post-neuron spike event would activate S_{post}, therefore allowing the evaluated output of the top comparator to see C_{Weight}. This explained sequence describes long term potentiation (LTP). The post-spiking before the pre-spiking would entail long term depression (LTD). To reduce area, the capacitors C_1 and C_2 were implemented with diode connected NMOS transistors operating in weak inversion. The voltage range between V_{charge} and V_Q is made to be about 100 mV. The decay shape of the voltages across C_1 and C_2 from V_{charge} to V_Q is a function of the difference between V_{charge} and V_Q. By reducing the voltage range, the decay appears more linear than exponential.

5.3.1.2 Memristor-MOS design description

The MMOST design will be delved into with more detail than the CMOS design. The design goal is to take advantage of the memristor crossbar, thereby simplifying the synapse and making it a fraction of the size of the CMOS synapse. The synapse itself is a simple memristor whose changes respond to pulses of equal widths provided through the neurons. STDP mechanism is moved from the synapse to the neuron. The neuron design utilizes a new way of realizing STDP by striking a tradeoff between neuron area and asynchrony. The neuron implementation of STDP is depicted graphically in Figure 5.6. The STDP behavior modeled is based on a linear approximation behavior observed in mushroom bodies as shown in [30].

Figure 5.6 shows the spike patterns between a pre-neuron's output and a postneuron's input (the memristor lies between these two terminals). In Figure 5.6, the pre-neuron spikes right before time t_0, so at time t_0, the pre-neuron's output is at 0 V. The 0 V level is held for 4 clock cycles (from t_0 to t_3) then pulses are allowed to pass for another 4 clock cycles (from t_4 to t_8). Afterwards, the pre-neuron's output rests at a reference voltage, V_{REFX}. The post-neuron's input exhibits a similar behavior as the pre-neuron's output, but instead of spiking before time t_0, it spikes sometime in the interval from t_2 to t_3. The post-neuron's input is pulled to 0V at time t_3, as opposed to time t_0 as the pre-neuron's output. The pre-neuron's output and the postneuron's input spiking patterns present a difference across the memristor's output, and this difference is shown in Figure 5.6 as "pre"-"post". As explained earlier, the memristor utilized is a threshold device, meaning its conductance experiences greater change when a voltage greater than its threshold voltage, *vth*, is met.

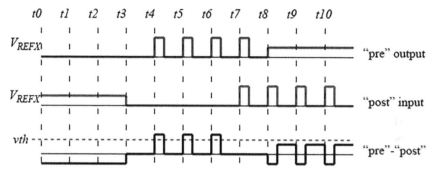

Figure 5.6 Pre-neuron and post-neuron spiking diagram showing three pulses above the memristor's threshold. The below threshold pulses do not greatly influence conductance.

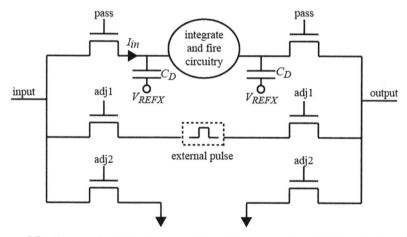

Figure 5.7 Neuron circuit that can provide spiking pattern for STDP realization with memristors.

The threshold is exceeded only by the three pulses shown in Figure 5.6. The neuron circuit that can implement the spiking patterns depicted in Figure 5.6 is shown in Figure 5.7.

The neuron in Figure 5.7 is composed of an integrate and fire circuitry, a path for passing an inhibitory current signal I_{in} to the integrate and fire circuitry (pass), paths for pulling the neuron's input and output nodes high (adj1), and paths for pulling both its inputs and output nodes low (adj2). The control signals (pass, adj1, and adj2) to turn each path on is controlled by the Finite State Machine (FSM) shown in Figure 5.8.

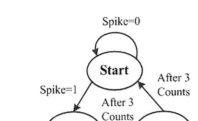

Figure 5.8   FSM showing control signal generation.

In Figure 5.8, Start is the default state — the neuron is not spiking, the neuron's input and output voltages are at reference voltage (V_{REFX}), pass is ON, adj1 is OFF, and adj2 is OFF. When the neuron receives excitatory inputs from the environment enough to cause a spike, then spike becomes 1, and in the next clock cycle, the neuron moves to the next state, Low. In the Low state, both the input and output ports of the neuron are pulled to 0 V — the neuron has spiked, pass is OFF, adj1 is OFF, and adj2 is ON. The neuron stays in this state for 4 clock cycles (a counting variable increments from 0 to 3) before moving to the Pulse state. The Pulse state is the state where the neuron passes the external pulse to both its input and output ports — pass is OFF, adj1 is ON, and adj2 is OFF. In order to move from Pulse to Start, a counting mechanism is employed for 4 clock cycles. This internal FSM resides within each neuron.

5.3.1.3 Comparison of CMOS design and memristor design
The CMOS design is an asynchronous design in which minor perturbations on a neuron's excitatory input can cause a spiking event. The MMOST design is a clocked design that synchronizes OFF-chip signals with the ON-chip logic. The MMOST design itself has asynchronous parts to it (neuron integration and signal input), but the timing of change in resistance of the memristor is a synchronous event. The WTA algorithm allows for spiking neurons to inhibit one another while changing synaptic weights to strengthen or weaken the inhibition. The change of synaptic weight for both the CMOS and memristor or MMOST design qualifies as the ability for the chip to learn. The advantage of choosing an STDP design is to take advantage of its noise handling capability. The lower the noise level, the lower the difference between signal and noise necessary for position detection. In comparing the CMOS and MMOST designs, the MMOST design has a higher potential

because consumes less area and requires less operating power. The quoted values in Table 5.2 for the MMOST design for both power and area are over-estimations, so the possibility of improving over CMOS with this technology is very appealing. This is without even considering potential synaptic and neuronal densities that can be achieved. The local connections adopted for this example are beneficial for the CMOS numbers but increasing the neighborhood connections will have a larger detrimental effect on CMOS density than on MMOST density.

Design complexity: For the current implementation, the timing of the CMOS circuitry is designed to perform STDP in the tens of microsecond range in order to conserve area. This value can be adjusted by using bigger capacitors (C_1 and C_2 in Figure 5.5) to extend the time constant or by putting the synaptic transistors (those that implement switches and comparators) even more into subthreshold. The CMOS design can become very complex when trying to design for its most dismaying feature: volatility. Currently, when the stimulus is removed, the weight decays exponentially to its DC steady state in about 100 ms, since synaptic weight is stored on capacitors. A way to improve this design would be to save these weights to memory and incorporate read, write and restore schemes which requires careful timing requirements.

The Chip area (5×5 array) for the CMOS design is about $2.9 \times 10^{-4} \mathrm{cm}^2$ from the CMOS layout, while that for the MMOST is about $6 \times 10^{-5} \mathrm{cm}^2$. The memristor design area is an over-estimation, so it is likely to be much less than the proposed value. From design automation, the current logic for the memristor design is expected to take about 488 minimum sized transistors. Since this automated design was not simulated for signal integrity, drive, etc., for a worst case scenario, we double this value by 2 in order to account for various signal buffering, clock signal regeneration, and via spaces to the crossbar structure. This gross estimation still shows that the memristor design consumes five times lower area than the CMOS design. This value can only improve, for a custom design would use fewer transistors. The area estimation assumes that the crossbar array area will be fully contained over the CMOS area.

Power: The CMOS design consumes less static power than the memristor design, mostly due to the fact that both designs are operating under different supply voltages (1 V for CMOS, 1.5 V for MMOST), and the memristor design has only a few transistors operating in the weak inversion region. The operating voltage difference is due to the fact that memristors will need to exceed a threshold voltage in order to change resistance, and the largest voltage across the memristor with under the 1.5 V power supply is about 0.9 V.

The static power can be reduced for later generations of the design by having a lower voltage supply and using charge pumps to achieve required threshold voltages. Although the static power consumption for CMOS is lower, its maximum dynamic power is higher than that of the memristor design. The memristor design consumes 15.6 μW, while the CMOS design consumes 55 μW. The memristor logic and comparators use most of the power due to heavy switching during spiking events. In the case of CMOS, as neurons begin to inhibit one another, they create or strengthen paths to ground allowing larger current draw especially when both excitatory and inhibitory inputs are activated. This current adds up pretty quickly as array size increases.

Noise: Both the CMOS and memristor designs were tested with a noise background between 0.1 V and 0.3 V. The conclusion for testing under CMOS is as noise level increases, the required signal level to counter this noise also increases. For example, at a noise level of 0.2 V, as long as the signal is at least 0.3 V, the neuron of interest will spike accordingly. This is a 100 mV difference between signal and noise. This value changes to 125 mV with the noise level increases to 0.3 V. In real world computing, we do not expect the noise to be quite that high, but as long as the signal level is above 0.425 V, the neural network will work as designed. For the memristor design, the noise level is actually used to randomly assert the memristors at different conductance states. Once the network is stabilized under a certain noise level, the signal input is capable of tuning the memristors around its signal level for the detecting purpose. The noise levels used for simulation are similar to that of the CMOS design (0.1 V, 0.2 V, and 0.3 V). At 0.3 V, as long as the input is about 200 mV greater than the noise level, then the signal is discernible.

5.3.2 Digital Example: Multi-function Chip Architecture

The previous example showed that through analog computation, localized signal detection can be computed. This section will show that digital functions can also be achieved with the proposed neuron design. The approach that will be presented may actually use more area than a digital approach would require, however the current approach can be reconfigured and can also interface well with other analog components. The multi-function chip architecture is shown in Figure 5.9. The neurons are shown in circle and implemented in CMOS while synapses are shown with arrowed lines and are implemented with memristors. Excitatory synapses are in red, and inhibitory synapses are in black. The architecture is amenable to STDP synapses whereby the spike timing between pre-neuron and post-neuron

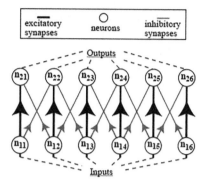

Figure 5.9 Neuromorphic architecture for the multi-function digital gate showing neurons and synapses.

determine how much the memristive synapses will adjust. The same approach to achieving STDP is used with this architecture as in the position detection architecture.

The neuromorphic architecture is composed of both input and output neurons, and based on the chosen structure of inhibitory to excitatory synapses, various functions can be obtained. The XOR and Edge Detector have the same synaptic weight profile, but perform different functions. An AND or an OR gate would have a different synaptic profile than both XOR and Edge Detection. The basic architecture shown needs pre- and post-processing circuits to interface with other systems. The postprocessing side may contain adders and integrators to convert the spiking outputs of the spiking neurons to leveled signals, while the pre-processing side would convert DC level signals to spiking inputs for the neurons. Each function will have different postprocessing requirements; the architecture is meant to be the barebones to allow for different functionality based on synaptic weight adjustments.

The training process involves using input patterns in order to adjust memristors to the desired relative values between excitatory and inhibitory synapses. On simulation startup, weights can either be initialized to a low value, or initialized to a random pattern and learned to low values. Using different input patterns, the memristors can be trained to predetermined weights or relative weights between excitatory and inhibitory synapses. For example, starting in a low weight state, n_{11}, n_{13}, and n_{15} can be made to spike at frequencies that cause n_{21}, n_{23}, and n_{25} to spike, thereby strengthening excitatory synaptic connection between these neurons according to STDP rules. This input pattern will not affect the inhibitory synapses due to the rules of STDP requiring pre-neuron and post-neuron to spike. After these synapses are trained to weights approximately twice the inhibitory synaptic weights, neurons n_{12}, n_{14}, and n_{16} are used to train the excitatory synapses between n_{22}, n_{24}, and n_{26}.

This training scheme is designed for the XOR and edge detection profile and allows the tuning of excitatory synapses without affecting inhibitory synapses as shown in Figure 5.10. The synapse naming follows the convention "pre-neuron post-neuron". In Figure 5.10, the XOR training is done for 30 ms to get a resistance profile for the excitatory neurons around 5.6 MΩ. The tuning of the memristors to exact resistance values is hard to accomplish, therefore, in a system, a timer would be used to stop training. This training scheme hints that the neurons have two different modes determined by a control signal deciding on either a training mode or running

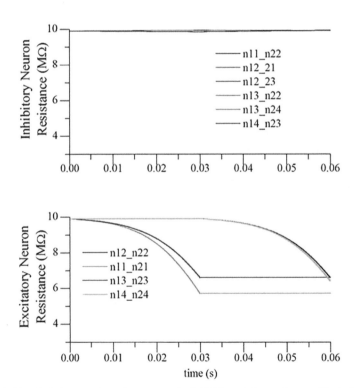

Figure 5.10 Training mode using prescribed XOR training scheme (Top) Inhibitory synapses unchanged during training (Bottom) Excitatory synapse trained using timestamps.

mode. The difference between the two modes lies in the voltage levels used for both. The training mode uses voltage levels that influence the memristors more than the running mode. The simulation results shown for the XOR and edge detection operation use the learned memristor resistance values of about 10 MΩ and between 5.6 and 6.8 MΩ for the inhibitory synapses and the excitatory synapses, respectively. The simulation results are shown in the run mode: learning has stabilized, and voltages adjusted so memristors are fairly static.

XOR Simulation: The neuromorphic architecture is simulated in the Cadence Analog Environment with the IBM 90 nm CMOS9RF process. The XOR simulation setup does not use all six input-output neuron pairs. Four neuron pairs are needed for the XOR operation. For example, to find the XOR between logic signals A and B, input A would be given to n_{11} and n_{12} while input B would be passed on to n_{13} and n_{14}. The outputs would be read from the sum of n_{22} and n_{23}. Figure 5.11 provides the results for the XOR operation for all cases. Figure 5.11a provides results for the case when both inputs A and B are Logic "0" thereby producing no spiking behavior at the outputs.

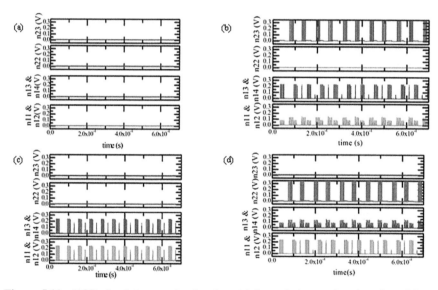

Figure 5.11 XOR simulation results showing: (a) Input A (n_{11} and n_{12}) = 0, and Input B (n_{13} and n_{14}) = 0, so Output (n_{22} or n_{23}) = 0. (b) Input A = 0, and Input B = 1, so Output = 1. (c) Input A = 1, and Input B = 1, so Output = 0. (d) Input A = 1, and Input B = 0, so Output = 1.

Figure 5.11b and Figure 5.11d provide the scenarios when one input is Logic "1" and the other input is Logic "0".

In Figure 5.11b, Input A is Logic "0" and Input B is Logic "1". The result from the simulation shows that n_{23} spikes in a pattern that signifies Logic "1" while n_{22} does not spike at all. The XOR post processing will integrate and add the results of n_{22} and n_{23} to obtain a final verdict. The spiking behavior of either n_{22} or n_{23} should be deciphered as a Logic "1" by the post-processing circuitry. Figure 5.11d provides results and works in a similar way to Figure 5.11b except this time, instead of n_{23} spiking and n_{22} not spiking, n_{23} does not spike but n_{22} spikes. The results from post-processing will be the same as the previous case.

Lastly, Figure 5.11c shows the case when both inputs A and B are Logic "1". The results show that neither n_{22} or n_{23} spikes therefore providing output results similar to Figure 5.11a. As expected, the XOR operation is verified with all test cases and shows that the neuromorphic architecture works as expected. Due to the bidirectional nature of the output node, Logic "0" when inputs do not induce spiking is different from Logic "0" when inputs induce spiking. For example, the Logic "0" seen for n_{11} and n_{12} in Figure 5.11d looks different from that of Figure 5.11a. The disturbance seen is directly related to the spiking behavior of the second layer of neurons. The pulses from this layer directly cause a disturbance in the output node of the input neurons.

Edge detection simulation: The edge detector operation is similar to XOR as shown in Figure 5.12. In Figure 5.12, the input neurons $n_{11},...,n_{16}$ receive '011110', respectively, and they cause the output neurons $n_{21},...,$ n_{26} to produce '010010', respectively. In the input pattern, there are two edges, i.e., between n_{11} and n_{12} and between n_{15} and n_{16}, and the neural network configuration was able to extract these edges in the output spiking pattern. The post-processing on the edge detector will integrate each output to determine output logic level. The verification of the edge detector is done by showing another pattern with input neurons $n_{11},...,n_{16}$ receiving '100110', respectively. This pattern clearly has two edges between n_{13} and n_{14} and between n_{15} and n_{16}. Another observation here is that since there is no wrap-around effect in the neural architecture, the neural network identifies Logic "1" values at the extremes as edges. This design decision is architecture dependent and behavior may be changed by modifying the synaptic weights of the synapses controlling neuron behavior at the extremes. The result for the input pattern "100110" turns out to be "100110" and the post-processing for the edge detection should be able to extract the position of the edges quite clearly.

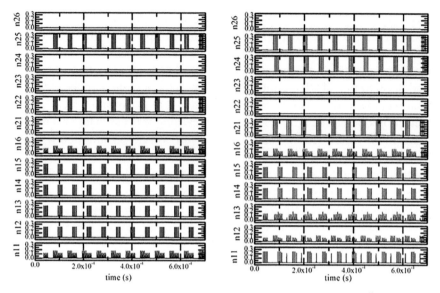

Figure 5.12 (Left) Edge detection simulation results for input pattern "011110" produces output pattern "010010". (Right) Edge detection simulation results for input pattern "100110" produces output pattern "100110".

The CMOS neuron from the position detector is scaled down from 130 nm CMOS process to 90 nm. In addition to process migration, the inclusion of synaptic weight dependent excitatory inputs was made. The neuron design was migrated from a mostly saturation design to a mostly subthreshold design to improve power efficiency. When conducting the simulation for edge detection and XOR, the average power consumption per neuron during the spiking is about 0.3 μW, which is more than an 8X saving over our position detector neuron.

5.4 Chapter Summary

We have explored the benefits of moving to an MMOST design for STDP circuit implementation on the bases of circuit area, power, and noise. The area considerations are implementation dependent, but scaling to denser networks favors the MMOST design, for a CMOS implementation will require more STDP synapses, which greatly limit connectivity. The power considerations show mixed results because moving to synchronous STDP for the MMOST implementation may actually waste more power in the idle state than the CMOS implementation. Dynamic power numbers are better for MMOST, so a more active circuit would take advantage of the MMOST design. The noise considerations show that both designs are comparable. However, this may change with device scaling, as both memristors and CMOS transistors become more susceptible to noise.

In addition to the STDP circuitry, a neuromorphic architecture for digital computation is proposed. The architecture is shown to perform the XOR and edge detection operations after a supervised learning process. The design is simulated in the 90 nm IBM CMOS process with power consumption while spiking at 0.3 μW. The amenable architecture is great for the memristor crossbar design, allowing the area savings possible with building crossbars above CMOS circuitry. The overall purpose of this work is to explore low level computing components that can utilize nanodevices in a manner that encourages parameter adjustment in order to facilitate on-site tuning when necessary.

References

[1] Treleaven, P. C. (1989), Neurocomputers, *Neurocomputing*, *1*(1), 4–31, doi:10.1016/S0925-2312(89)80014-1.

[2] Türel, O., J. H. Lee, X. Ma, and K. K. Likharev (2004), Neuromorphic architectures for nanoelectronic circuits, *International Journal of Circuit Theory and Applications*, *32*(5), 277–302, doi:10.1002/cta.282.

[3] Zhao, W. S., G. Agnus, V. Derycke, A. Filoramo, J.-P. Bourgoin, and C. Gamrat (2010), Nanotube devices based crossbar architecture: toward neuromorphic computing, *Nanotechnology*, *21*(17), 175–202.

[4] Strukov, D. B., and R. S. Williams (2009), Four-dimensional address topology for circuits with stacked multilayer crossbar arrays, *Proceedings of the National Academy of Sciences*, *106*(48), 20155–20158, doi:10.1073/pnas.0906949106.

[5] Zaveri, M. S., and D. Hammerstrom (2011), Performance/price estimates for cortex-scale hardware: A design space exploration, *Neural Networks*, *24*(3), 291–304, doi:10.1016/j.neunet.2010.12.003.

[6] Zaveri, M. S., and D. Hammerstrom (2008), Cmol/cmos implementations of bayesian polytree inference: Digital and mixed-signal architectures and performance/price, *Nanotechnology, IEEE Transactions on*, *9*(2), 194–211.

[7] Cauwenberghs, G. (1998), Neuromorphic learning vlsi systems: A survey, in *Neuromorphic Systems Engineering, The Kluwer International Series in Engineering and Computer Science*, vol. 447, edited by T. S. Lande, 381–408, Springer US.

[8] von Békésy, G. (1968), Mach- and hering-type lateral inhibition in vision, *Vision Research*, *8*(12), 1483–1499, doi:10.1016/0042-6989(68)90123-5.

[9] Blakemore, C., and E. A. Tobin (1972), Lateral inhibition between orientation detectors in the cat's visual cortex, *Experimental Brain Research*, *15*(4), 439–440, doi:10.1007/BF00234129.

[10] Meinhardt, H., and A. Gierer (2000), Pattern formation by local self-activation and lateral inhibition, *BioEssays*, *22*(8), 753–760.

[11] McCulloch, W., and W. Pitts (1943), A logical calculus of the ideas immanent in nervous activity, *Bulletin of Mathematical Biology*, *5*(4), 115–133, doi:10.1007/BF02478259.

[12] Rosenblatt, F. (1958), The perceptron: A probabilistic model for information storage and organization in the brain, *Psychological Review*, *65*(6), 386–408, doi:10.1037/h0042519.

[13] Wolpert, S., and E. Micheli-Tzanakou (1993), Silicon models of lateral inhibition, *Neural Networks, IEEE Transactions on*, *4*(6), 955–961, doi:10.1109/72.286890.

[14] Indiveri, G. (2001), A current-mode hysteretic winner-take-all network, with excitatory and inhibitory coupling, *Analog Integrated Circuits and Signal Processing*, *28*(3), 279–291.

[15] Pedroni, V. A. (1995), Inhibitory mechanism analysis of complexity o(n) mos winner-take-all networks, *Circuits and Systems I: Fundamental Theory and Applications, IEEE Transactions on*, *42*(3), 172–175.

[16] Pouliquen, P. O., A. G. Andreou, and K. Strohbehn (1997), Winner-takes-all associative memory: A hamming distance vector quantizer, *Analog Integrated Circuits and Signal Processing*, *13*(1), 211–222.

[17] Serrano-Gotarredona, R., et al. (2009), Caviar: A 45k neuron, 5m synapse, 12g connects/s aer hardware sensory-processing-learning-actuating system for high-speed visual object recognition and tracking, *Ieee Transactions on Neural Networks*, *20*(9), 1417–1438.

[18] Urahama, K., and T. Nagao (1995), K-winners-take-all circuit with o(n) complexity, *Neural Networks, IEEE Transactions on*, *6*(3), 776–778.

[19] Choi, J., and B. Sheu (1993), A high-precision vlsi winner-take-all circuit for self-organizing neural networks, *Solid-State Circuits, IEEE Journal of*, *28*(5), 576–584, doi:10.1109/4.229397.

[20] Klein, R. M. (2000), Inhibition of return, *Trends in Cognitive Sciences*, *4*(4), 138–147, doi:10.1016/S1364-6613(00)01452-2.

[21] Morris, T. G., and S. P. DeWeerth (1996), Analog vlsi circuits for covert attentional shifts, in *Microelectronics for Neural Networks, 1996., Proceedings of Fifth International Conference on*, 30–37.

[22] Dan, Y., and M.-m. Poo (2004), Spike timing-dependent plasticity of neural circuits, *Neuron*, *44*(1), 23–30, doi:10.1016/j.neuron.2004. 09.007.

[23] Afifi, A., A. Ayatollahi, and F. Raissi (2009), Implementation of biologically plausible spiking neural network models on the memristor crossbar-based cmos/nano circuits, in *European Conference on Circuit Theory and Design*, ECCTD '09, IEEE, New York, doi:10.1109/ECCTD.2009.5275035.

[24] Linares-Barranco, B., and T. Serrano-Gotarredona (2009), Memristance can explain spike-time-dependent-plasticity in neural synapses, *Nature precedings*, 1–4.

[25] Snider, G. S. (2008), Spike-timing-dependent learning in memristive nanodevices, in *IEEE International Symposium on Nanoscale Architect ures*, 85–92.

[26] Bofill-i Petit, A., and A. F. Murray (2004), Synchrony detection and amplification by silicon neurons with stdp synapses, *Neural Networks, IEEE Transactions on*, 15(5), 1296–1304.

[27] Ishikawa, M., K. Doya, H. Miyamoto, T. Yamakawa, G. Tovar, E. Fukuda, T. Asai, T. Hirose, and Y. Amemiya (2008), Analog cmos circuits implementing neural segmentation model based on symmetric stdp learning, in *Neural Information Processing, Lecture Notes in Computer Science*, vol. 4985, 117–126, Springer Berlin/Heidelberg.

[28] Tanaka, H., T. Morie, and K. Aihara (2009), A cmos spiking neural network circuit with symmetric/asymmetric stdp function, *IEICE TRANS-ACTIONS on Fundamentals of Electronics, Communications and Computer Sciences*, 92(7), 1690–1698.

[29] Koickal, T. J., A. Hamilton, S. L. Tan, J. A. Covington, J. W. Gardner, and T. C. Pearce (2007), Analog vlsi circuit implementation of an adaptive neuromorphic olfaction chip, *Circuits and Systems I: Regular Papers, IEEE Transactions on*, 54(1), 60–73.

[30] Cutsuridis, V., S. Cobb, and B. P. Graham (2008), A ca2 + dynamics model of the stdp symmetry-to-asymmetry transition in the ca1 pyramidal cell of the hippocampus, in *Proceedings of the 18th international conference on Artificial Neural Networks, Part II*, ICANN '08, 627–635, Springer-Verlag, Berlin, Heidelberg.

[31] Cassenaer, S., and G. Laurent (2007), Hebbian stdp in mushroom bodies facilitates the synchronous flow of olfactory information in locusts, *Nature*, *448*(7154), 709–713.

6

Value Iteration with Memristors

Idongesit Ebong and Pinaki Mazumder

This chapter presents an attempt to bridge higher level learning to a memristor crossbar, therefore paving the way to realizing self-configurable circuits. The approach, or training methodology, is compared to Q-Learning in order to re-emphasize that reliably using memristors may require not knowing the precise resistance of each device, but instead working with relative magnitudes of one device to another.

6.1 Introduction

Memristors [1, 2] have been proposed for use in different applications in both an evolutionary and revolutionary manner with respect to hardware complexity. In the evolutionary sense, memristors have been proposed for FPGA, cellular neural networks, digital memory and programmable analog resistors. From the revolutionary perspective, memristors are offered for applications that bring together higher-level algorithms, usually implemented in software, down to the hardware level. These include the proposed architectures including the hardware that will utilize MoNETA [3], instar and outstar training [4], optimal control [5], visual cortex [6], etc. The use of memristors in the evolutionary sense has several limitations [7] due to the nature of the memristors being used in a Boolean logic computation when direct control of memristors is very imprecise.

A plausible area to use memristors would be in applications whereby precise resistance values are not required, but the relative values between memristors in the crossbar are maintained. This approach has been shown successful in simulating a maze problem [8], but the maze hardware architecture seems harder to fabricate and realize since memristors are not used in a crossbar configuration. The access transistors inhibit a crossbar structure, and

169

even if this drawback were ignored, the approach described in [8] requires probing voltages in the memristor network that may be inaccessible due to the spacing requirements for vias and contacts. The work presented in this chapter seeks to solve the maze problem in using a different approach that may be realized with current fabrication methods.

The approach whereby precise resistance values on memristors are not required has been shown successfully through simulation to implement a fuzzy system [9] and extended to an edge detector learning [10]. Fuzzy systems have multiple applications, but this paper focuses on ways to bridge the gap between successful algorithms in artificial intelligence (AI) and memristors. AI algorithms are grounded deeply in mathematical formulations that breed reproducibility. If memristor crossbars can implement AI algorithms, then software interface of AI hardware will become less intricate since hardware will handle more complex computation. This work strives to link both memristor properties and AI through a basic learning tool, value iteration through Q-learning [11].

Q-learning is used as an example because of its memory requirement. Q-Learning learns state-action values (dubbed Q-values) and storing these Q-values in a tabular form for the entire state-action space is shown to reach optimal solutions even under exploration. The drawback associated with Q-learning stems from the prohibitive nature of the memory requirement for the tabular form of the algorithm. In order to circumvent this problem, function approximators have been used to reduce the memory size required. Function approximators though need careful design because poor design may lead to divergence.

This chapter is organized as follows: Section 6.2 deals with the mathematical details concerning Q-learning and an extension of the memristor equations; Section 6.3 introduces the maze application and reconciles Q-learning equation and memristor equation; Section 6.4 provides simulation results and discussion; and Section 6.5 relays concluding remarks.

6.2 Q-Learning and Memristor Modeling

Equation (6.1) provides the update for the estimated Q-value ($\tilde{Q}$) at the current state (s_t) and action taken at the current state (α_t). a_t is a learning parameter, r_t is the reward. In this form of Q-learning, the model of the environment or Markov Decision Process (MDP) does not need to be accurate. After learning, the exact reward values do not affect the overall behavior of the network [12].

$$\tilde{Q}(s_t, a_t) \leftarrow \tilde{Q}(s_t, a_t) + \alpha_t(s_t, a_t)$$
$$\times [r_t + \max_{at+1}[\tilde{Q}(s_{t+1}, a_{t+1}) - \tilde{Q}(s_t, a_t)]] \quad (6.1)$$

Examination of (6.1) shows that the learning parameter scales the reward and a difference between $\tilde{Q}(s_t, a_t)$ and $\tilde{Q}(s_{t+1}, a_{t+1})$. This difference lands this form of Q-learning under the temporal difference (TD) category, whereby the TD error in learning the value function is used to update the Q-values.

As mentioned earlier, one of the drawbacks to Q-learning is the memory required to store Q-values. By discretizing the time steps, the Q-values for every admissible state-action pair should be stored in memory. While calculating (6.1), multiple readouts from memory need to occur in order to compute the MAX function before generating the TD error and updating $\tilde{Q}(s_t, a_t)$. Our approach tries to bypass mass readouts from memory by utilizing the memristor crossbar in a neuromorphic manner, therefore reducing the number of operations required to update $\tilde{Q}(s_t, a_t)$. The apparent drawback of this approach is a reduction in accuracy of an analog memory compared to a digital. The analog method is suitable in this case because it allows for a direct comparison of values utilizing a neural network approach.

By further expanding (6.1), the following equation can be obtained:

$$\tilde{Q}(s_t, a_t) \leftarrow \tilde{Q}(s_t, a_t)(1 - \alpha_t(s_t, a_t)) + \alpha_t(s_t, a_t)$$
$$\times r_t + \alpha_t(s_t, a_t) \times \max_{at+1} \tilde{Q}(s_{t+1}, a_{t+1}) \quad (6.2)$$

The takeaway from (6.2) is that the learning rate places importance on which parameter is more important by adjusting the contribution of each to the value update. The MAX function in (6.2) will produce a value, $\tilde{Q}_{max}(s_{t+1}, a_{t+1})$ that can be seen as a linear combination between $\tilde{Q}(s_t, a_t)$ and another value δ_t thereby giving the relationship in (6.3).

$$\max_{at+1}[\tilde{Q}(s_{t+1}, a_{t+1})] = \tilde{Q}_{max}(s_{t+1}, a_{t+1}) = \tilde{Q}(s_t, a_t) + \delta_t \quad (6.3)$$

The value of δ_t can be zero, positive or negative. It is a correcting factor that discerns how far apart the Q-value of the current state-action pair is from the Q-value of the next state-action pair. By substituting (6.3) back into (6.2) and eliminating some terms we obtain (6.4), the final equation describing the targeted value function updates.

$$\tilde{Q}(s_t, a_t) \leftarrow \tilde{Q}(s_t, a_t) + \alpha_t(s_t, a_t) \times (r_t + \delta_t) \quad (6.4)$$

Neural network inspired approaches have been shown to efficiently perform maximizing and minimizing functions [13]. Memristors in the crossbar configuration are not only used as memory but also as processing elements, i.e., synapses in a neural network exhibit both functions. By monitoring the current through selected memory devices, the MAX function can be evaluated in parallel. The next step is to cast (6.4) in a form that is readily applied to the memristor crossbar, so memristor modeling is discussed next.

Referring back to total memristance given by (3.1 or A.11), the flux term $\varphi(t)$ is the independent variable and hence the control for the memristor's resistance. By choosing a constant voltage pulse V_{app} and applying this constant pulse for a specified time t_{spec}, (3.1 or A.11) can be discretized into n different applications of V_{app} for t_{spec} thereby producing:

$$M_T = R_0 \sqrt{1 - \beta \sum_n V_{app} \cdot t_{spec}} = R_0 \sqrt{1 - \beta \cdot V_{app} \cdot t_{spec} \cdot n} \quad (6.5)$$

where β is defined as $(2 \cdot \eta \cdot \Delta R)/(Q_0 R_0^2)$ and n is an integer. The goal in value iteration is to update the value function, and since we strive to use M_T to store the value function, then the updates to M_T will depend on the value of n. The change between the previous value and the updated value ΔM_T can be described by

$$M_T = R_0 \sqrt{1 - \beta \cdot V_{app} \cdot t_{spec} \cdot n} - R_0 \sqrt{1 - \beta \cdot V_{app} \cdot t_{spec} \cdot (n - 1)}$$
$$(6.6)$$

Furthermore defining $a = 1 - \beta \cdot V_{app} \cdot t_{spec} \cdot n$ and $b = \beta \cdot V_{app} \cdot t_{spec}$, (6.6) can be rewritten as:

$$\Delta M_T = R_0 \sqrt{a} \left(1 - \sqrt{1 + \frac{b}{a}} \right) \quad (6.7)$$

Since $|b/a| < 1$, (6.7) can be approximated with the Taylor expansion as:

$$\Delta M_T \cong R_0 \sqrt{a} \left[-\frac{1}{2} \left(\frac{b}{a} \right) + \frac{1}{8} \left(\frac{b}{a} \right)^2 - \frac{1}{16} \left(\frac{b}{a} \right)^3 + \cdots \right] \quad (6.8)$$

For the intended application, Hebbian learning is envisioned. Therefore, if updating the memristor in one direction, whereby the resistance is always increasing with increasing n, then the piecewise relationship in (6.9) (keeping two terms of the Taylor expansion) describes an approximate discretized memristor used in this application.

$$
M_T \cong \begin{cases} R_0 & n = 0 \\ R_0 + \sum_n R_0 \sqrt{a} \left[-\frac{1}{2} \left(\frac{b}{2} \right) + \frac{1}{8} \left(\frac{b}{a} \right)^2 \right] & n > 0 \end{cases} \tag{6.9}
$$

The next section will explain the maze application in detail, thereby reconciling the derived memristor behavior in (6.9) and the value iteration equation in (6.4).

6.3 Maze Search Application

6.3.1 Introduction

Given a test maze (Figure 6.1) we would like to train through value iteration, the generation of optimal actions to reach the target (RED) from the start position (GREEN). The 16×16 maze shows admissible states in white

Figure 6.1 Maze example showing starting position (green square) and ending position (red square).

and inadmissible states in black. Our approach to solving this maze using memristors is to store the value of each state (admissible or inadmissible) in a memristor crossbar. A 16×16 memristor crossbar array is therefore needed to store all values. The maze pattern can be preprogrammed to the crossbar array whereby inadmissible states are programmed to R_{OFF} and the admissible states are programmed to values around an initial resistance R_0. The search space is discretized into time periods where one move is made per unit time. Each move made must either progress to adjacent states or stay at the current state. For example, if at time period $p = 1$ and the current state is the green square, then the three valid states for transition are the two adjacent white squares and the green square.

Decisions regarding state transition are made by obtaining the stored values of valid states in reference to the present state. The drawback to this one-step lookahead approach is its limited depth search that takes longer for training to converge to an approximation of the optimal path from start to end. The advantage of this approach is a less complex hardware implementation when using the crossbar.

6.3.2 Hardware Architecture

The memristor crossbar is used to store state values. To reduce hardware complexity with respect to accessing the crossbar, two crossbars are used whereby one stores the values in order (Network 1 in Figure 6.2) as prescribed by the state order in Figure 6.1, while the other crossbar stores values of Figure 6.1 mirrored about the diagonal from the top left corner to the bottom right corner (Network 2 in Figure 6.2). The top level system in Figure 6.2 shows an agent acting on the environment (the maze). The components of the system are: controller, memristor network, comparators (C blocks), and actor. The actor performs chosen actions, the comparators compare two values within the memristor crossbar, the memristor network performs the MAX function and generates next state information, and the controller coordinates communication between all components. The two memristor networks have the same components and a detailed network schematic is also provided in Figure 6.2.

The network blocks have two sets of neurons. Each set contains 16 neurons, allowing access to the value of each state on the memristor crossbar array. This architecture is chosen to approximate a recurrent neural network. Network 1 may be viewed as the forward path and Network 2 may be viewed as the feedback path. Neurons correspond to horizontal and vertical

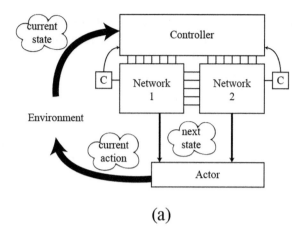

(a)

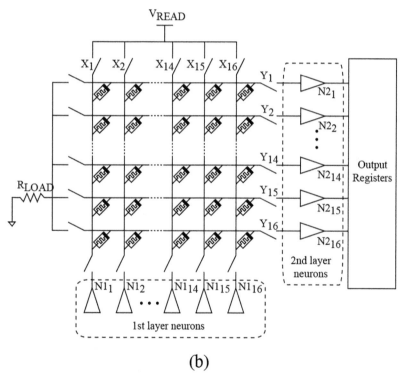

(b)

Figure 6.2 (a) Top Level system showing information flow (b) Network schematic showing analog and digital components.

coordinates in Figure 6.1. At any given time the admissible actions are: stay at current state, move one space in any diagonal direction, any horizontal direction, or any vertical direction. Network 1 determines the next Y position, while Network 2 determines the next X position. The controller coordinates the actions of the networks using four control phases: *Start, Run, Check,* and *Train.*

That *Start* phase is a wait phase whereby the crossbar network is not accessed. All the switches in Figure 6.2 are open, all input and output neurons disabled, and output registers are zeros. In the *Run* phase, the network obtains the next position; the first neuron to spike will have its corresponding output register latch a "1" while the others are "0," and will provide a signal to the controller that this phase is complete. In the *Check* phase, the digital network asserts V_{READ} and connects R_{LOAD} to decipher the values stored at two locations (the value of current state vs. that of the next state). If the current state's value is greater than or equal to the next state's then a punish signal is generated. In the *Train* phase, the punish signal is used to reduce the weight of the current state. The neural network is used to translate the time to spike to approximate the environment. The architecture is a hybrid architecture that combines both analog processing with digital controls. The next section makes a case as to why this architecture is suitable for value iteration and the maze problem.

6.3.3 Hardware Connection to Q-Learning

For the maze application, value iteration updates based on (6.4), but the exact nature of the update term, $\alpha_t(s_t, a_t) \times (r_t + \delta_t)$, has not clearly been defined. In the maze problem, $\alpha_t(s_t, a_t)$ will be limited to take on a value of either -1 or 0, and the sum of r_t and δ_t can be cast to take on the value of ΔM_T in (6.8), $R_0 \sqrt{a}[-\frac{1}{2}(\frac{b}{a}) + \frac{1}{8}(\frac{b}{a})^2]$.

This proposed matching works in this application because the envisioned system has memristors initialized around R_0 and any memristor updates will adjust resistance by $\Delta M_T \cdot \alpha_t(s_t, a_t)$ is -1 if $\tilde{Q}(s_t, a_t)$ is greater than or equal to $\tilde{Q}_{max}(s_{t+1}, a_{t+1})$, otherwise $\alpha_t(s_t, a_t)$ is 0. The punish signal generated in the *Check* phase determines which value $\alpha_t(s_t, a_t)$ takes. This restriction on $\alpha_t(s_t, a_t)$ ensures $\tilde{Q}(s_t, a_t)$ is always decreased when updated since $r_t + \delta_t$ is always a positive number.

The value for $\alpha_t(s_t, a_t)$ depends on $\tilde{Q}_{max}(s_{t+1}, a_{t+1})$, and $\tilde{Q}_{max}(s_{t+1}, a_{t+1})$ is obtained from the neuromorphic side of the circuit. A simple leaky integrate and fire neuron should work for this purpose. The schematic in

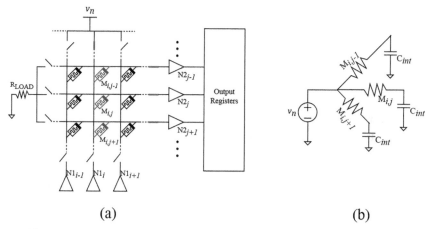

Figure 6.3 (a) Activation of neurons (b) Equivalent circuit of activated devices.

Figure 6.3(a) is used to explain the nearest neighbor concept. From a current X position and a current Y position, switch corresponding to X_j is activated and Y_{i-1}, Y_i, and Y_{i+1} are enabled. Using RC integrators to model neuron internal state, the equivalent circuit for these activated devices is shown in Figure 6.3(b). The first order RC circuit shows that the internal state of the neurons take on the form $v_n(1 - e^{-t/(M_{ij}C_{int})})$. By choosing a spiking threshold v_{thresh} for the neurons less than v_n, neuron j can spike whenever $v_n(1 - e^{-t/(M_{ij}C_{int})})$ reaches v_{thresh}. The difference between the activated neurons lies in $t^j{}_{spike}$, how long it takes for neuron j to spike:

$$t^j_{spike} > -M_{ij} \cdot C_{int} \cdot \ln\left(1 - \frac{v_{thresh}}{v_n}\right) \tag{6.10}$$

According to (6.10), each memristor allows each neuron to spike at a different time. If three memristors were chosen, then the memristor with the lowest value will cause its neuron to spike sooner than the other memristors, thereby guaranteeing the highest conductance memristor will be chosen when trying to determine the MAX function. This disparity in charging activated neurons' internal state capacitors is therefore used to determine $\tilde{Q}_{max}(s_{t+1}, a_{t+1})$.

6.4 Results and Discussion

MATLAB simulations were performed on the derived models. The parameters used to evaluate performance are: $v_{thresh}/v_n = 0.75$, $V_{app} = 1.2$ V,

$C_{int} = 1$ pF, $t_{spec} = 2$ ms, $\beta = -199.84$ V$^{-1}\cdot$s^{-1}, $R_0 = 2$ MΩ, $R_{ON} = 20$ kΩ, and $R_{OFF}=20$ MΩ. Figure 6.4(a) compares the effect of keeping more terms of the Taylor expansion in (6.8), showing that preserving at least two terms provides enough accuracy for the current modeling. When two terms are kept, the error quickly reduces to less than one percent at $n = 3$.

Figure 6.4(b) shows the graph of (6.10) and how the choice of v_{thresh} can affect circuit operation. Since the MAX function depends on the comparison of spike times of different neurons, separation of these spike times for different n values is critical for correct circuit operation. By increasing v_{thresh}, while other parameters are kept at their previous levels, there is a wider change in spike time. The quoted times are in μs, and if transistors used for implementation are sensitive to the hundreds of nanosecond range,

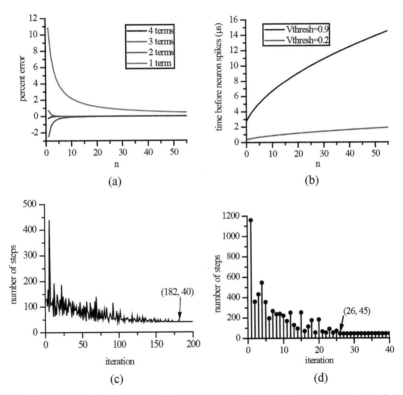

Figure 6.4   (a) Number of terms vs. the percent error (b) Effect of v_{thresh} on the charging time to spike (c) Number of steps before convergence using the baseline value function (d) Number of steps before convergence using memristors.

then there should be minimal problem detecting the larger of $n = 50$ and $n = 51$.

Figure 6.4(d) shows the relationship between the number of steps to reaching target and the number of training stages for convergence. The outlined process in this paper prefers exploration in the first iteration. During the second iteration, the number of steps is drastically reduced. To show that learning converges, after 26 iterations, the network stops updating since $\alpha_t(s_t, a_t)$ does not reach the value of -1 due to the stable path chosen by the neural network. The results in Figure 6.4(d) are juxtaposed with that in Figure 6.4(c) using conventional methods where $\tilde{Q}(s_t, a_t)$ is updated using the relationship in (6.1) with $\alpha_t(s_t, a_t) = 0.2$. This value function is dubbed the baseline value function and is used as a point of comparison. The results show that conventional method converges to one of the optimal paths, taking much more effort with designing the value function to limit the number of steps to convergence.

Figure 6.5 shows two solutions: Figure 6.5(a) shows the path obtained through the baseline value function while Figure 6.5(b) shows results using the memristor modeled network. The first path is an optimal path, while the second is near-optimal. The discrepancy between the two lies in the current method being inefficient at diagonal moves. All the sub-optimal moves made in the memristor implementation were due to making a vertical

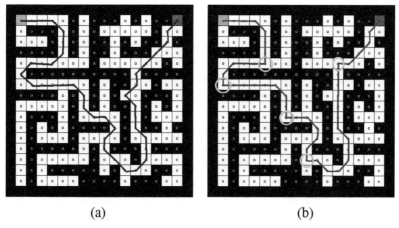

(a)   (b)

Figure 6.5   (a) Optimal path using the baseline value function (b) Near optimal path using the memristor crossbar (suboptimal moves circled).

and a horizontal move instead of one diagonal move. The discrepancy is due to the 2-step process in obtaining the next state location.

6.5 Chapter Conclusion

We have shown the concept of value iteration being applied to the memristor crossbar in a way that is realizable with the aid of CMOS hardware. We have shown how maze learning can be implemented using the crossbar. We have dissected the memristor modeling equation to show that the neural network model whereby state information can be translated to delayed spike timing is shown. The goal behind this work was the mapping of a higher-level algorithm to the memristor crossbar, and the simulation results in the chapter have proven this is possible.

References

[1] Chua, L. O. (1971), Memristor – missing circuit element, *IEEE Transactions on Circuit Theory*, *CT18*(5), 507–519.

[2] Strukov, D. B., G. S. Snider, D. R. Stewart, and R. S. Williams (2008), The missing memristor found, *Nature*, *453*(7191), 80–83.

[3] Versace, M., and B. Chandler (2010), The brain of a new machine, *Spectrum, IEEE*, *47*(12), 30–37, doi:10.1109/MSPEC.2010.5644776.

[4] Snider, G. (2011), Instar and outstar learning with memristive nanodevices, *Nanotechnology*, *22*(1), 015,201.

[5] Werbos, P. J. (2012), Memristors for more than just memory: How to use learning to expand applications, in *Advances in Neuromorphic Memristor Science and Applications*, *Springer Series in Cognitive and Neural Systems*, vol. 4, edited by R. Kozma, R. E. Pino, G. E. Pazienza, J. G. Taylor, and V. Cutsuridis, 63–73, Springer Netherlands.

[6] Zamarreno-Ramos, C., L. A. Camunas-Mesa, J. A. Perez-Carrasco, T. Masquelier, T. Serrano-Gotarredona, and B. Linares-Barranco (2011), On spike-timingdependent-plasticity, memristive devices, and building a self-learning visual cortex, *Frontiers in Neuroscience*, *5*, 26.

[7] Ebong, I. E., and P. Mazumder (2011), Self-controlled writing and erasing in a memristor crossbar memory, *Nanotechnology, IEEE Transactions on*, *10*(6), 1454–1463.

[8] Pershin, Y. V., and M. Di Ventra (2011), Solving mazes with memristors: A massively parallel approach, *Phys. Rev. E*, *84*(4), 046,703, doi: 10.1103/PhysRevE.84.046703.

[9] Merrikh-Bayat, F., and S. Bagheri Shouraki (2012), Memristive neuro-fuzzy system, *Systems, Man, and Cybernetics, Part B: Cybernetics, IEEE Transactions on*, *PP*(99), 1–17.

[10] Merrikh-Bayat, F., S. Bagheri Shouraki, and F. Merrikh-Bayat (2011), Memristive fuzzy edge detector, *Journal of Real-Time Image Processing*, 1–11.

[11] Watkins, C. J. C. H. (1989), Learning from delayed rewards, ph.D. thesis, Cambridge University.

[12] Balleine, B., N. Daw, and J. ODoherty (2009), Multiple forms of value learning and the function of dopamine, in *Neuroeconomics: decision making and the brain*, edited by P. W. Glimcher, 367–385, Academic Press.

[13] Hopfield, J. J., and D. W. Tank (1985), "neural" computation of decisions in optimization problems, *Biological Cybernetics*, *52*(3), 141–152, doi:10.1007/BF00339943.

7

Tunneling-Based Cellular Nonlinear Network Architectures for Image Processing

Pinaki Mazumder, Sing-Rong Li and Idongesit Ebong

In this chapter, an RTD-based CNN architecture is presented and its operation through driving-point-plot analysis, stability and settling time study, and circuit simulation is investigated. Full-array simulation of a 128×128 RTD-based CNN for several image processing functions is performed using the Quantum Spice simulator designed at the University of Michigan, where the RTD is represented in SPICE simulator by a physics-based model derived by solving Schrödinger's and Poisson's equations self-consistently. A comparative study between different CNN implementations reveals that the RTD-based CNN can be designed superior to conventional CMOS technologies in terms of integration density, operating speed, and functionality.

7.1 Introduction

Since its invention by Chua and Yang in 1988 [1, 2], the cellular neural/ nonlinear network (CNN) has been much acclaimed as a powerful back-end analog array processor, capable of accelerating various computation-intensive tasks in image processing, pattern formation and recognition, motion detection, robotics, and various other real-time problem solving that requires complex computation [3]. In such real-world applications, massively parallel computation of spatial data over a 2-D surface is needed to process data in real-time, albeit computational functions are rather simple algebraic operations and each array element concurrently performs identical operation. The features of CNN that make it an easily implementable parallel computing architecture relative to fully connected neural networks are mentioned in [4] but reiterated here:

(1) local interconnection between nodal components, with each node component called a cell or processing element; (2) regular placement of all cells in space; (3) identical cell configuration as well as space invariant interconnection; (4) real-time signal processing capability due to continuous-time dynamics and concurrent operation of cells; and (5) huge amount of templates to be exploited for various image processing algorithms. Numerous CNN implementations with versatile embedded applications have been developed using CMOS technology due to its low cost and high integration capability [5–8]. However, during the last two decades, CMOS technology has advanced by leaps and bounds so much so that it will encounter physical and manufacturing limitations, thereby ending the era of scaling down the transistor. To sustain the exponential growth of the integrated circuits as espoused by Moore's Law, several meso- and nano-scale technologies have been investigated to overcome the limitations of CMOS technology. Among several proposed nano-electronics devices, the resonant tunneling diode (RTD) has been explored for sometime due to its relatively easy fabrication process along with its unique folded-back negative differential resistance (NDR) current-voltage (I-V) characteristics [9–11]. The RTD has found several applications in both digital and analog circuits [12, 13].

Another advantage of RTD's include the possibility of cointegration with InP [14] or GaAs [15] three terminal devices such as HFETs, HEMTs, and HBTs that have one or two orders of higher electron mobility than CMOS. Previous work has shown that circuits implemented with RTDs have very fast operating speed, compact integration density, and rich functionality [12, 16]. These promising advantages the RTD possesses have spurred interest in its development for massively parallel architectures such as the CNN.

A compact bistable CNN architecture comprising the well-known monostable-bistable logic element (MOBILE) circuit models, first proposed by Maezawa et al. [16], were studied by Hanggi et al. [17]. Their preliminary study of RTD-based CNN array with 10 or more cells can be integrated in a standard CMOS chip. Dogaru et al. further extended the previous work by proposing an RTD based CNN cell configuration capable of performing various types of Boolean functions [18–20]. The operation principle of the RTD-based cell was clearly explained by them; however, no full-array simulation of a 2-D CNN array was presented in that work, and no image processing algorithms were simulated on a full array. Although subsequently Itoh et al. reported a full-array simulation for RTD-based CNNs [21], a simplistic piecewise linear model was used to represent the tunneling I-V characteristic of the RTD, thereby failing to precisely estimate the dynamics of the CNN in the real case.

7.2 CNN Operation Principle

7.2.1 CNN Based on Chua and Yang's Model

The cell model for the conventional CNN proposed by Chua and Yang (see Figure 7.1) consists of one linear resistor, one linear capacitor, several linear voltage-controlled current sources representing the feedback and feed-forward currents supplied by the neighboring cells, one independent current source and one nonlinear voltage-controlled voltage source.

According to this model, every cell acts as a nonlinear dynamic system, with its transient behavior governed by the following nonlinear ordinary differential equation:

$$C\frac{dx_{ij}(t)}{dt} = -\frac{x_{ij}(t)}{R}$$
$$+ \sum_{C(k,l)\in N_r(i,j)} (a_{ij,kl}f(x_{kl}(t)) + b_{ij,kl}u_{kt}) + I$$
$$1 < i < M; \quad 1 < j < N \tag{7.1}$$
$$f(x_{ij}) = 0.5 \times (|x_{ij} + 1| - |x_{ij} - 1|) \tag{7.2}$$

where $N_r(i,j)$ represents the neighborhood of cell (i,j) in an $M \times N$ CNN array; $x_{ij(t)}$, u_{ij} and $f(x_{ij})$ are state, input, and output variables of cell (i,j), respectively. $a_{ij,kl}$ and $b_{ij,kl}$ are the space invariant feedback and feed-forward parameters, providing the weighting for the feedback and feed-forward currents from cell (k,l) to cell (i,j). The number of elements in the set formed by feedback parameters $a_{ij,kl}$ and in the set formed by feed-forward parameters $b_{ij,kl}$ depends on how far a central cell, cell (i,j), is

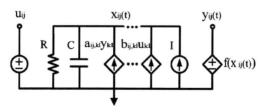

Figure 7.1   Circuit model for the conventional CNN.

connected to its adjacent cells, which is usually formulated as $(2r + 1)^2$, where r is an integer starting from 1. For example, if is r equal to 1, there are nine elements in each of the feedback and feed-forward parameter set: each cell communicates with its nearest eight neighboring cells through feedback and feed-forward branches whereas it also contains a self-feedback loop and a self-feed-forward branch. Due to the space invariant connection between cells in the CNNs, $a_{ij, kl}$ and $b_{ij, kl}$ are usually denoted by a pair of matrices of synaptic elements (e.g., 3×3 matrix for $r = 1$; 5×5 matrix for $r = 2$), called feedback template, A, and feed-forward template, B, respectively.

In this work, we consider only the case where $r = 1$ since it generates the simplest structure. When the CNN is employed in image processing applications, its templates A and B merely act as image filters in order to map the input image into the desired output image. As a result, by designing templates of different matrix coefficients, the wide gamut of image processing algorithms can be executed by running an ordered sequence of different A and B matrices.

In conventional CNNs, the input and state variables are continuous values (analog signal) whereas the output variable is a binary value at steady state, either $+1$ or -1, according to the state-to-output transfer function $f(x_{ij})$. Figure 7.2 graphically depicts this function as a piecewise linear function that saturates at either $+1$ or -1 when the state variable is greater than $+1$ or less than -1. The last component in the model presented in Figure 7.1 is the independent current bias I; it is associated with each node as externally injected current which provides design flexibility for the CNN design.

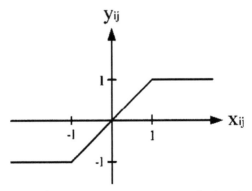

Figure 7.2   Sigmoid state-to-output transfer function.

7.2.2 CNN Equations Based on RTD's Model

The RTD is a symmetric two-terminal meso-scale device with non-monotonic I-V characteristics and extremely small intrinsic capacitance, which render its suitability in compact and highspeed CNN implementations. The RTD-based CNN cell model described in this paper is illustrated in Figure 7.3, where one RTD replaces the linear resistor in the conventional CNN cell model.

Contrasting Figures 7.1 and 7.3, the RTD introduces a nonlinearity that the resistor did not have. With the CNN in Figure 7.1, the sigmoid representation in Figure 7.2 is necessary for correct operation in representing the relationship between the input and output states, while with the RTD representation, the Figure 7.2 transfer function is no longer necessary. The RTD would guarantee saturation due to its nonlinearity. This simplifies the relationship between the state variable and output variable, making them equivalent to each other. Equation (7.1) is then modified to

$$C\frac{dx_{ij}(t)}{dt} = -h(x_{ij}(t))$$
$$+ \sum_{C(k,l)\in N_r(i,j)} (a_{ij,kl}x_{kl}(t) + b_{ij,kl}u_{kl}) + I. \quad (7.3)$$

where $h(x)$ represents the RTD's I-V characteristics. An accurate physics model developed by Schulman et al. [24] is used to model $h(x)$:

$$h(x_{ij}(t)) = A \cdot \ln\left[\frac{1 + e^{(B-C+n_1 x_{ij}(t))q/kT}}{1 + e^{(B-C-n_1 x_{ij}(t))q/kT}}\right]$$
$$\cdot \left[\frac{\pi}{2} + \tan^{-1}\left(\frac{C - n_1 x_{ij}}{D}\right)\right]$$
$$+ H \cdot (e^{n_2 x_{ij}q/kT} - 1). \quad (7.4)$$

Equation (7.4) is current per unit area, and the parameters A, B, C, D, H, n_1 and n_2 depend on the physical model of the RTD. Equation (7.3)

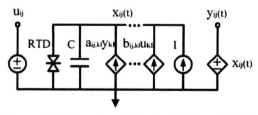

Figure 7.3 Circuit model for the RTD-based CNN.

states that the current flowing into the capacitor is equal to the sum of the currents flowing out of the RTD, the feedback branches, the feed-forward branches, and the constant current bias. Referring to Figure 7.3, this would be a Kirchhoff's Current Law (KCL) at the node denoted by the current state.

The RTD-based CNN's ability to perform the same functions as the conventional cell without the sigmoid function is explained through driving point plots in Figure 7.4. The plots in Figure 7.4 were generated for uncoupled CNNs $- a_{ij, kl} = 0$ for $l \neq ij$. The x-axis indicates the state variable presented as a voltage while the y-axis indicates the derivative of the state variable with respect to time, which can be regarded as the normalized current flowing through the capacitor C in Figure 7.3.

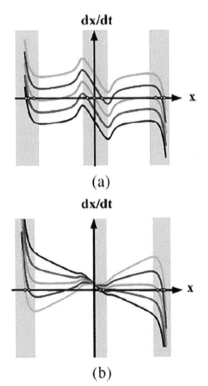

(a)

(b)

Figure 7.4 Driving-point plots for (a) $a_{ij, ij} = 20\ \mu A/V$, $w = 100\ \mu A$, $50\ \mu A$, $0\ \mu A$, $-50\ \mu A$ $-100\ \mu A$; (b) $w = 100\ \mu A$, $a_{ij, ij} = 400\ \mu A/V$, $200\ \mu A/V$, $0\ \mu A/V$, $200\ \mu A/V$, $400\ \mu A/V$.

A parameter introduced in Figure 7.4, w, is defined as

$$w = \sum_{C(k,l) \in N_r(i,j)} b_{ij,kl} u_{kl} + I. \tag{7.5}$$

This definition along with the uncoupling assumption thereby lets us write (7.3) as

$$C\frac{dx_{ij}(t)}{dt} = -h(x_{ij}(t)) + a_{ij,ij} x_{ij}(t) + w. \tag{7.6}$$

Figure 7.4(a) shows the plots for different w – multiple combinations of input variables u_{kl}, feed-forward parameter $b_{ij,kl}$ and constant bias I – with a fixed self-feedback parameter $a_{ij,ij}$. Figure 7.4(b) shows how the I–V relationship changes with different $a_{ij,ij}$, while other parameters stay the same.

The intersection points of the driving-point curves and the axis are the equilibrium states of the cell. The stable equilibrium states – marked with open circles – are distributed in three domains separated by the NDR regions of the RTD. Therefore the steady state of each cell can be assigned to three levels – 1, 0, or +1 – by an analog-to-digital converter. Such conversion can be done using RTD-based quantizers whose threshold voltages are designed in the NDR regions [25].

With an RTD-based CNN, a need for a quantizer may not be necessary. From Figure 7.4(a), when $w = 50$ uA, we only have two equilibrium points. With proper design, a binary output (two stable points) can be obtained without the need of a quantizer.

7.2.3 Comparison Between Different CNN Models

To demonstrate the superior advantages of the RTD-based CNN cell, a comparison is made with a two diode model as well as conventional cell based on CMOS technology. Figure 7.5 shows the three circuit models used in the comparison.

The two-diode model is introduced because the antiparallel configuration of the diodes – D_1 and D_2 in Figure 7.5(c) – provides a similar saturating effect on the state variable as the RTD does. The assumptions for comparison are as follows:

• The circuit implementation for the RTD and the two diodes is the same for these three models in terms of device counts, area, power consumption, and the capacitance being driven.

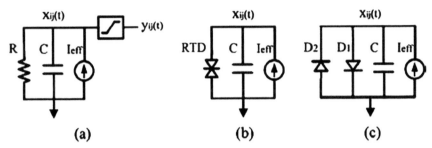

Figure 7.5   Simplified circuit model for: (a) the conventional CNN; (b) the RTD-based CNN; and (c) the two-diode-based CNN.

Table 7.1   Comparison between the three CNN models

	Conventional	RTD based	Two diode based
C		1 pF	
R	5 kΩ–10 kΩ	Negligible	Negligible
Intrinsic Capacitance	Negligible	0.75 aF–0.75 fF	10 pF–110 pF
Estimated additional area	$>10\ \mu m^2$	$0.01–1\ \mu m^2$	$0.18–2\ \mu m^2$
Setting time (90%) zero initial condition with $I_{\text{eff}} = 200$ mA	20 ns	6.9 ns–12.5 ns	77.3 ns–716 ns
Saturated output voltage	−2 V 2 V	−1.8 V to −1.4 V 1.4 V to 1.8 V	−0.8 V to −0.7 V 0.7 V to 0.8 V
# of steady states (Design flexibility)	2	3	2

Assumption: Same design, occupied area, power consumption for I_{eff} based on 02.5 μm CMOS technology.

- The total injected current from the feedback, the feed-forward, and the constant current bias is the same—represented as I_{eff} in Figure 7.5.

Table 7.1 lists the comparison of circuit performance based on the simulation results, where the resistor, sigmoid circuits and diodes are designed using 250-nm CMOS technology. In the conventional CNN cell, the sigmoid circuit would occupy a large amount of area to compensate for device mismatch issues. With continual CMOS scaling, the device mismatch issue is worsened since small atomic displacements render the operation of identical devices different. On the other hand, although the two-diode model holds similar nonlinearity as the RTD-based model, the intrinsic capacitance of the diodes is much larger than that of the RTD, resulting in slower operating speed. Moreover, the RTD-based model is the only one to have three output states due to its NDR property. This characteristic provides richer design flexibility than the other two methods. Table 7.1 summarizes the RTD-based model possesses the smallest area, smallest settling time, and gives the most design flexibility.

7.3 Circuit Analysis

The stability criteria and settling time analysis are two important factors regarding the design of a CNN architecture. Due to the complex interaction existing between cells, CNNs can be classified according to the feedback connection conditions: coupled CNNs and uncoupled CNNs ($a_{ij, kl} = 0$ for $kl \neq ij$). For uncoupled CNNs, the dynamic equation of one cell is represented as (7.6); this will enable our analysis henceforth.

7.3.1 Stability

Circuit stability analyses tend to use two methods, namely, Lyapunov theorem and graphing methods. According to the Lyapunov theorem, the energy function of the RTD-based CNN can be defined as [1]

$$E(t) = -\frac{1}{2} \sum_{(i,j)} \sum_{(k,l)} a_{ij, kl} x_{kl}(t) x_{ij}(t)$$
$$- \sum_{(i,j)} \sum_{(k,l)} b_{ij, kl} u_{kl}(t) x_{ij}(t)$$
$$- \sum_{(i,j)} I x_{ij}(t) + \sum_{(i,j)} \int_0^{x_{ij}} h(s) ds. \tag{7.7}$$

If a feedback template is symmetric $a_{ij, kl} = a_{kl, ij}$, the derivative of the energy function with respect to time t is

$$\frac{dE(t)}{dt} = -\sum_{(i,j)} \sum_{(k,l)} a_{ij, kl} x_{kl}(t) \frac{dx_{ij}(t)}{dt}$$
$$- \sum_{(i,j)} \sum_{(k,l)} b_{ij, kl} u_{kl}(t) \frac{dx_{ij}(t)}{dt}$$
$$- \sum_{(i,j)} I \frac{dx_{ij}(t)}{dt}$$
$$+ \sum_{(i,j)} \frac{dx_{ij}(t)}{dt} \frac{d}{dx_{ij}(t)} \int_0^{x_{ij}} h(s) ds \tag{7.8}$$

$$\frac{dE(t)}{dt} = -\sum_{(i,j)} \frac{dx_{ij}(t)}{dt}$$

$$\times \left\{ \sum_{(k,l)} a_{ij,kl} x_{kl}(t) + \sum_{(k,l)} b_{ij,kl} u_{kl}(t) \right.$$

$$\left. + I - \frac{d}{dx_{ij}(t)} \int_0^{x_{ij}} h(s) ds \right\} \tag{7.9}$$

$$\frac{dE(t)}{dt} = -\sum_{(i,j)} C \left(\frac{dx_{ij}(t)}{dt} \right)^2 \leq 0. \tag{7.10}$$

Equation (7.10) shows that the energy function is monotonically decreasing $|u_{ij}| < c_1$, $|x_{ij}(t = 0)| < c_2$, $|x_{ij}(t)| < c$. Also, the energy function is bounded with certain constraints—c_1, c_2, and c are constants. As a result, the state variable will also be bounded; dc output is always generated in the RTD-based CNN.

The graphical representation in Figure 7.4 of the driving plots supports this conclusion. In Figure 7.4(a), when w was changed, the plots shifted vertically, but each plot crossed the axis at least once. In Figure 7.4(b), when $a_{ij,ij}$ was changed, the shape of the plots changed, but there is always an intersection with the axis. There exists at least one stable equilibrium state for every cell in the RTD-based CNN. Design flexibility comes into play because based on the number of equilibriums to design for, the method prescribed in [26] can be used to obtain a unary, binary, or ternary output.

7.3.2 Settling Time

CNNs prevail over sequential signal processors due to their real-time functional characteristics. Consequently, the operating speed is considered a critical performance index in CNN system design. To determine the operational speed of a dynamic system, the concept of settling time (t_s) is introduced. t_s of a cell is defined as the time required to reach a steady state (e.g., a stable equilibrium point) from the initial condition. The settling time of a CNN system is determined by the settling time of the slowest cell. The value of the settling time for a cell depends on several factors, such as initial condition, input, output, feedback and feed-forward templates, constant bias, capacitance and the size of the RTD (i.e., the current magnitude). Hence, analyzing the dependence of t_s on circuit parameters is the first step to design a high-speed CNN processor.

In this section, the settling time of one RTD-based CNN cell is investigated through SPICE simulation and mathematical modeling. Since the dynamics is very difficult to predict for propagated type CNN, this work focuses on the uncoupled RTD-based CNN. Without the loss of generality, zero initial condition is assumed in our analysis.

(1) *Simulation Results*: From (7.6), the settling time is dependent on w, $a_{ij,ij}$, and the area of the RTD, which govern the peak current of the RTD. Three cases are simulated to explore the relationship between the settling time and these three parameters. For the first case, we will investigate the dependence of settling time on the size of the RTD for five w values, 180, 270, 360, 540 and 900 uA, with $a_{ij,ij} = 0$. From the conclusion drawn in [26], we want the absolute value of w to be greater than the RTDs peak current (e.g., $I_{peak} = 178$ uA for area $= 1$ um^2) for binary output. The simulation results illustrated in Figure 7.6 show the settling time results for the first case.

In Figure 7.6, the settling time has a weak dependence on peak current for the other w cases except for when $w = 180$ uA. For this case, there is a drastic increase in settling time as peak current increases. This phenomenon may result from a very small current experienced by the capacitor during the trek from the initial state to the stable equilibrium state due to the small difference between the values of w and I_{peak}. Here we will define γ as the ratio of w/I_{peak}. When $1 < \gamma < 2$, the settling time greatly depends on I_{peak}. On the other hand, the settling time has a weak dependence on I_{peak} when $\gamma > 2$. From Figure 7.6, cells with larger take shorter time to achieve steady state.

In the second case, the settling time is simulated for various $a_{ij,ij}$ for the same five w values with an RTD area of 1 um^2 ($I_{peak} = 178$ A). The results are illustrated in Figure 7.7.

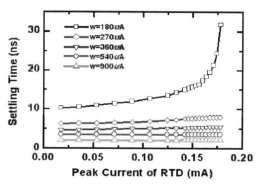

Figure 7.6 Simulated settling time for various I_{peak} of RTD with $a_{ij,ij}$ for five different w.

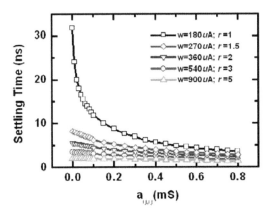

Figure 7.7 Simulated settling time for various $a_{ij,ij}$ with I_{peak} of RTD $= 178 \ \mu$A for five different w.

From Figure 7.7, when $1 < \gamma < 2$, the settling time decreases greatly as $a_{ij,ij}$ increases from 0 to 0.8 ms, whereas the decreasing rate is not drastic when $\gamma > 2$. Note that as $a_{ij,ij}$ increases, the shape of the driving point plot changes, which increases the minimum current experienced by the capacitor, thus reducing the settling time.

The last case investigates the dependence of the settling time on the values of w. Different RTD sizes are used to modulate the peak current; the results are shown in Figure 7.9.

From Figure 7.8, the settling time seems to saturate when $w > 540$ uA no matter what I_{peak} and $a_{ij,ij}$ are. Moreover, the settling time is reduced if gamma increases (from the black curve to the blue one) or $a_{ij,ij}$ increases (from the black curve to the violet one). This observation agrees with the previous experiments due to the same reason that w and $a_{ij,ij}$ have better control on the settling time as the effect of the RTD decreases. $a_{ij,ij}$ might not be controllable but since w depends on the constant current source I, we have more design control over this than feedback/feedforward parameters.

(2) *Mathematical Modeling*: Since the dynamic behavior of a single RTD-based CNN is determined by a nonlinear differential equation, it is very difficult to find an analytic solution for the transient response of the state variable. Even though a complex analytic solution can be determined, it may not make any sense to us since we may not be familiar with the complex function. Hence, this work simplifies the settling time analysis by

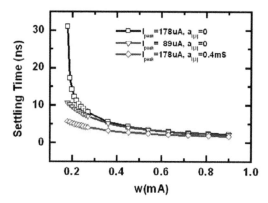

Figure 7.8 Simulated settling time for various w with different combinations of $a_{ij, ij}$ and I_{peak}.

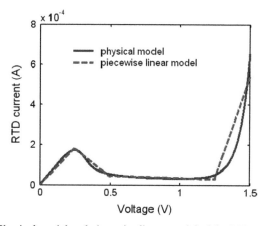

Figure 7.9 Physical model and piecewise linear model of the I-V curve of the RTD.

using a piecewise linear model for the I-V curve of the RTD as illustrated in Figure 7.9. This method can provide an intuitive understanding for the transient response of a cell from the initial condition to the steady state as well as determine which parameters significantly affect the settling time.

From Figure 7.9, the physical model of the RTD used in the simulation has a peak around 0.25 V and a valley current starting at 0.5 V followed by a wide valley region. In addition, the diodelike current activates at around 1.25 V. Therefore, the I-V curve of the RTD is modeled with a piecewise

Table 7.2   Parameters for the PWL modeling

	A		A/V
a_0	0	a_1	6.96e-4
b_0	2.95e-4	b_1	−4.84e-4
c_0	5.41e-4	c_1	−2.27e-6
d_0	−2.93e-3	d_1	2.41e-3

linear (PWL) function, with each section following:

$$h(x) = \begin{cases} a_0 + a_1 x & 0 \text{ V} \leq x < 0.25 \text{ V} & a_0 = 0, & a_1 > 0 \\ b_0 + b_1 x & 0.25 \text{ V} \leq x < 0.5 \text{ V} & b_0 > 0, & b_1 < 0 \\ c_0 + c_1 x & 0.5 \text{ V} \leq x < 1.25 \text{ V} & c_0 > 0, & c_1 < 0 \\ d_0 + d_1 x & 1.25 \text{ V} \leq x & d_0 > 0, & d_1 > 0 \end{cases} \qquad (7.11)$$

The fitting parameters for (7.11) are summarized in Table 7.2. The nonlinear differential equation describing the dynamics of one cell, thus, is decomposed into four linear ordinary differential equations (ODE), with each representing some period of the trek from the zero initial state to the steady state. Therefore, the settling time can be solved sequentially from one ODE to another ODE to obtain four solutions to (7.6) depending on initial and final conditions. The results of each of the four sections were combined linearly to get an overall approximation of settling time as depicted in (7.12):

$$t_s = \left(\frac{C}{a_{ij,\,ij} - d_1} \right) \cdot \ln \left[\frac{x_{ij,\,\text{steady}} - \left(\frac{d_0 - w}{a_{ij,\,ij} - d_1} \right)}{1.25 - \left(\frac{d_0 - w}{a_{ij,\,ij} - d_1} \right)} \right]$$

$$+ \left(\frac{C}{a_{ij,\,ij} - c_1} \right) \cdot \ln \left[\frac{1.25 - \left(\frac{c_0 - w}{a_{ij,\,ij} - c_1} \right)}{0.5 - \left(\frac{c_0 - w}{a_{ij,\,ij} - c_1} \right)} \right]$$

$$+ \left(\frac{C}{a_{ij,\,ij} - b_1} \right) \cdot \ln \left[\frac{0.5 - \left(\frac{b_0 - w}{a_{ij} - b_1} \right)}{0.25 - \left(\frac{b_0 - w}{a_{ij,\,ij} - b_1} \right)} \right]$$

$$+ \left(\frac{C}{a_{ij,\,ij} - a_1} \right) \cdot \ln \left[\frac{0.25 - \left(\frac{-w}{a_{ij,\,ij} - d_1} \right)}{\frac{w}{a_{ij,\,ij} - a_1}} \right]. \qquad (7.12)$$

As can be seen from (7.12), the settling time depends on the self-feedback parameter, steady state target, $a_{ij,ij}$, w, and the size of the RTD (related to the parameters a_1, b_0, b_1, c_0, c_1, d_0, d_1). The settling time is proportional to the capacitance but is inversely proportional to $a_{ij,ij} - a_1$, $a_{ij,ij} - b_1$, $a_{ij,ij} - c_1$ and $a_{ij,ij} - d_1$. Another interesting phenomenon is that the impact of w is reduced due to the likely cancellation (if w dominates) of the numerator and denominator together with the log function. This result in conjunction with the results from Figures 7.7 and 7.8 show that when w is large enough (greater than the peak current), additional increases in w does not affect the settling time tremendously.

7.4 Simulation Results

In this work, a method of quantifying the effects of parameters that influence stability and settling time were presented. A simplification to uncouple the effects of neighboring cells were made with knowledge this method will introduce a percentage error between the values obtained during design and those acquired from simulation. A small 12×12 RTD-based CNN was designed in order to verify and quantify the errors associated with the assumptions made during analysis. The expression in (7.12) was used to appropriately design settling times in the nanosecond range. The calculated results were on the same order as what was observed during simulation. The results are shown in Figure 7.10.

Figure 7.10 was obtained through simulation results performed through SPICE simulation tools, where RTD's are represented by current sources describing the physics-based models and the feedback and feed-forward branches are modeled as ideal voltage controlled current sources. The boundary condition for the cells at the edge is zero, i.e., no feedback and feed-forward loops from outside of the array. In Figure 7.11, the input voltages of black and white cells are $+1$ and -1 V, respectively, where as the output of each cell converges to values between $+1.5$ and -1.8 V or -1.5 and -1.8 V, representing $+1$ (black) or -1 (white) after A/D conversion.

The dynamics of the state variable of each cell in a full row [e.g., the fifth row for Figure 7.10(a); the third row for Figure 7.10(b); the fifth row for Figure 7.10(c)] is demonstrated in the waveform. As can be seen, for the uncoupled image processing functions (e.g., the horizontal line detection

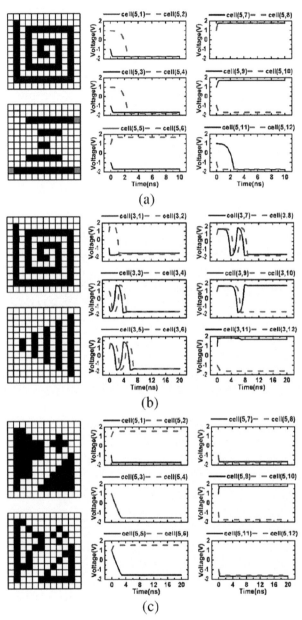

Figure 7.10   Simulation results of 12×12 RTD based CNN area with the input and output image patterns for: (a) horizontal line detection; (b) horizontal physical model and piecewise linear model of the I-V curve of the RTD; (c) edge extraction.

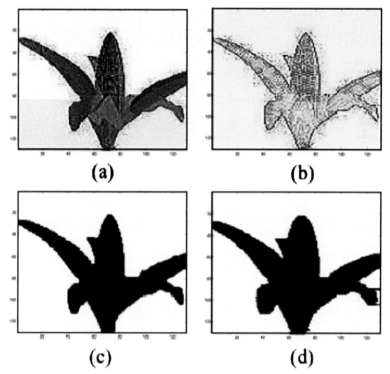

Figure 7.11 Simulation results for 128 × 128 RTD-based CNN array. (a) Input image. (b) Output image for EE. (c) Output image for HF. (d) Output image for shadowing.

and the edge extraction), the settling time is around a few nanoseconds with 1 pF capacitance. On the other hand, for the propagate type algorithm such as the horizontal connected component detection, the interaction between neighboring cells can be obviously observed from the transient responses of state variables, which contributes to a much longer settling time.

For image processing on a grander scale, a full array for a 128 128 RTD-based CNN with grayscale input images was simulated. Figure 7.11 demonstrates the results of different functions—edge extraction (EE), hole filling (HF), and shadowing— on the same input image.

In addition to the input image in Figure 7.11, another image shown in Figures 7.12 and 7.13 are used to show the intermediate process between the input and output images. These are different snapshot images at different timestamps of the transient response of the full array simulation.

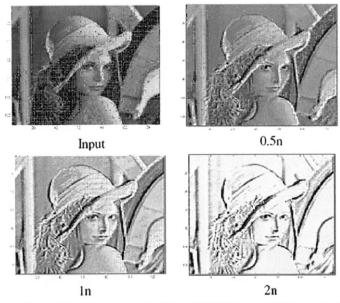

Figure 7.12   Simulation results for 128×128 RTD-based CNN array for EE.

Figure 7.13   Simulation results for 128×128 RTD-based CNN array for averaging.

7.5 Conclusion

Several image processing functions—EE, HF, shadowing, HLD, HCCD—are successfully demonstrated for a 12×12 an a 128×128 RTD-based CNN array using the QSPICE simulator developed at the University of Michigan. The RTD-based CNN is proven to always be stable if the feedback template is symmetric. A method and expression for approximating settling time of RTD-based CNN cells for binary outputs is presented. The settling time analysis shows that when the ratio of the total injected or extracted current in a cell to the RTD's peak current approaches 1, settling time increases exponentially. Simulation results on the comparison of the RTD-based CNN with the conventional CNN and the two-diode implementation show that the RTD presents unique design advantages in terms of compactness, speed, and design flexibility.

References

[1] L. O. Chua and L. Yang, "Cellular neural networks: Theory," IEEE Trans. Circuits Syst., vol. CAS-35, no. 10, 1257–1272, Oct. 1988.

[2] L. O. Chua and L. Yang, "Cellular neural networks: Applications," IEEE Trans. Circuits Syst., vol. CAS-35, no. 10, 1273–1290, Oct. 1988.

[3] L. O. Chua and T. Roska, "The CNN paradigm," IEEE Trans. Circuits Syst. I, Fundam. Theory Appl., vol. 40, no. 3, 147–156, Mar. 1993.

[4] K. R. Crounse and L. O. Chua, "Methods for image processing and pattern formation in cellular neural networks: A tutorial," IEEE Trans. Circuits Syst. I, Fundam. Theory Appl., vol. 42, no. 10, 583–601, Oct. 1995.

[5] T. Roska and A. Rodriguez-Vazquez, "Toward visual microprocessors," Proc. IEEE, vol. 90, no. 7, 1244–1257, Jul. 2002.

[6] K. Karahaliloglu and S. Balkir, "Bio-inspired compact cell circuit for reaction-diffusion systems," IEEE Trans. Circuits Syst. II, Express Briefs, vol. 52, no. 9, 558–562, Sep. 2005.

[7] J. Kowalski, "0.8 m CMOS implementation of weighted-order statistic image filter based on cellular neural network architecture," IEEE Trans. Neural Netw., vol. 14, no. 5, 1366–1374, May 2003.

[8] P. Kinget and M. S. J. Steyaert, "A programmable analog cellular neural network CMOS chip for high speed image processing," IEEE J. Solid-State Circuits, vol. 30, no. 3, 235–243, Mar. 1995.

[9] K. Karahaliloglu and S. Balkir, "Nanostructure array of coupled RTDs as cellular neural networks," Int. J. Circuit Theory Appl., vol. 31, no. 6, 571–589, 2003.

[10] P. Julian, R. Dogaru, M. Itoh, M. Hanggi, and L. O. Chua, "Simplicial RTD-based cellular nonlinear networks," IEEE Trans. Circuits Syst. I, Fundam. Theory Appl., vol. 50, no. 4, 500–509, Apr. 2003.

[11] M. Hanggi and L. O. Chua, "Cellular neural networks based on resonant tunnelling diodes," Int. J. Circuit Theory Appl., vol. 29, no. 5, 487–504, 2001.

[12] P. Mazumder, S. Kulkarni, M. Bhattacharya, J. P. Sun, and G. I. Haddad, "Digital circuit applications of resonant tunneling devices," Proc. IEEE, vol. 86, no. 4, 664–686, Apr. 1998.

[13] Y. Tsuji and T. Waho, "Design of flash analog-to-digital converters using resonant-tunneling circuits," IEICE Trans. Electron., vol. E87C, no. 11, 1863–1868, 2004.

[14] J. I. Bergman, J. Chang, Y. Joo, B. Matinpour, J. Laskar, N. M. Jokerst, M. A. Brooke, B. Brar, and E. Beam III, "RTD/CMOS nanoelectronic circuits: Thin-film InP-based resonant tunneling diodes integrated with CMOS circuits," IEEE Electron Device Lett., vol. 20, no. 3, 119–122, Mar. 1999.

[15] Y. L. Huang, L. Ma, F. H. Yang, L. C. Wang, and Y. P. Zeng, "Resonant tunnelling diodes and high electron mobility transistors integrated on GaAs substrates," Chinese Phys. Lett., vol. 23, no. 3, 697–700, 2006.

[16] K. Maezawa, T. Akeyoshi, and T. Mizutani, "Functions and applications of monostable-bistable transition logic elements (MOBILE's) having multiple-input terminals," IEEE Trans. Electron Devices, vol. 41, no. 2, 148–154, Feb. 1994.

[17] M. Hanggi and L. O. Chua, "Compact bistable CNNs based on resonant tunneling diodes," in Proc. IEEE Int. Symp. Circuits Syst. (ISCAS), 2001, 93–96.

[18] M. Hanggi, R. Dogaru, and L. O. Chua, "Physical modeling of RTD-based CNN cells," in Proc. 6th IEEE Int. Workshop Cellular Neural Networks Their Appl. (CNNA), 2000, 177–182.

[19] M. Hanggi, L. O. Chua, and R. Dogaru, "A simple RTD-based circuit for Boolean CNN cells," in Proc. 6th IEEE Int. Workshop Cellular Neural Netw. Their Appl. (CNNA), 2000, 189–194.

[20] R. Dogaru, M. Hanggi, and L. O. Chua, "A compact and universal cellular neural network cell based on resonant tunneling diodes: Circuit,

model, and functional capabilities," in Proc. 6th IEEE Int. Workshop Cellular Neural Netw. Their Appl. (CNNA), 2000, 183–188.

[21] M. Itoh, P. Julián, and L. O. Chua, "RTD-based cellular neural networks with multiple steady states," Int. J. Bifurcation Chaos, vol. 11, no. 12, 2913–2959, 2001.

[22] C. P. Gerousis and S. M. G. W. P., "Nanoelectronic single-electron transistor circuits and architectures," Int. J. Circuit Theory Appl., vol. 32, no. 5, 323–338, 2004.

[23] S. Bandyopadhyay, K. Karahaliloglu, S. Balkir, and S. Pramanik, "Computational paradigm for nanoelectronics: Self-assembled quantum dot cellular neural networks," IEE Proc. Circuits, Devices Syst., vol. 152, no. 2, 85–92, 2005.

[24] J. N. Schulman, H. J. De Los Santos, and D. H. Chow, "Physics-based RTD current-voltage equation," IEEE Electron Device Lett., vol. 17, no. 5, 220–222, May 1996.

[25] S.-R. Li, P. Mazumder, and L. O. Chua, "On the implementation of RTD based CNNs," in Proc. Int. Symp. Circuits Syst. (ISCAS), 2004, III-25–III-28.

[26] S.-R. Li, P. Mazumder, and L. O. Chua, "Cellular neural/nonlinear networks using resonant tunneling diode," in Proc. 4th IEEE Conf. Nanotechnol., 2004, 164–167.

8

Color Image Processing with Multi-peak Resonant Tunneling Diodes

Woo Hyung Lee and Pinaki Mazumder

This chapter introduces a novel approach to color image processing that utilizes multi-peak resonant tunneling diodes for encoding color information in quantized states of the diodes. The multi-peak resonant tunneling diodes (MPRTDs) are organized as a two-dimensional array of vertical pillars which are locally connected by programmable passive and active elements with a view to realizing a wide variety of color image processing functions such as quantization, color extraction, image smoothing, edge detection and line detection. In order to process color information in the input images, two different methods for color representation schemes have been used: one using color mapping and the other using direct RGB representation.

8.1 Introduction

Since 1987, when the Cellular Neural Networks (CNN) paradigm of massively parallel computing through local interactions between ensemble of simple processing elements capable of aggregating the weighted analog stimuli from their neighboring processors and generating digital outputs was first developed by Chua and Yang [1, 20], it has rapidly morphed into a general-purpose powerful backend hardware substrate to accelerate several classes of non-number crunching computations much faster than conventional digital computers can perform within the constraint of energy budget as well as Silicon real estate. Several types of two- and multidimensional spatial data processing used in color image analysis [25], video motion detection [18], DNA microarray pattern classification [24], real-time brain waves study in epileptic episodes [26] and in many other real-world applications have been successfully demonstrated by both analog and digital implementations of the CNN array embedded within a single VLSI chip. However, ultimately heat dissipation and Silicon areas of conventional CNN cells limit the size of CMOS based CNN arrays to 256×256 processing elements or so, thereby severely limiting the resolution of the input and the output data. The goal of this paper is to describe the design of a new quantum tunneling device (called multi-peak resonant tunneling diode, MPRTD)-based nonlinear dynamical system capable of performing various types of color image processing tasks with higher precisions. Though nanoscale MPRTD-based neurons (processing elements) along with their programmable synapses comprising both passive and active devices strikingly differ from CMOS-based analog and digital CNNs, the overall behavior of the nonlinear system has been shown in this paper to mimic the CN paradigm of computation.

The resonant tunneling diode (RTD) was invented by Chang, Esaki and Tsu [2] in the early seventies by demonstrating that when a bias voltage is applied across its two terminals, the *I-V* characteristic of the RTD has a folded-back shape owing to the tunneling through the mesoscale double-barrier quantum well of the RTD. This negative differential resistance (NDR) characteristic of the RTD has been cleverly exploited by a number of researchers to design several high-speed and high-density integrated circuits [3–5, 8] as well as utilized to build CNNs with improved speed and integration density [20]. Vertically integrated RTDs have sub-pico second switching speed as opposed to planar and sluggish PMOS and NMOS devices, thereby enabling the RTD-based digital and CNN systems to have superior power-delay performance than their CMOS counterparts. To show the advantages of RTD's in CNN architectures, Hänggi and Chua [20] incorporated RTDs in CNN processing elements and implemented monochromatic image processing using self-latching bi-stable (1-bit storage) property of the RTD.

However, in order to perform color image processing, simple RTD configuration of [20] is not adequate since color images need more than one bit per pixel. In this paper, we discuss a new type of two-dimensional array consisting of the MPRTDs and programmable synapses that may consist of resistive, capacitive and active devices. Since MPRTDs are fabricated by vertically stacking double-barrier (one-peak) RTDs, the resulting 3-D architecture requires very small Silicon area. Waho, Chen and Yamamoto [19] implemented multiple thresholds and outputs using the planar RTDs configured into MOBILE (Mono stable to bi-stable transition logic elements) mode. In this paper, we have realized color image processing by implementing multiple output levels with vertically stacked RTDs grown under the metallic islands as shown in Figure 8.1. To demonstrate simulation results, we used a piece-wise continuous model of an eight-peak RTD with nine stable states which was originally fabricated at the Texas Instruments by stacking serially eight pseudo-morphic ALAs/In0.53Ga0.47As/InAs RTDs and the existence of nine states were experimentally verified by Seabaugh, Kao and Yuan [7]. To exploit the nine stable states for color image processing, we introduce two different methods for the color representation: Color mapping and RGB encoding. While the first method uses a color map to index the color value, the latter method uses three cells to represent RGB (Red, Green and Blue) colors for a single pixel. Using the color depiction methods and characteristics of the multi-peak RTDs, we describe several color image processing functions such as quantization, smoothing and color extraction.

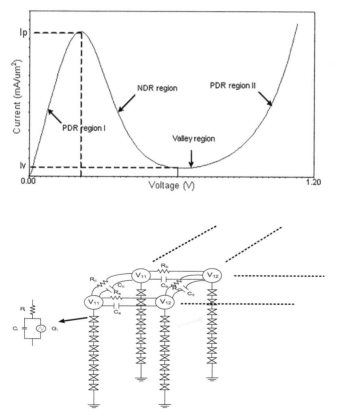

Figure 8.1 *I–V* characteristic of a resonant tunneling diode (top) and side view and top view of the multi-peak RTD based image processor (bottom).

8.2 The Multi-Peak RTD-Based Color Image Processor

Silicon-based Resonant Tunneling Diodes (RTDs) are touted as a strong candidate to augment the performance of conventional CMOS circuits by largely replacing bulky and sluggish PMOS transistors generally used in the pull-up network. The *I-V* device characteristics of the RTD comprise three distinct regions, namely, two positive differential resistance (PDR) regions and one intervening negative differential resistance (NDR) region, as shown in Figure 8.1(top). The resulting non-monotonic tunneling characteristics of the RTD can be cleverly exploited in key circuit nodes to significantly improve the device performance in terms of power and delay, thereby yielding low power-delay product in comparison to other form of CMOS circuits [16–18]. A definite advantage of RTD-based circuits is that they are generally more process-variation tolerant since the operation of the RTD remains unaffected as long as its peak current, I_P is higher than its valley current, I_V.

RTD-based image processors published in the literature utilized the bistable state of the resonant tunneling diode [16, 17, 23] to decide the output image and processing methods. By using parallel processing and small intrinsic capacitance, fast image processing is possible with this structure. The disadvantage is that it is impossible to present various functions, to generate color images, and to process those color images because three stable states are required to display color information. Multiple stable states are essential to represent the color information due to the various color values. Each RTD can be modeled as the R, C and voltage-controlled current source in the circuit simulation. Multi-peak RTDs are built by stacking single-peak RTDs to obtain nine states in this paper, though in practice any number of states can be realized as long as the process of RTD stacking is possible. As we discussed before, we used nine states because it was proven RTD stats by experimental results [7]. If the process evolves more and allows more stacking the RTDs, then we can use more stable states for our color image processor.

In Figure 8.1, the interconnection between two multi-peak RTDs is implemented with the RC models. This RC models change based on the image function that the processor realizes. In the CNN perspective, the changes of the RC model parameters correspond to the change of the templates in the CNNs. For implementation of the conductance network, [16] showed the feasibility of implementation and initial guidance for more complex conductance network.

The *I-V* characteristics of the multi-peak RTD in the vertical direction can be simply piecewise-linearly modeled as shown in Figure 8.2(a). Since the eight-peak RTD is used, nine stable states exist when using constant current source that delivers current in the range between the lowest peak current and the highest valley current. However, the *I-V* curve of the experimental data shows a DC shift as shown in Figure 8.2(b). To be valid in functionality, the lowest peak should be higher than the highest valley. These peak or valley values also change based on the size of the metallic islands. Since the size of RTDs varies depending on the process variation, we need to consider the shift of *I-V* curve to up and down in the direction. Based on the experimental result from [7], if the size deviation of the metallic islands is 67% to bigger size or 33% to smaller size, the functionality is guaranteed. The process variation also affects the power dissipation and speed. However, the overall speed and power dissipation are not much changed if we assume the randomly distributed inputs. The worst pixel speed is degraded around 30 ps from Equation (8.9) and the power dissipation of the worst pixel is increased around 0.2 nW from Equation (8.14). The overall speed and power dissipation is more dependent on the initial value of output and templates as shown in Figures 8.5, 8.9 and 8.15. In the case of the smoothing function, the stabilization time and power dissipation is proportional to the difference between the initial color and the smoothed output color. The discrepancy between the initial color and the background color affects the stabilization time and power dissipation proportionally for the color extraction function.

The standard CNN state equation can be written as follows [1, 17, 18, 20].

$$\frac{dx_{ij}}{dt} = -x_{ij} + \sum_{k,l \in N_{i,j}} (a_{k-i,l-j} f(x_{kl}) + b_{k-i,l-j} u_{kl}) + I_{ij}. \quad (8.1)$$

where $X_{ij} \in R$, $f(X_{ij}) \in R$, $u_{kl} \in R$, and $I_{ij} \in R$ denote the state, output, input and threshold. $a(i, j; k, l)$ and $b(i, j; k, l)$ are called the feedback and the feed-forward operators or templates, respectively. Considering the network connection between cells and neglecting the feed-forward effect, Equation (8.1) can be transformed into Equation (8.2) below which will be used for quantization, smoothing, and color extraction.

$$\frac{C dv_{ij}}{dt} = -J(v_{ij}) + \sum_{k,l \in N_{i,j}} (a_{k-i,l-j} v_{kl}) + I_{ij}. \quad (8.2)$$

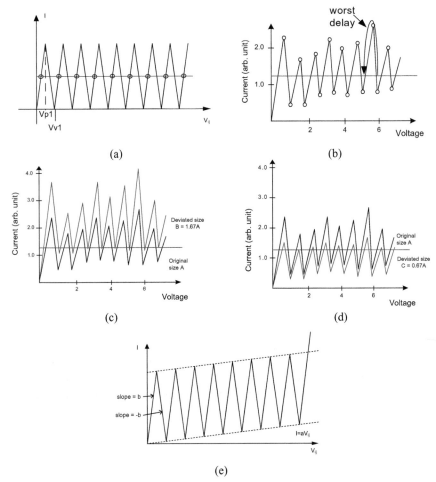

Figure 8.2   Simplified piecewise-linear model of *I-V* characteristics for an eight-peak RTD in the vertical direction (a), piecewise-linear model based on experimental data (b), maximum size deviation for valid functionality (c), minimum size deviation for valid functionality (d) and piecewise-linear model with a linear DC shift (e).

where the template

$$a = \begin{bmatrix} 0 & q & 0 \\ q & -4q & q \\ 0 & q & 0 \end{bmatrix},$$

I_{ij} is the driving current, and C is the contact capacitance of each RTD.

In the cases of quantization and smoothing function, the template for feedback is represented as a_{ij} in Equation (8.2). The template for the quantization function and smoothing function is decided by the conduction from the neighboring cells to the target cell. In the above example, there is a target cell which has connections with four neighboring cells with q conduction value. The center cell has $-4q$ conduction value which means that it reversely dissipates the incoming current. By the value of q in the template, the quantization function or smoothing function is decided. The quantization and smoothing function are realized with resistive network connection with the defined neighboring cells in the template. So, the current flows from the four neighboring cells to the center cell or reverse direction. The center cell circulates the received currents through itself to satisfy the KCL. Also, this current flow depends on the outputs of the neighboring cells. Hence, the template a_{ij} is represented as a matrix which has q in the left, right, up and down current flow and $-4q$ for the center cell. The value of q is a conductance value in a real number which changes depending on the image functions.

The *I-V* curves of Figure 8.2(a) and (e) can be represented as Equations (8.3) and (8.4), respectively. Even if Equation (8.4) demonstrates more realistic curve form based on the experimental data, Equation (8.3) is chosen for a model for the eight-peak RTD since the functionality is not affected with simple implementation and the overall processing time changes less than 30 ps (5%) from Equation (8.9) and Figure 8.5. This is because the initial voltage and the templates of each cell more affects the processing time than the eight-peak RTD models.

$$J(V_{ij}) = \begin{cases} \alpha(V_{ij}) & \text{if } 0 < V_{ij} \leq V_{p1} \\ \alpha(|V_{ij} - n \cdot V_{p1}|) & \text{if } (2n-1)V_{p1} < V_{ij} \leq (2n+1)V_{p1} \end{cases} \tag{8.3}$$

$$J(V_{ij}) = \begin{cases} b((V_{ij} - n) \cdot k, & 0 < V_{ij} \leq \dfrac{nc_1}{2b} \\ -b((V_{ij} - n) \cdot k) + c_1, & \dfrac{nc_1}{2b} < V_{ij} \leq \dfrac{nc_1}{a+b} \end{cases} \tag{8.4}$$

where n is an integer from 1 to 8 and k is the real number of the shift amount.

8.3 Color Representation Method

As shown in Figure 8.2(a) and (b) and Equation (8.3), an eight-peak resonant tunneling diode has nine stable states. The nine states can be used to indicate color values. That can be used for processing the color images.

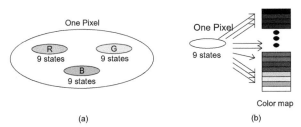

Figure 8.3   The color representation methods with metallic islands on the eight-peak RTDs. (a) RGB value method and (b) color map method.

Two methods are suggested to store color images using the array of the eight-peak RTDs. One method is to match the voltages of the array of the eight-peak RTDs to color index values. In this method, each index value designates each color in the color map as shown in Figure 8.3(b). This method, called the color map method, is used for color images which have color maps.

However, there is a limitation of the color number in this method. The maximum number of colors to represent is 9 because the eight-peak RTD has the nine stable states. Despite this drawback, the color map method is easier to perform in fabrication than the other method. The second method uses three metallic islands to represent one pixel, and each metallic island corresponds to the red, green, and blue (RGB) colors. The voltage value of the metallic island indicates the intensity of the RGB color. Since each metallic island can express the nine different intensities, a total of 729 colors can be represented, and the number of colors can be extended by increasing the number of metallic islands per pixel. This method is called RGB color method as shown in Figure 8.3(a) and is used to process bitmap images.

The multiple quantization levels can also be used to provide gray scale to images. If we assume three multi-peak RTDs represent one gray scale, a total of 729 gray levels can be represented for images. A fine resolution can be achievable in the gray images due to the fact above. This will be useful when we implement the smooth function which requires polishing the edges of gray levels in the neighboring pixels.

8.4 Color Quantization

8.4.1 Implementation and Results

Using the discrete stable states of the multi-peak RTDs, the color quantization is implemented. Figure 8.4 represents the changes of colors of a 4×4 pixel

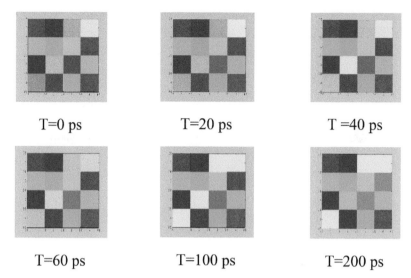

T=0 ps   T=20 ps   T =40 ps

T=60 ps   T=100 ps   T=200 ps

Figure 8.4   Transient representation of 4×4 color image quantization from 0 ps to 200 ps.

image. This change of the color image arises from the *I-V* characteristics of the multi-peak RTDs. From the initial color value, the quantized value is obtained by the minimum Euclidean distance. From Equation (8.3), the quantized color value can be obtained as a function of I_{ij}.

$$J(v_{ij}) = I_{ij} \tag{8.5}$$
$$v_{ijk} = J^{-1}(I_{ij}) \tag{8.6}$$

where v_{ijk} is the solution of Equation (8.3) and $k = 1, 2, \ldots, 9$, where 1 corresponds to the lowest value of the solution and 9 corresponds to the highest value of the solution in the Figure 8.3. If we assume that the input image value is u_{ij}, the minimum Euclidean distance can be derived as follows.

$$Dist_{\min} = \min_{k=1,..,9} (|u_{ij} - v_{ijk}|) \tag{8.7}$$

The solution of Equation (8.3) which has the minimum Euclidean distance is the quantization value. The change from the unstable state to the stable state is depicted in Figure 8.5. In Figure 8.5, the time to get to the final stable state is confirmed by monitoring the tracks of colors in the images. As shown in the HSPICE simulation results, the color values are stabilized after 200 ps. The images representing the HSPICE simulation results show the same characteristics. Based on the simulation, the minimum

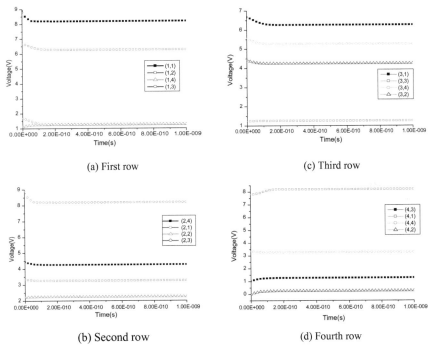

(a) First row

(c) Third row

(b) Second row

(d) Fourth row

Figure 8.5   HSPICE simulation results divided into rows with time from 0 ns to 1 ns.

connection resistance between the metallic islands is 250 MΩ to implement the quantization function.

The feedback template and feed-forward template for the quantization function are represented as

$$a = \begin{pmatrix} 0 & k & 0 \\ k & -4k & k \\ 0 & k & 0 \end{pmatrix}, \quad b = \begin{pmatrix} 0 & 0 & 0 \\ 0 & 0 & 0 \\ 0 & 0 & 0 \end{pmatrix}$$

where k is the conductance between the neighboring cells and center cell. The typical value for k is greater than 0.01 μmho. External current value should be chosen to be between the lowest peak and the highest valley in the *I-V* curve.

The trace of 50×50 color image that is originally from Looney Tunes Wiki with time from 0 ns to 1 ns is depicted in Figure 8.6. Each pixel represents a color using the image value, and this color value is quantized due to the multi-peak RTD. The color change of the eyes and the mouth of the duck in the Figure 8.6 originate from the limitation of the numbers of stable states of the multi-stable state resonant tunneling diode.

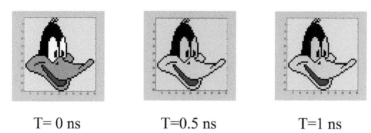

T= 0 ns   T=0.5 ns   T=1 ns

Figure 8.6 50×50 pixel color image quantization from 0 ns to 1 ns.

8.4.2 Settling Time Analysis

The settling time of the array of the multi-peak RTD structure is defined as the time needed to reach the stable state from the initial condition value. The stabilization time is determined by the slowest cell in the CNN. In Equation (8.2), if we assume the initial voltage is v_0, then the solution of the first order differential equation is given as

$$v(t) = \left[v_0 - \left(\frac{I_{ij}}{\alpha + \sum\limits_{k,l \in N_{i,j}} a_{k-i,l-j}} \right) \right] e^{-t(\alpha + \sum\limits_{k,l \in N_{i,j}} a_{k-i,l-j})/C}$$

$$+ \frac{I_{ij}}{\alpha + \sum\limits_{k,l \in N_{i,j}} a_{k-i,l-j}}. \tag{8.8}$$

From Equation (8.8), the output of the center cell is changed when a set of neighboring cells is interconnected by any template. If the template is changed, a vector $a_{k-i,l-j}$ is also changed. The changed vector affects the conductance between the neighboring cells and the center cell from Equation (8.8), and the output of the center cell is stabilized when there is no change in the current flow which is changed by the template and the outputs of the neighboring cells.

The current from the feedback operator in the quantization function is negligible, so Equation (8.8) can be approximated as

$$v(t) = \left(v_0 - \frac{I_{ij}}{\alpha} \right) e^{-t\alpha/C} + \frac{I_{ij}}{\alpha}. \tag{8.9}$$

From Equation (8.9), the difference between the initial voltage and the quantized voltage is proportional to the stabilization time. The stabilization time is maximal in the initial voltages where the red line meets the negative slope of the *I-V* curve as in Figure 8.2.

Table 8.1 ·Comparison of the stabilization time among the image processing functions

	Worst	Best	Avg.
Quantization	120 ps	50 ps	83 ps
Smoothing	0.05 ns	0.7 ns	0.35 ns
Color extraction	0.05 ns	1.3 ns	0.7 ns

In Table 8.1, the stabilization time according to image processing functions is demonstrated. The stabilization time depends on the image processing function. This is attributed to the change of the template between the image processing functions. The changed template (α) in Equation (8.9) affects the output $v(t)$ leading to the deviation of stabilization time.

8.4.3 Power Consumption Analysis

The information on the power consumption of the multi-peak RTDs is needed to determine whether the array of the multi-peak RTDs is energy efficient. For quantization, the equivalent circuit model is shown in Figure 8.7, where we assume the current flows vertically from the metallic islands to the substrate. The *I-V* characteristic of the G1 is defined by the multi-peak resonant tunneling diode. To use the negative differential resistance characteristic of the device, we need to flow current which is between the peak value and the valley value in the IV curve of the device. We also assume the current between the

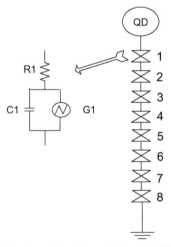

Figure 8.7   Equivalent circuit model of the multi-peak RTDs for color quantization.

metallic islands for quantization is negligible. In Figure 8.7, each one-peak RTD can be modeled with a resistance serially connected to a capacitor and nonlinear voltage controlled current source with a parallel connection. The supplied input energy stored on the capacitor is expressed by the following equation

$$
\begin{aligned}
E_{v_0} &= \int_0^\infty i_0(t) \cdot v_0 \, dt \\
&= v_0 \int_0^\infty C_{total} \frac{dv_0}{dt} \, dt \\
&= C_{total} \, v_0 \int_0^\infty dv_0 \\
&= C_{total} \, v_0^2
\end{aligned}
\tag{8.10}
$$

where C_{total} is the sum of the serially-connected capacitances. Considering the external current source, the energy supplied to the quantum dot equals

$$
\begin{aligned}
E_{total} &= E_{v_0} + E_{ext} \\
&= C_{total} \, v_0^2 + I^2 R_{total}
\end{aligned}
\tag{8.11}
$$

where R_{total} is the summation of the serially connected resistances in the vertical direction.

The energy of the capacitor after the quantization is given as

$$
\begin{aligned}
E_{Voutput} &= \int_0^\infty i_{Vinput}(t) \cdot v_{output}(t) \, dt \\
&= \int_0^\infty C_{total} \frac{dV_{output}}{dt} \cdot v_{output}(t) \, dt \\
&= \frac{C_{total} \, v_{output}^2(t)}{2}
\end{aligned}
\tag{8.12}
$$

In Equation (8.11), the output voltage changes with time in Equation (8.8). From Equation (8.8), the energy change with time is expressed by the following equation

$$
E_{Voutput} = \frac{C_{total}}{2} \left[\left(v_0 - \frac{I}{\alpha} \right) e^{-t\alpha/C} + \frac{I}{\alpha} \right]^2
\tag{8.13}
$$

Therefore, the energy consumption after quantization is given as

$$
\begin{aligned}
E_{diss} &= E_{total} - E_{Voutput} \\
&= C_{total}\, v_0^2 + I^2 R_{total} \\
&\quad - \frac{C_{total}}{2}\left[\left(v_0 - \frac{I}{\alpha}\right)e^{-t\alpha/C} + \frac{I}{\alpha}\right]^2
\end{aligned}
\tag{8.14}
$$

8.5 Smoothing Function

8.5.1 Implementation and Results

As shown in Figure 8.8, the changes in the pixel values due to the multi-peak RTD and resistance between metallic islands are depicted. These color changes arise from the current change in each metallic island that is connected with resistors. In this simulation, we chose 50 MΩ for the connection resistors. As shown in the Figure 8.8, the colors in the neighboring positions change such that the color values between neighbor cells are similar. This phenomenon provides a way of detecting similarity of the detected colors nearby. The color similarity can be detected when the similar color pixels are merged to the same color in the long run.

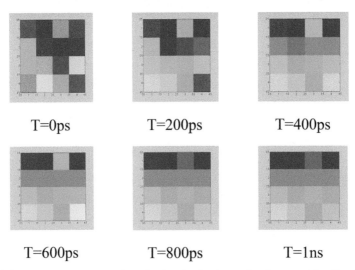

Figure 8.8   Transient representation of a 4 × 4 color image for the smoothing function from 0 ps to 200 ps.

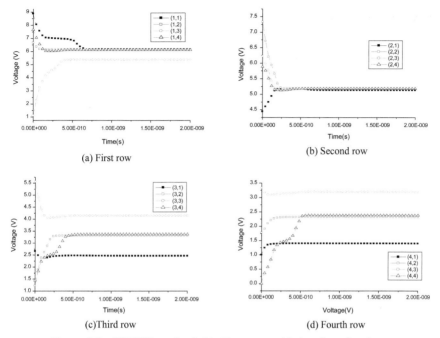

Figure 8.9 HSPICE results divided into rows with time from 0 to 2 ns.

Also, the traces of the image color that change with time arise from the negative differential resistance which is one of the characteristics of the resonant tunneling diodes. Therefore, the simulation results reflect both the effects of resistances which connect the nearest neighbor cells and the negative differential resistance characteristic. In Figure 8.9, the change in the color images shows that the colors of the first row and the second row are quite different from those of the other rows, which are also shown in the simulation results as the affinity of colors. In addition, the color value change by the neighboring cells can be applied to the noise removal of an image.

The feedback template and feed-forward template for the smoothing function is represented as

$$a = \begin{pmatrix} 0 & l & 0 \\ l & -4l & l \\ 0 & l & 0 \end{pmatrix}, \quad b = \begin{pmatrix} 0 & 0 & 0 \\ 0 & 0 & 0 \\ 0 & 0 & 0 \end{pmatrix}$$

where l is conductance between the neighboring cells and center cell. The typical value for l is between 0.04 and 0.02 μmho. Then, the output changes based on Equation (8.8).

8.5.2 Stabilization time

As shown in Figure 8.9, the settling time of the smoothing function is longer than that of the quantization function. This difference is attributed to the feedback operation in Eqns. (8.2) and (8.8). In Equation (8.8), the feedback operation term works as a function increasing the settling time. After the system is settled, the feedback operation term becomes zero. Hence, Equation (8.9) describes the smoothing function after settling.

8.5.3 Power Consumption Analysis

To analyze the smoothing function, we simplify the array of the multi-peak RTD structure as two cells as shown in Figure 8.10. We then expand the two cells to an $N \times N$ array. If the voltages of the two cells are equal, the power consumption will be the same as that in the quantization function. When the voltages of the two cells are different and $V_{QDj} > V_{QDi}$, the supplied input energy is expressed by the following equation.

$$
\begin{aligned}
E_{vo} &= \int_0^\infty i_0(t) \cdot (v_{0i} + v_{0j})\, dt \\
&= v_{0i} \int_0^\infty C_t \frac{dv_{0i}}{dt}\, dt \\
&\quad + v_{0j} \int_0^\infty C_t \frac{dv_{0j}}{dt}\, dt \\
&= C_t v_{0i} \int_0^\infty dv_{0i} + C_t v_{0j} \int_0^\infty dv_{0j} \\
&= C_t(v_{0i}^2 + v_{0j}^2)
\end{aligned}
\tag{8.15}
$$

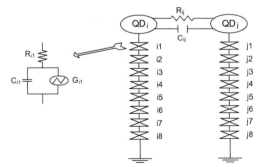

Figure 8.10   Equivalent circuit model of the array of the multi-peak RTD structure for the smoothing function.

where C_t is the sum of the capacitances surrounding the cells. Including the energy from the external current source, the total energy supplied is given as

$$
\begin{aligned}
E_{total} &= E_{v0i} + E_{v0j} + 2E_{ext} \\
&= C_t v_{0i}^2 + C_t v_{0j}^2 + 2I_{ij}^2 R_{total}
\end{aligned}
\tag{8.16}
$$

Considering current from QD_i to QD_j, the energy after the smoothing function is expressed by the following equation.

$$
\begin{aligned}
E_{Voutput} &= \int_0^\infty i_{QDi}(t) \cdot v_{QDi}(t)\, dt \\
&\quad + \int_0^\infty i_{QDj}(t) \cdot v_{QDj}(t)\, dt \\
&= \int_0^\infty C_t \frac{dV_{QDi}}{dt} \cdot v_{QDi}(t)\, dt \\
&\quad + \int_0^\infty C_t \frac{dV_{QDj}}{dt} \cdot v_{QDj}(t)\, dt \\
&= \frac{C_t v_{QDi}^2(t)}{2} + \frac{C_t v_{QDj}^2(t)}{2}
\end{aligned}
\tag{8.17}
$$

$$
\begin{aligned}
v_{QDi}(t) &= \left[v_0 - \left(\frac{I_{ij}}{\alpha - (1/R_{ij})} \right) \right] e^{-t(\alpha - (1/R_{ij}))/C} \\
&\quad + \frac{I_{ij}}{\alpha - \frac{1}{R_{ij}}}
\end{aligned}
\tag{8.18}
$$

$$
\begin{aligned}
v_{QDj}(t) &= \left[v_0 - \left(\frac{I_{ij}}{\alpha + (1/R_{ij})} \right) \right] e^{-t(\alpha + (1/R_{ij}))/C} \\
&\quad + \frac{I_{ij}}{\alpha + \frac{1}{R_{ij}}}
\end{aligned}
\tag{8.19}
$$

The total energy consumption used for the smoothing function is given as

$$
\begin{aligned}
E_{diss} &= E_{total} - E_{Voutput} \\
&= C_t(v_{0i}^2 + v_{0j}^2) + 2I_{ij}^2 R_{total} \\
&\quad - \frac{C_t}{2}(v_{QD_i}^2(t) + v_{QDj}^2(t))
\end{aligned}
\tag{8.20}
$$

If we expand the two dots to an $N \times N$ array, the total energy consumption is modified as follows.

$$E_{diss} = E_{total} - E_{Voutput}$$

$$= C_t \sum_{n=1}^{N^2} v_{0n}^2 + 4N^2 \cdot I_{ij}^2 R_{total} - \frac{C_t}{2} \sum_{n=1}^{N^2} V_{QDn}^2(t) \qquad (8.21)$$

8.6 Color Extraction

We need three cells each pixel output for color extraction function. Nine cells are required if we represent the color image with RGB method. The relationship between three cells is depicted as shown in Figure 8.11.

However, if we want to represent the relationship with template methods, the feedback and feed-forward template can be represented as

$$a = \begin{pmatrix} 0 & m & 0 \\ 0 & 0 & 0 \\ 0 & 0 & 0 \end{pmatrix}, \quad b = \begin{pmatrix} 0 & 0 & 0 \\ n & 0 & 0 \\ 0 & 0 & 0 \end{pmatrix}.$$

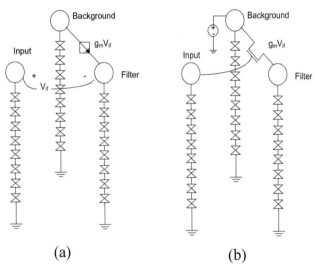

(a)   (b)

Figure 8.11   Circuit configurations for color extraction. (a) using voltage controlled current source; (b) using MOSFET.

The relationship between the inputs to outputs is represented as n, and the relationship between the background and output as m. From these templates, we expect the input image affects the conductance between the background output and the output image. The values of m and n are decided according to the devices for implementation.

The circuits shown in Figure 8.11 are designed to extract the color in the array of the multi-peak RTD structure. Figure 8.11(a) is the circuit which uses a voltage controlled current source for color extraction. The input in Figure 8.11(a) is the RGB color value of the image to be processed. The background color after the color image processing is determined by the background cell value. Also, the filtering cell value is the color value of the image to be extracted.

In Figure 8.11(a), the change of the voltage in the filtering cell can be obtained by

$$C\frac{dV_{filter}}{dt} = -J(V_{filter}) + g_m V_{if} \qquad (8.22)$$

where g_m, V_{if}, and V_{filter} are conductance, voltage difference between the filtering cell and input cell, and voltage in the filtering cell, respectively.

Two methods can be used to assign the initial state in Figure 8.11. One is to initialize the filtering cells with one color value. The other method is to initialize the filter cells with the input image color value. These two methods can be used in both of the circuits in Figure 8.11.

As shown in Figure 8.11(a), the circuit implementation with the voltage controlled current source is described. In this circuit, the color extraction is possible using the principle that when V_{if} is greater than threshold voltage, the result is a current flow from the background cell to the filter cell. Figure 8.11(b) shows the circuit implementation with a MOSFET instead of the voltage controlled current source. The feasibility of implementation with the MOSFET hybrid with III–V material is shown in patents and papers [21, 22]. Even if it is not simple to fabricate this structure in these days, as the fabrication technology evolves, it will be getting easier in the future. In this paper, we want to demonstrate the proto-type of work which will be realized in the near future.

When the MOSFET is in the triode region, the change of the voltage in the filtering cell can be obtained by

$$C\frac{dV_{filter}}{dt} = -J(V_{filter}) + \mu_n C_{ox}\frac{W}{L}[(V_{if} - V_{TH})V_{bf} - \frac{1}{2}V_{bf}^2 \qquad (8.23)$$

where $V_{if} - V_{bf} \leq V_{TH}$. In the saturation region, the change of the voltage in the filtering cell is given by

$$C\frac{dV_{filter}}{dt} = -J(V_{filter}) + \frac{1}{2}\mu_n C_{ox}\frac{W}{L}(V_{if} - V_{TH})^2(1 + \lambda V_{DS}) \quad (8.24)$$

where $V_{if} - V_{bf} > V_{TH}$ and λ is the channel-length modulation which is an empirical constant parameter.

The circuits in Figure 8.11(a) and (b) are limited in terms of the number of colors that can be extracted because the current controllers with the critical voltage allow only the extraction of the highest or lowest color value. Therefore, we suggest an advanced circuit which can extract any color in the image as shown in Figure 8.12. The color to be extracted is controlled by changing the four resistor values. The output voltage is given as follows.

$$V_{out} = V_{in} \begin{cases} V_{in} \cdot \dfrac{R_2}{R_1 + R_2} > \dfrac{V_{DD}}{2} \quad or \\[2ex] \dfrac{R_3}{R_3 + R_4}V_{DD} + \dfrac{R_4}{R_3 + R_4}V_{in} < \dfrac{V_{DD}}{2} \end{cases}$$

$$\quad (8.25)$$

$$V_{out} = 0 \begin{cases} V_{in} \cdot \dfrac{R_2}{R_1 + R_2} < \dfrac{V_{DD}}{2} \quad or \\[2ex] \dfrac{R_3}{R_3 + R_4}V_{DD} + \dfrac{R_4}{R_3 + R_4}V_{in} > \dfrac{V_{DD}}{2} \end{cases}$$

Using the circuit in Figure 8.12, red color extraction of the 4×4 image is shown for time from 0 ns to 1 ns in Figure 8.13. The HSPICE simulation results with time are shown in Figure 8.14. The HSPICE simulation

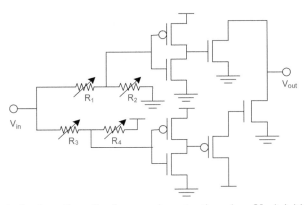

Figure 8.12　A circuit configuration for any color extraction where V_{out} is initialized into V_{in}.

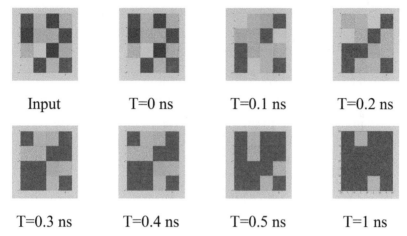

Figure 8.13 Cyan color extraction of 4×4 pixel image using the circuit shown in Figure 8.12.

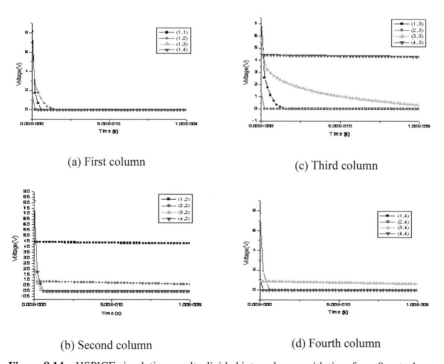

Figure 8.14 HSPICE simulation results divided into columns with time from 0 ns to 1 ns.

Table 8.2 Comparison of properties of image functions

Parameters	Quantization (50 × 50)	Smoothing (50 × 50)	Color Extraction (Figure 8.9(b)) (4 × 4)	Color Extraction (Figure 8.12) (50 × 50)
Delay (worst case)	0.13 ns	0.6 ns	0.3 ns (w/o interconnect delay)	1.3 ns (w/o interconnect delay)
Power (worst case)	10 nW	10 nW	0.038 mW	0.015 W
Power-Delay Product	1.3×10^{-18} J	6×10^{-18} J	1.14×10^{-14} J	1.95×10^{-11} J
Resistor	250 MΩ	50 MΩ	NA	NA
Area	$\sim 50 \times 50$ (nm)^2	$\sim 50 \times 50$ (nm)^2	~ 2 (um)^2	~ 30 (um)^2

result of the changes of the color values confirms the final output image in Figure 8.13.

In Table 8.2, comparisons in the performance and property of the image functions are described. The image processor shows better performance in the functions of the quantization and smoothing than color extraction. This originates from the devices for the function implementation. Since the conventional CMOS was used for the color extraction, the performance was dominated by the CMOS. However, this degradation can be overcome by using nano-electronic devices. This will be the focus of our further study.

8.7 Comparison with DSP Chips

In Table 8.3, the performance comparison among commercial DSPs is described [9–14]. The operating frequencies range from 150 to 600 MHz. In the relationship between IPC and power dissipation, the power dissipation is reversely proportional to the IPC because the more processing instructions per clock cycles results in more power dissipation.

In Table 8.4, the processing time and power dissipation comparison among different DSPs for the smoothing function are demonstrated. Based

Table 8.3 Operational specifications for different processors [9–14]

Parameters	DM 642	C641x	C6711	C55x
Frequency (MHz)	400~600	500~600	150	144~200
Power	1~1.7	0.64~1.04	1.1	0.065~0.16
CPI	1.67~2.5	1.67~2	6.7	2.5~6.94

Table 8.4 Performance comparison among different processors for smoothing function (500 × 500 pixels)

Parameters	DM 642	C641x	C6711	C55x	Suggested Processor
Smoothing (instructions)	20 × (1.67∼2.5)	20 × (1.67∼2)	20 × (6.7)	20 × (2.5∼6.94)	1 (*)
Smoothing (ns)	55.6∼125	66.8∼80	893	250∼964	0.6
Power (W)	1∼1.7	0.64∼1.04	1.1	0.065∼0.16	1×10^{-6}
Power-delay product (nJ)	∼212.5	∼83.2	982.3	∼154	6×10^{-7}
Area (μm^2)	∼25 × 10^6	∼25 × 10^6	∼25 × 10^6	∼25 × 10^6	25 × 10

(*)The MPRTD processor data is based on HSPICE simulation as opposed to actual fabrication data of commercial microprocessors.

on [13], the smoothing function requires around 20 instructions per pixel for averaging the neighboring pixel values and divided by the number of pixels. The processing time is calculated considering the CPI and frequency of the DSPs. The result shows that the DSPs are slower than the suggested image processor by 2 to 4 orders of magnitude. However, since we did not include interconnect delay and I/O delay for the suggested image processor, the realistic speed of the suggested image processor will be degraded by several nano seconds, while the energy dissipation of MPRTD based processor is expected to be a few orders of magnitude lower than the DSPs power consumption. Also, the area overhead of DSPs is bigger than the suggested processor by 5 orders of magnitude. Even if we consider the speed degradation by the interconnect delay, I/O delay and area overhead of I/O pad, the suggested image processor shows better performance than DSPs by a considerable margin.

Implementing parallel processing with the conventional DSPs, one can reduce the processing time comparable to the MPRTD image processor. However, this will increase the power dissipation dramatically so that the implementation will become unrealistic. For instance, if we use 1000 DSPs for the parallel processing, the power dissipation will be around 1 KW. Therefore, this implementation will not be promising approach to reduce the processing time. Therefore, under the tight energy and Silicon area budget, the MPRTD based image processor will provide far superior performance than existing digital computers.

Using the structure with the single RTD plate as discussed in [16, 17] cannot process the color image algorithm such as the functions shown Table 8.2 as it has two states by itself. To process the color image algorithm, the structure needs to be changed by adding heavy interconnection between the cells. In this way, the gain of the compact size of the structure will be disappeared.

8.8 Stability

However, we must ensure that the networks of MPRTDs must stabilize when an image function is performed by changing the programmable interconnects between the MPRTDs. To examine the stability of the non-linear dynamic system describing the color image processor architecture, the classical Lyapunov theorem will be applied here. We consider an array of $M \times N$ MPRTD cells, where the cell on the ith row and jth column is denoted by $C(i, j)$. The output of the MPRTD on the ith row and jth column is represented by V_{ij}. We can then define the Lyapunov function, $E(t)$, of the MPRTD image processor by Equation (8.26):

$$
\begin{aligned}
E(t) \;=\; & -\frac{1}{2} \sum_{n,\,m} \sum_{k,\,l} A(n, m; k, l) \cdot F_{n,\,m}(t) \cdot F_{k,\,l}(t) \\
& + \frac{1}{R(V_{n,\,m})} \sum_{n,\,m} \int_{0}^{F_{n,\,m}} G^{-1}(F_{n,\,m})\, dF_{n,\,m} \\
& - \sum_{n,\,m} \sum_{k,\,l} B(n, m; k, l) \cdot F_{n,\,m}(t) \cdot U_{k,\,l}(t) \\
& - \sum_{n,\,m} I_{n,\,m} F_{n,\,m}(t).
\end{aligned}
\tag{8.26}
$$

where $U_{k,\,l}(t)$, $F_{n,\,m}(t)$ and G is feed-forward inputs, outputs from neighboring cells and relationship between the neighboring cell output and the target cell output. The stability of the image processor can be established by the following two theorems that claim the Lyapunov energy function is both bounded for a given power supply for the MPRTD image processor and the energy function monotonically reduces to a minimum value.

1) *Theorem* 1: The Lyapunov function, $E(t)$, of the multi-peak RTD based color image processor is bounded by $E_{\max}$ when a supplied voltage source is bounded.

Proof:

If the supplied voltage source is bounded, $V_{n,\,m}$, $F_{k,\,l}$, $V_{k,\,l}$, and $U_{k,\,l}$ are bounded. If $V_{n,\,m}$ is bounded, the differential resistance which is the function of $V_{n,m}$ is also bounded. The current flow between the neighboring cells and the target cell which is defined as $A(i,j;k,l)$ and $B(i,j;k,l)$ is bounded as the supplied voltage source is bounded. Considering a bounded external current source, $E(t)$ is bounded.

2) *Theorem 2*: The differential of the Lyapunov function, $E(t)$, of the multi-peak-based color image processor is less than or equal to zero in the region where $dF_{n,m}/dV_{n,m} \geq 0$, that is

$$\frac{dE(t)}{dt} \leq 0 \quad where \quad \frac{dF_{n,m}}{dV_{n,m}} \geq 0. \tag{8.27}$$

Proof:

From Equation (8.26), the differential of $E(t)$ with respect to time t can be described by

$$\frac{dE(t)}{dt} = -\sum_{n,m}\sum_{k,l} A(n,m;k,l)\frac{dF_{n,m}}{dV_{n,m}} \cdot \frac{dV_{n,m}}{dt} \cdot F_{k,l}(t)$$

$$+ \frac{1}{R(V_{n,m})} \sum_{n,m} \frac{dF_{n,m}}{dV_{n,m}} \cdot \frac{dV_{n,m}}{dt} \cdot G^{-1}(F_{n,m})$$

$$- \sum_{n,m}\sum_{k,l} B(n,m;k,l)\frac{dF_{n,m}}{dV_{n,m}} \cdot \frac{dV_{n,m}}{dt} \cdot U_{k,l}(t)$$

$$- \sum_{n,m} I_{n,m} \cdot \frac{dF_{n,m}}{dV_{n,m}} \cdot \frac{dV_{n,m}}{dt} \tag{8.28}$$

$$= -\sum_{n,m} \frac{dF_{n,m}}{dV_{n,m}} \cdot \frac{dV_{n,m}}{dt} \left(\sum_{k,l} A(n,m;k,l) \cdot F_{k,l}(t) - \frac{G^{-1}(F_{n,m})}{R(V_{n,m})} \right.$$

$$\left. + \sum_{k,l} B(n,m;k,l) \cdot U_{k,l}(t) + I_{n,m} \right)$$

$$= -\sum_{n,m} \frac{dF_{n,m}}{dV_{n,m}} \cdot \left[\frac{dV_{n,m}}{dt} \right]^2 \cdot C$$

Since we can assume C is positive in physical meaning, the polarity of $E(t)/dt$ depends on $dF_{n,m}/dV_{n,m}$. From this theorem, the multi-peak

RTD based color image processor is stable in a limited region where $dF_{n,m}/dV_{n,m} \geq 0$. The differential of $F_{n,m}$ with respect to $V_{n,m}$ is positive when the differential resistance of multi-peak RTD is positive. When the functions of the color image processor are the quantization or smoothing function, the template A is changed and has all positive values. In the case of color extraction, the template A has positive values which are the variable with respect to the $F_{k,l}$ and template B also has positive values which are the variable with respect to the $U_{k,l}$. The changed template A and B do not affect the stability of the image processor. Based on Equation (8.28), the multi-peak RTD based color image processor is stable regardless of the functions of the image processor.

8.9 Conclusion

In this chapter, we have introduced a new architecture to process color images by using a spatially distributed array of multi-peak RTDs. We have demonstrated through simulation various color image-processing functions such as the quantization, smoothing and color extraction by programming the interconnecting patterns (synapses) between the MPRTDs which act as neurons. We implemented quantization function and smoothing function through changing the conductance value between the MPRTDs. In the case of the color extraction function, we suggested three different methods to extract the selected colors. We demonstrate the HSPICE simulation results of those functions. Based on the simulation, we demonstrate that 130 ps processing time is required for quantization, 600 ps for smoothing, and 1.3 ns for color extraction. Evidently, these performance data affirm that the MPRTD-based color image processor provides faster processing speed and lower power dissipation than the conventional digital signal processors (DSPs) as shown in Table 8.4.

References

[1] L. O. Chua and L. Yang, "Cellular Neural Networks: Applications," IEEE Trans. Circuits and Systems, vol. 35, Oct. 1988, 1273–1290.
[2] L. L. Chang, L. Esaki, and R. Tsu, "Resonant Tunneling Diode in Semiconductor Double Barriers," Appl. Phys. Lett., vol. 24, Jun. 1974, 593–595.

[3] P. Mazumder, S. Kulkarni, G. I. Haddad, and J. P. Sun, "Digital Applications of Quantum Tunneling Devices," Proceedings of the IEEE, Apr. 1998, 664–688.

[4] A. Seabaugh and P. Mazumder, "Quanmtum Devices and Their Applications," Proceedings of the IEEE, vol. 7, no. 4, April 1999.

[5] G. I. Haddad and P. Mazumder, "Tunneling Devices and Their Applications in High-Functionality/Speed Digital Circuits," Journal of Solid State Electronics, vol. 41, no. 10, Oct. 1997, 1515–1524.

[6] J. P. Sun, G. I. Haddad, P. Mazumder and J. N. Schulman, "Resonant Tunneling Diodes: Models and Properties", Proceedings of the IEEE, 1998.

[7] A. C. Seabaugh, Y. C. Kao and H. T. Yuan, "Nine-state Resonant Tunneling Diode Memory," IEEE Electron Device Lett., vol. 13, no. 9, 1992, 479–481.

[8] Li Ding and Pinaki Mazumder, "Noise- Tolerant Quantum MOS Circuits Using Resonant Tunneling Devices," IEEE Trans. Nanotechnology, vol. 3, 2004, 134–146.

[9] Texas Instruments, TMS320C6711 DSK: http://focus.ti.com/ docs/ toolsw/folders/print/tmds320006711.html

[10] Texas Instruments, TMS320DM642 Product folder: http://focus.ti.com/ docs/toolsw/folders/print/tmds320dm642.html

[11] Texas Instruments, TMS320DM642 Power Consumption: http://focus.ti. com/lit/an/spra962a/spra962a.pdf

[12] Texas Instruments, TMS320C6416 Product folder: http://focus.ti.com/ docs/prod/folders/print/tms320c6416.html

[13] Texas Instruments, TMS320C6416 Power Consumption: http://focus.ti. com/lit/an/spra811c/spra811c.pdf

[14] http://focus.ti.com/dsp/docs/dsphome.tsp?sectionId=46&DCMP=TI HeaderTracking&HQS=Other+OT+hdr_p_dsp

[15] W. Pratt, Digital Image Processing, 3rd Edition, Wiley 2001.

[16] V. P. Roychowdhury, D. B. Janes, and S. Bandyopadhyay, "Collective Computational Activity in Self-Assembled Arrays of Quantum Dots: A Novel Neuromorphic Architecture for Nanoelectronics," IEEE Trans. on Electron Devices, vol. 43, no.10, 1996, 1688–1699.

[17] K. Karahaliloglu, S. Balkir, S, "Image processing with quantum dot nanostructures," International Symposium on Circuits and Systems, vol. 5, 2002, 217–220.

[18] Woo Hyung Lee and Pinaki Mazumder, "Motion detection by quantum dots based velocity tuned filter," IEEE Trans. on Nanotechnology, vol. 7, 2008.

[19] Takao Waho, Kevin J. Chen and Masafumi Yamamoto, "Resonant-Tunneling Diode and HEMT Logic Circuits with Multiple Thresholds and Multilevel Output," IEEE Journal of solid-state circuits, vol. 33, no. 2, 1998.

[20] Martin Hänggi and Leon O. Chua, "Cellular neural networks based on resonant tunnelling diodes," International Journal of Circuit Theory and Applications, vol 29 issue 5, 487–504, 2001.

[21] W. Huang, et al., "Enhancement-mode GaN Hybrid MOS-HEMTs with Ron,sp of 20 mΩ-cm2," Proceedings of the 20th ISPSD, 2008.

[22] Bong-Hoon Lee and Yoon-Ha Jeong, "A Novel SET/MOSFET Hybrid Static Memory Cell Design," IEEE Trans. on Nanotechnology, vol. 3, 2004.

[23] K. Karahaliloglu, S. Balkir, S. Pramanik, and S. Bandyopadhyay, "A Quantum Dot Image Processor," IEEE Trans. on Electron Devices, vol. 50, no. 7, July 2003.

[24] Arena, P., Fortuna, L., Occhipinti, L, "A CNN algorithm for real time analysis of DNA microarrays," IEEE Transactions on Circuits and Systems I: Fundamental Theory and Applications, vol. 49 no. 3, Mar 2002, 335–340.

[25] Wang, L., De Gyvez, J.P. and Sanchez- Sinencio, E, "Time multiplexed color image processing based on a CNN with cell-state outputs," IEEE Trans. VLSI Systems, vol. 6, no. 2, June 1998, 314–322.

[26] Tetzlaff, R., R. Kunz, C. Ames, D. Wolf, "Analysis of Brain Electrical Activity in Epilepsy with Cellular Neural Networks (CNN)," Proc. on European Conference on Circuit Theory and Design, 1999.

9

Design of a Velocity-Tuned Filter Using a Matrix of Resonant Tunneling Diodes

Woo Hyung Lee and Pinaki Mazumder

In this chapter, a nanoscale velocity-tuned filter that employs resonant tunneling diodes to perform temporal filtering to track moving and stationary objects is presented. The new velocity-tuned filter is not only amenable for nanocomputing, but also superior to other approaches in terms of area, power and speed. It is shown that the proposed nanoarchitecture for velocity-tuned filter is asymptotically stable in the specific region.

9.1 Introduction

It is well known that real-time vision machine application tasks are computationally intensive and require complex and costly resources. In addition, certain specific tasks such as bio-robots and biomedical applications put additional constraints on the overall system in terms of its size, power consumption, shock resistance, and manufacturing cost. A real-time vision machine requires motion computation involving a large number of computations and many computational resources. An attractive solution to the problems is to use parallel image processing architectures [1, 2].

Since velocity-tuned filters are one of the main parts of motion computation, they require a compact area and low power consumption. The velocity-tuned filter can be combined with pattern recognizers, optical flow sensors and noise removers to create real-time vision machines. Even though velocity-tuned filters using spatiotemporal derivatives and Reichardt correlation detectors have been studied, they cannot provide sufficient area compactness, low power consumption and/or, speed [3, 4]. These problems can be attributed to limitations of the conventional analog circuits that cannot achieve high performance for the real-time vision machine.

Among nanoelectronic devices, an array of resonant tunneling diode structure shows good performance in terms of area, power consumption and speed. Since being studied by several researchers [1, 2], their applications have been confined to Boolean logic or image processing.

Using the bi-stable states of the resonant tunneling diode, it can be also applied to amplify output signal difference in a system. In the case of velocity tuned filter, it is required to implement a circuit which magnifies the output difference to differentiate an object with a specific speed. Hence, the resonant tunneling diode is used to magnify the output signal difference based on speeds in the velocity tuned filter.

In this chapter, a new type of velocity-tuned filter using nanoelectronic devices such as resonant tunneling diodes is studied. The velocity-tuned filter achieves performance in area, power, and speed superior to other conventional velocity-tuned filters by taking advantage of efficient parallel processing capability of an array of resonant tunneling diodes (RTDs).

9.2 Array of RTD-based Velocity-Tuned Filters

9.2.1 Conventional Velocity-tuned Filter

Figure 9.1 shows a conventional velocity-tuned filter. The conventional velocity-tuned filter consists of pre-amplifiers, filters, multipliers and differential amplifiers. Pre-amplifiers consist of a pre-filtering part and an amplification part. In pre-filtering part, the input signal In_n comes from photo detectors which are not shown in Figure 9.1. After the input signal is amplified by the pre-amplifiers, the input signal passes through the two layered filters, the receptor layer and the horizontal cell layer respectively. The main function of the receptor layer is to improve the signal to noise ratio. The second layer calculates a spatiotemporal average of the receptor output. The combined function of the two layered filter is a spatiotemporal band-pass filter. In [5], Torralba used the CNN structure to implement the spatiotemporal band-pass filter and active and passive devices to connect the CNN cells.

After pre-filtering the signal from the optical flow, the filtered signal is amplified for providing inputs to two layered filters. The signal, then, is filtered with the two layered filters which allow the signals to have different values based on the velocities. The filtered signal is then magnified with local energy integration and the output is linearized with the difference between signals using shunting inhibition circuitry.

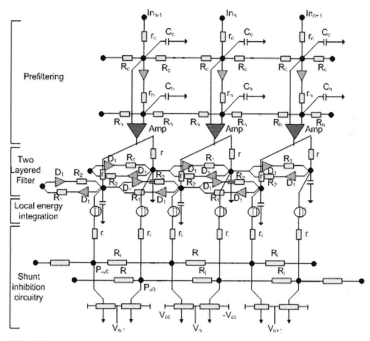

Figure 9.1 Conventional velocity-tuned filter that consists of pre-filtering, two layered filters, local energy integration, and shunting inhibition circuitry [5].

The local energy integration can be written in a mathematical form as:

$$
\begin{aligned}
P(\Delta v) &= \int \Gamma_s(f_x) df_x \\
&= \int \Gamma_e(f_x) |G(f_x)|^2 df_x \\
&= \Gamma \int |G(f_x)|^2 df_x \\
&= \frac{\Gamma}{2\gamma \sqrt{1 + (v_x - v_0)^2 / \Delta v_0^2}}
\end{aligned}
\tag{9.1}
$$

where v_x and $G(f_s)$ are input velocity and frequency response, and Γ, γ, and Δv_0 are constants [5]. The velocity-tuned filter needs the shunting inhibition circuitry to linearize the output. The function of the shunting inhibition is written mathematically as

$$
V_{out} = \frac{G_+ - G_-}{G_+ + G_-} V_{cc} = \frac{P_{V0} - P_{-V0}}{P_{V0} + P_{-V0}} V_{cc}
\tag{9.2}
$$

From (9.2), the output voltage is proportional to the difference between G_+ and G_-. The difference can be represented as the difference between P_{v0} and P_{-v0}. Therefore, an analog amplifier was implemented to magnify the difference in Torralba's work [5].

Since the conventional velocity-tuned filters use an analogue circuitry to compute Equations (9.1) and (9.2), the conventional velocity-tuned filters are not well suited for real-time motion computation circuits for bio-robots or biomedical applications, which require high performance in area, power, and speed. Hence, the conventional velocity-tuned filters hardly meet the condition for real-time motion estimation for those applications.

To implement the real-time motion estimation, an array of resonant tunneling diodes that are implemented by metallic islands on the resonant tunneling diodes is used. Since the incident light is pre-filtered and then amplified as an electric signal in the first part in Figure 9.1, the remaining part to improve the performance of the velocity tuned filter is focused. The array of the resonant tunneling diodes is used to replace the filtering part, the local energy integration part, and the shunt inhibition part. Those functions can be implemented using the negative resistance characteristic and area compactness of the array of the resonant tunneling diodes.

9.2.2 Resonant Tunneling Diode

Since the resonant tunneling diode (RTD) was introduced by Esaki et al. [6,7], it has been applied to various types of circuitry [8]. The main characteristic of the resonant tunneling diode is negative differential resistance (NDR). This characteristic originates from its heterostructure with a low-bandgap quantum well between high-bandgap materials. The thickness and width of the resonant tunneling diode are required to be fabricated in the order of several nanometers with epitaxial deposition techniques. The low-bandgap quantum well is quantized resulting in discretized energy levels in the quantum well. Figure 9.2(a) shows the modeling of the RTD I-V characteristic from experimental results. The fabricated RTD describes a peak-to-valley current ratio (PVCR) of 13 with a peak voltage (VP) of 0.28 V at room temperature. To model this I-V curve, factors that affect the conduction of RTD need to be examined.

However, the size of RTDs deviates with the process variation. The I-V curve also shifts up and down since the current flow proportional to the size of RTDs. This will affect the functionality of the system since the functions are valid when the NDR region is guaranteed. This means that if the lowest

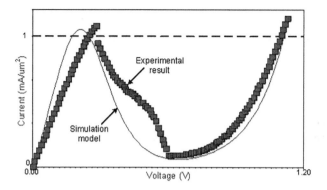

(a) Experimental result and simulation model

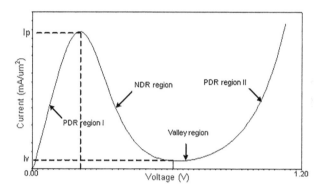

(b) Simulation model of resonant tunneling diode with several regions

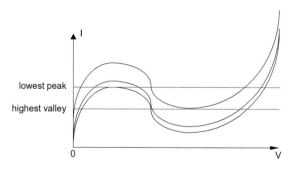

(c) I-V curve deviation with the size of metallic islands

Figure 9.2　Modeling of resonant tunneling diode based on experimental result.

peak is higher than highest valley, the functionality is valid from the deviation of the process. From the experimental result, a PVCR of 13 is assumed. Therefore, functionality is still valid as long as the size deviation of RTDs is less than 13.

However, it is possible to make a RTD having a PVCR of 46 with cutting edge technology [9]. In this case, the maximum cell size deviation from the process to guarantee the functionality of the system is 46. The situation will become even better as the process technology is developed.

Conduction of the resonant tunneling diode consists of two parts. One is conduction due to resonant tunneling and the other is conduction from diode conduction. The negative differential resistance results from the resonant tunneling conduction effect [10–12]. Using a physics-based model suggested by Schulman et al., the resonant tunneling current of RTD is modeled as a summation of resonant tunneling effect, J_1, and diode conduction effect, J_2:

$$J(V) = J_1(V) + J_2(V), \tag{9.3}$$

$$J_1(V) = \frac{qm \cdot kT\Gamma}{4\pi^2\hbar^3} \ln\left(\frac{1 + e^{\frac{E_F - E_r + \frac{n_1 qV}{2}}{kT}}}{1 + e^{\frac{E_F - E_r - \frac{n_1 qV}{2}}{kT}}}\right)$$

$$\cdot \left(\frac{\pi}{2} + \arctan\left(\frac{E_r - \frac{n_1 qV}{2}}{\frac{\Gamma}{2}}\right)\right), \tag{9.4}$$

$$J_2(V) = H(e^{\frac{n_2 qV}{kT}} - 1) \tag{9.5}$$

where E_F is the Fermi energy, E_r is the energy of the resonant level, Γ is the resonant width, and n_1 and n_2 are model parameters. Those parameters are often obtained empirically and affect the slope of the curve in Figure 9.2(b).

The resonant tunneling occurs when the applied voltage across the diode is aligned to one of the quantized energy levels in the quantum well as shown in PDR region I in Figure 9.2(b). However, when the applied voltage is increased to be misaligned to the quantized energy level, the conduction is decreased as shown in the NDR region in Figure 9.2(b). The current subsequently increases as conduction through higher energy states becomes possible as shown in the PDR region in Figure 9.2(b). This characteristic enables the circuit switch fast and self-latching or bi-stable. Using this characteristic, a wide class of circuit applications, for instance high speed circuits, low power-delay product circuits, and multi-valued logic, can be implemented.

9.2.3 Velocity-tuned Filter

The studied velocity-tuned filter consists of an array of filters and RTDs. The RTDs are connected to filters with diodes. Each filter cell is connected to four neighboring cells and one output cell. The output cell consists of RTDs, and needs a static current source which is vertically connected to each RTD. Conventional velocity-tuned filters require local energy integration and shunting inhibition circuits to magnify the output difference. The filter structure does not employ these circuits and instead, uses the RTD to magnify the filtered output differences. Through the differential resistance characteristic of the RTD, the final value is determined by the RTD's *I-V* characteristic and the external current source. In the investigated velocity-tuned filter, the output of the filter makes the RTD device operate when the output of the filter is higher than 1.4 V. This threshold voltage can be given by

$$V_{th} = V_d + \frac{V_p + V_u}{2} \tag{9.6}$$

where V_d is the diode junction voltage and V_p and V_v are the peak voltage and the valley voltage respectively in RTD *I-V* curve.

Figure 9.3 shows the schematic diagrams of the velocity-tuned filter's side view and top view. Analysis of the velocity-tuned filter requires a new state equation. This new state equation of the structure is obtained by modifying the state equation of diffusion circuit in Equation (9.7). In Equation (9.7), $X_{n,m}$, $S_{n,m}$, and v represent input voltage, output voltage, and velocity respectively. Also, γ and τ are defined in Equations (9.8) and (9.9). Here, γ is related with conductance between cells in the filtering part and τ affects the processing time of the filter. Also, v_{x0} and v_{y0} represent tuning velocity in the x direction and the y direction, respectively:

$$\tau \frac{dS_{n,m}(t)}{dt} = X_{n,m}(t) - S_{n,m}(t) + \left(\gamma^2 + \frac{v_{x0}\tau}{2}\right)(S_{n-1,m}(t) - S_{n,m}(t))$$

$$+ \left(\gamma^2 + \frac{v_{x0}\tau}{2}\right)(S_{n,m-1}(t) - S_{n,m}(t))$$

$$+ \left(\gamma^2 - \frac{v_{x0}\tau}{2}\right)(S_{n+1,m}(t) - S_{n,m}(t))$$

$$+ \left(\gamma^2 - \frac{v_{x0}\tau}{2}\right)(S_{n,m+1}(t) - S_{n,m}(t)) \tag{9.7}$$

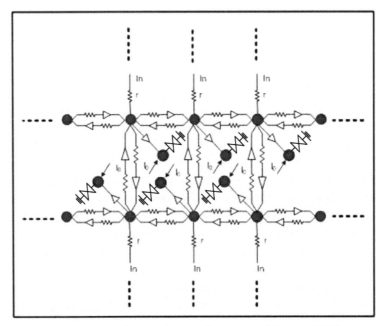

Figure 9.3   Velocity-tuned filter with an array of filters and RTDs.

$$\gamma^2 = \frac{r}{\frac{kT}{qI_s(e^{qV_d/kT}-1)} + R},$$ (9.8)

$$\tau = r \cdot C.$$ (9.9)

Considering the resonant tunneling diode, the state equation can be obtained as follows:

$$\frac{d}{dt}q_{n,m} = C_{n,m}\frac{dS_{n,m}}{dt} = -f(S_{n,m}(t)) + h(S_{n,m}),$$ (9.10)

$$f(S_{n,m}(t)) = F \cdot (\exp(n_2 S_{n,m}q/kT) - 1)$$
$$+ A \cdot \ln\left[\frac{1 + e^{((B-C+n_1 S_{n,m})q/kT)}}{1 + e^{((B-C-n_1 S_{n,m})q/kT)}}\right]$$
$$\cdot \left[\frac{\pi}{2} + arctan\left(\frac{C - n_1 S_{n,m}}{D}\right)\right] + H(e^{\frac{n2eV}{kT}} - 1),$$ (9.11)

$$h(S_{n,m}(t)) = \sum_{Cell(k,l)\in N_r(n,m)} G(Nr(n, m), \overrightarrow{v}(t, x, y))$$
$$\cdot (e^{q(N_r(n,m)-I_r R-S_{n,m})/kT} - 1) \cdot I_s$$ (9.12)

where

$$G(N_r(n,m), \overrightarrow{v(t,x,y)})$$
$$= \begin{cases} \gamma^2 + \frac{\overrightarrow{v(t,x,y)} \cdot \tau}{2} & \text{when } \cos\theta(\overrightarrow{v(t,x,y)}, \overrightarrow{N_r(n,m)S_{n,m}}) = 0; \\ \gamma^2 - \frac{\overrightarrow{v(t,x,y)} \cdot \tau}{2} & \text{when } \cos\theta(\overrightarrow{v(t,x,y)}, \overrightarrow{N_r(n,m)S_{n,m}}) \neq 0; \end{cases}$$

$$(9.13)$$

and $N_r(n,m)$ is a r-neighborhood of a cell *Cell(n,m)* and is defined by

$$N_r(n,m) = \begin{cases} Cell(k,l) | \max\{|k-n|,|l-m|\} \leq r \\ 1 \leq k \leq N; \quad 1 \leq l \leq M \end{cases}, \qquad (9.14)$$

where $I_{n,m}$ is the bias current and $f(S_{n,m})$ describes the *I-V* characteristic of the resonant tunneling diode with physical model parameters, *A, B, C, D, F, n_1,* and *n_2*; $h(S_{n,m})$ represents the current from neighboring cells to $S_{n,m}$ where I_s is the saturation current in the diode.

The relationship between $S_{n,m}$ and $O_{n,m}$ can be described by combining Equations (9.10) and (9.15):

$$g(O_{n,m}(t)) = (e^{q(S_{n,m} - O_{n,m})/kT} - 1) \cdot I_s. \qquad (9.15)$$

Using these equations, a new state equation is obtained as given by

$$C_{n,m} \frac{dO_{n,m}}{dt} = -f(O_{n,m}(t)) + g(O_{n,m}(t)) + I_{n,m}(t), \qquad (9.16)$$

which represents the velocity-tuned filter. The schematic diagram of the velocity-tuned filter for a basic cell is described in Figure 9.4. Figures 9.5 and 9.6 describe experimental results of the velocity-tuned filter. In Figure 9.5 and Figure 9.6, the dark round shaped dots represent input impulse signals. Observing signals based on the time sequence, the moving object and the standing object are divided with time, T. Hence, $T = 2$ represents the next time to $T = 1$ in the sampling clock sequence. In Figure 9.5, the standing object (S_4) is amplified in the output, O_4, while other moving objects from S_2 to S_1 hold low output values in O_1. In Figure 9.6, the moving object from S_2 to S_1 in the left direction is latched to the high value in O_1, while the standing object, S_4 is latched to the low output value in O_4. In Figure 9.7, the left figure is the moving input. The first row represents an object moving with a velocity of 2 pixels/sec to the right. The second row shows an object with a velocity of 1 pixel/second to the right. The third row represents a non-moving

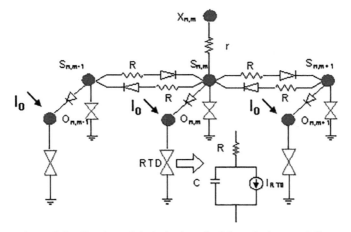

Figure 9.4   Circuit model of a basic cell of the velocity-tuned filter.

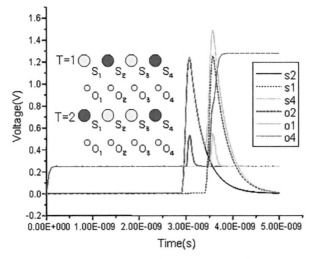

Figure 9.5   HSPICE simulation result for velocity-tuned filter with velocity 0.

object. The fourth row shows an object with a velocity of 1 pixel/second to the left. Finally, the fifth row represents an object with a velocity of 2 pixels/second to the left. The right sides of Figures 9.7 and 9.8 show the output of the filter. In Figure 9.7, only the target object, which does not move has a high output value shown with colors, red and yellow. Therefore, the non-moving object can be detected as shown in Figure 9.7(b). In Figure 9.8,

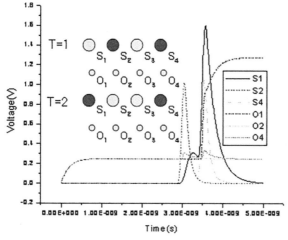

Figure 9.6 HSPICE simulation result for velocity-tuned filter with velocity 1.

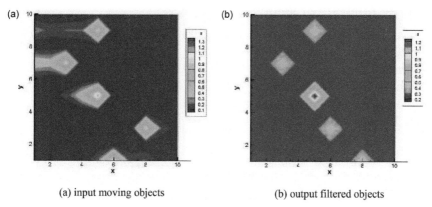

(a) input moving objects   (b) output filtered objects

Figure 9.7 Experimental result of velocity-tuned filter for a velocity of 0 pixel/second. (a) input moving objects (b) output filtered objects.

the moving object with a velocity of 1 pixel/second to the left has a high output value. Hence, the object with a velocity of 1 pixel/second to the left can be detected as shown in Figure 9.8(b). The velocity to be filtered can be chosen by controlling the resistance values between the cells. Based on Equation (9.7), the resistance value to filter objects with a certain velocity is controlled. The output results in Figure 9.8(b) can be explained using Equation (9.16).

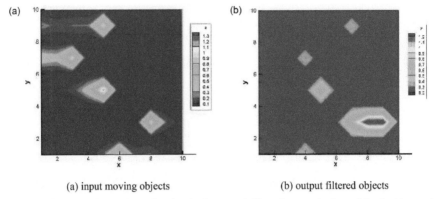

(a) input moving objects   (b) output filtered objects

Figure 9.8 Experimental result of velocity-tuned filter for a velocity of 1 pixel/second. (a) input moving objects (b) output filtered objects.

9.3 System Analysis

9.3.1 Delay Analysis of Velocity-tuned Filter

In this section, the delay of the input signal to the output node is investigated. To analyze the signal delay, the velocity-tuned filter is regarded as an RC network in the form of a tree. To examine the system as an RC tree, the velocity-tuned filter is simplified as shown in Figure 9.9. In Figure 9.9, the velocity-tuned filter only with capacitors and resistors is described. Hence the following Equation (9.17) always holds for the output node $O_{n,m}$:

$$\int_0^\infty V_{O_{n,n}}(t)dt = \sum_{k=1}^{N} C_k S_k(0) R_{O_{n,m},k}, \tag{9.17}$$

where

$$R_{O_{n,m},k} = \sum R_j \Rightarrow (R_j \in [path(O_{n,m} \to X_{n,m}) \cap path(k \to X_{n,m})]. \tag{9.18}$$

Assuming that the array of filter nodes is a $2n \times 2m$ matrix, same as that of the output nodes, the delay of the signal from the input node to the output node in the position of n, m can be represented by

$$r \sum_{l=1}^{2m} \sum_{kl=1}^{2m} (C_{S_{k,l}} + C_{O_{k,l}}) - rC_{O_{n,m}} + (r + R_d(V) + 5R_{rtd}(V))C_{O_{n,m}}, \tag{9.19}$$

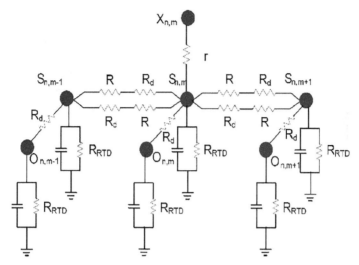

Figure 9.9 RC tree representation of the velocity-tuned filter.

where $R_{rtd}(V)$ and $R_d(V)$ are the resistances with functions of applied voltages across each device. Since it is assumed that the output is stabilized after 5τ (time constant) due to the RTD, $5R_{rtd}(V)$ for the RTD stabilization time on the output is added. Based on the simulation results in Figures 9.5 and 9.6, the RTD stabilization takes time from 200 ps to 500 ps in the simulation. Assuming the capacitances are same for all nodes, Equation (9.19) can be approximated as

$$C(4rn \cdot 4rm + R_d(V) + 5R_{rtd}(V)). \tag{9.20}$$

Given the size of the array, the delay is governed by the input resistance, the diode resistance, and the RTD resistance from Equation (9.20).

However, since the conventional velocity-tuned filter uses local energy integration and shunt inhibition circuitry after filtering, the velocity-tuned filter is much faster than the conventional velocity-tuned filter. To examine this in detail, the propagation delay of local energy integration and shunt inhibition circuitry needs to be investigated. In circuit implementation, the integration logic consists of multipliers and adders. Using cutting-edge technology, 0.11 *um* CMOS standard cell library, the whole processing of those combinational logics takes 4.5 *ns* to 5 *ns* for 4 bits (1 digit) multipliers [13]. Including adders, the processing time will be increased until 7 *ns* to 8 *ns*, which is around 100 times slower than that of the nanoelectronic circuitry as shown in Figures 9.5 and 9.6.

9.3.2 Power Consumption Analysis of Velocity-tuned Filter

Since the investigated velocity-tuned filter has fewer device components than conventional velocity-tuned filters, the velocity-tuned filter is assumed to consume less power. To analyze the power consumption, it is needed to partition the velocity-tuned filter into two components. The first part corresponds to the filter part, which includes filtering connections, input connections, metallic islands, and the resonant tunneling diode in Figure 9.4 and the second part consists of output interconnections, output metallic islands, and the output resonant tunneling diode in Figure 9.4. To calculate the power dissipation, Figure 9.4 is redrawn as shown in Figure 9.10 which excludes the right hand side of the filters for circuit analysis, assuming that the current flow in the filter area is unidirectional. Assuming the input voltage to node $X_{n,m}$ in Figure 9.9 is V_{supply} which is a pulse signal with a rising and a falling time, the average power consumption of resistor r is given as

$$P_r = r \cdot I^2 = r \cdot (YV_{supply})^2 = r \cdot \left|\frac{1}{Z}\right|^2 \cdot |V_{supply}|^2, \qquad (9.21)$$

$$Y = \frac{1}{Z} \qquad (9.22)$$

where

$$Z = r + \cfrac{(R + R_d(V) + \frac{A \cdot (R_d(V) + A)}{2A + R_d(V)}) \cdot \frac{A \cdot (R_d(V) + A)}{2A + R_d(V)}}{R + R_d(V) + \frac{2A \cdot (R_d(V) + A)}{2A + R_d(V)}} \qquad (9.23)$$

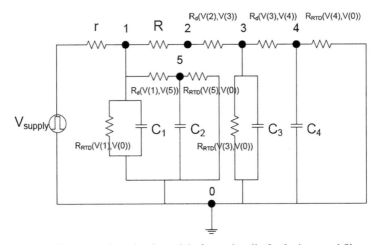

Figure 9.10 Equivalent circuit model of an unit cell of velocity-tuned filters.

where

$$A = \frac{\frac{1}{j\omega C} \cdot R_{RTD}(V)}{\frac{1}{j\omega C} + R_{RTD}(V)}, \tag{9.24}$$

where $R_{RTD}(V)$ and $R_d(V)$ are functions of voltage, V, applied across the devices. $R_{RTD}(V)$ and $R_d(V)$ can be obtained from *I-V* curves when voltages are applied across the devices. The *I-V* curve of the diode is described by

$$I = I_s(e^{\frac{qV}{kT}} - 1). \tag{9.25}$$

Also, the *I-V* curve of RTDs is described by Equations (9.3)–(9.5). Since V_{supply} is not a DC source, the voltages across the devices change with time. This calculation can be processed by partitioning the simulation time and applying the equations with each interval repeatedly. The average power consumption of resistor R is given as

$$P_R = R \cdot I_1^2 = R \cdot \left| \frac{\frac{A \cdot (R_d(V)+A)}{2A+R_d(V)}}{R + R_d(V) + \frac{2A \cdot (R_d(V)+A)}{2A+R_d(V)}} \right|^2 \cdot \left| \frac{1}{Z} \right|^2 \cdot |V|^2. \tag{9.26}$$

In the case of a diode in the filter parts, the average power consumption is written as

$$P_{R_d(V)} = R_d(V) \cdot \left| \frac{\frac{A \cdot (R_d(V)+A)}{2A+R_d(V)}}{R + R_d(V) + \frac{2A \cdot (R_d(V)+A)}{2A+R_d(V)}} \right|^2 \cdot \left| \frac{1}{Z} \right|^2 \cdot |V|^2. \tag{9.27}$$

The output which includes the diode, metallic islands, and resonant tunneling diode consumes power described by

$$P_{output} = \frac{\left| \frac{A \cdot (R_d(V)+A)}{2A+R_d(V)} \right|^3}{\left| R + R_d(V) + \frac{2A \cdot (R_d(V)+A)}{2A+R_d(V)} \right|^2} \cdot \left| \frac{1}{Z} \right|^2 \cdot |V|^2. \tag{9.28}$$

The total power dissipation can be obtained by measuring current from the power source and multiplying with the supplied voltage. The total power dissipation could be obtained by using HSPICE simulation which includes the above calculations. However, since the output part include the external current into the output cell, the power consumption due to the external current needs to be added to the output power consumption. The modified power consumption including the contribution from the external current is given as

$$P_{output\text{-}mod} = P_{output} + I_{ext}^2 \cdot R_{RTD}(V). \tag{9.29}$$

In detail, the external current can be minimized to minimize the static power consumption by satisfying the following condition,

$$I_v + I_{leak} < I_{ext} < I_{peak} + I_{leak}, \tag{9.30}$$

where the I_{leak}, I_{peak} and I_v are the leakage current from the capacitance components of the RTDs, the peak current and the valley current, respectively as shown in Figure 9.2(b). Applying 0.5 mA/um^2 of the current density, two stable voltage points, 0.15 V and 1.1 V, exist. Therefore, the corresponding power dissipations from the external current source are 15 nW and 110 nW. Applying the experimental values of the resonant tunneling diodes' capacitance of 10^{-15} F and V_{supply} of 1.5 V to the HSPICE simulation, the total average power dissipation is obtained in the range from 29 nW to 114 nW depending on output voltages. Hence, if high output voltages is considered, the external current is dominant in the total power dissipation; otherwise, it is comparable with dynamic power dissipation. However, the dynamic power dissipation is dominated by the power from the input resistance r and the resistances, R in the filter, each of which has much higher value than the other resistive components.

Based on the above calculations, Table 9.1 gives the performance comparison with the velocity-tuned filters in [6, 15]. In terms of area, the VTF is 10 to 100 times smaller than the other VTFs. As shown in Table 9.1, the processing time of the VTF is 3 to 6 orders of magnitude faster than the other VTFs and consumes around 100 times less power than the other VTFs.

9.3.3 Stability of Velocity-tuned Filter

Stability is an important issue in designing a system. Since the velocity-tuned filter consists of nonlinear elements, namely an array of RTDs, it is needed to use appropriate methods for examining the stability of the nonlinear system. One such method is the Lyapunov theorem. To apply the Lyapunov method, the Lyapunov function, $E(t)$, of the velocity-tuned filter is defined as in

Table 9.1   Performance comparison of VTFs with 20×20 pixel images

	Sequential VTF	Conventional network VTF	New VTF
Area	$\sim 1000\ \mu m^2$	$\sim 100\ \mu m^2$	$\sim 10\ \mu m^2$
Processing Time	0.1~10 ms (1 GHz)	0.1~10 μs	0.3~2 ns
Power	0.1~1 mW	0.1~1 mW	1~10 μW

Equation (9.31):

$$E(t) = -\frac{1}{2}\sum_{n,m}h(S_{n,m}(t))\cdot h(S_{k,l}(t)) + \frac{1}{2}\sum_{n,m}f^2(S_{n,m}(t))$$

$$-\sum_{n,m}I_{n,m}(S_{n,m}(t)). \tag{9.31}$$

Considering $f(S_{n,m})$ is a nonlinear equation, $E(t)$ can be described by

$$E(t) = -\frac{1}{2}\sum_{n,m}h(S_{n,m}(t))\cdot h(S_{k,l}(t)) + \sum_{n,m}\int_0^{S_{n,m}}f(S_{n,m})dS_{n,m}$$

$$-\sum_{n,m}I_{n,m}(S_{n,m}(t)). \tag{9.32}$$

1) *Theorem* 1: The Lyapunov function, $E(t)$, of the velocity-tuned filter is bounded by $E_{\max}$ when a supplied voltage source is bounded.

Proof:

To prove $E(t)$ is bounded, $h(S_{n,m})$ is firstly examined. Assuming that the supplied voltage source is bounded, if it is proven that the differential of $h(S_{n,m})$ with respect to $S_{n,m}$ is bounded, it is concluded that $h(S_{n,m})$ is bounded. From Equation (4.10), a differential of $h(S_{n,m})$ is obtained with respect to $S_{n,m}$ as

$$h'(S_{n,m}(t)) = \sum_{C(k,l)\in N_r(n,m)}G(N_r(n,m),\overrightarrow{v(t,x,y)})\cdot(q/kT)$$

$$\cdot e^{q(N_r(n,m)-I_rR-S_{n,m})/kT}\cdot I_s. \tag{9.33}$$

In Equation (9.33), $h'(S_{n,m}(t))$ is bounded with given bounded $S_{n,m}$. Therefore, it is obvious that $h(S_{n,m})\cdot h(S_{k,l})$ is bounded. To prove $f(S_{n,m})$ is bounded, $f'(S_{n,m})$ from Equation (9.9) is obtained as

$$f'(S_{n,m}(t)) = b\exp(bn_2S_{n,m})Fn_2 + \frac{\alpha\beta(\frac{\pi}{2} + \arctan(\frac{C-n_2S_{n,m}}{D}))]}{1+b\exp(B-C+n_2S_{n,m})} - \gamma,$$

$$\tag{9.34}$$

where

$$\alpha = A[1 + b\cdot\exp(B-C-n_2S_{n,m})], \tag{9.35}$$

$$\beta = \frac{b\cdot\exp(B-C+n_2S_{n,m})}{1+b\cdot\exp(B-C-n_2S_{n,m})}$$

$$+ b \cdot \exp(B - C - n_2 S_{n,m})$$
$$\cdot \frac{[1 + b \cdot \exp(B - C + n_2 S_{n,m})]n_2}{[1 + b \cdot \exp(B - C - n_2 S_{n,m})]^2}, \tag{9.36}$$

$$\gamma = \frac{A \cdot n_2 \cdot \ln\left[\frac{1 + b \cdot \exp(B - C + n_2 S_{n,m})}{1 + b \cdot \exp(B - C - n_2 S_{n,m})}\right]}{D\left[1 + \frac{(c - n_2 S_{n,m})^2}{D^2}\right]}$$

$$- H \cdot n_2 \cdot b \cdot \exp(n_2 b S_{n,m}), \tag{9.37}$$

where b is q/kT. In Equation (9.34), all the exponential functions are bounded with the given function, $S_{n,m}$, which is bounded. Therefore, it is proven that $f(S_{n,m})$ is bounded. In addition, having the bounded function, $S_{n,m}$, the current source is bounded. Hence, $I_{n,m}$ is bounded. From the fact that every element of $E(t)$ is bounded, it is concluded that $E(t)$ is bounded.

2) **Theorem 2**: The differential of the Lyapunov function, $E(t)$, of velocity-tuned filter is less than or equal to zero in the region where $f'(S_{n,m}) \geq 0$, that is

$$\frac{dE(t)}{dt} \leq 0, \ f'(S_{n,m}) \geq 0. \tag{9.38}$$

Proof:
From Equation (9.31), the differential of $E(t)$ with respect to time t can be described by

$$\frac{dE(t)}{dt} = -\sum_{n,m} \frac{df(S_{n,m}(t))}{dS_{n,m}(t)} \cdot \frac{dS_{n,m}(t)}{dt} \cdot h(S_{k,l}(t))$$

$$+ \sum_{n,m} \frac{df(S_{n,m}(t))}{dS_{n,m}(t)} \cdot \frac{dS_{n,m}(t)}{dt} \cdot f(S_{n,m}(t))$$

$$- \sum_{n,m} I_{n,m} \cdot \frac{df(S_{n,m}(t))}{dS_{n,m}(t)} \cdot \frac{dS_{n,m}(t)}{dt} \cdot f(S_{n,m}(t))$$

$$= -\sum_{n,m} \frac{df(S_{n,m}(t))}{dS_{n,m}(t)} \cdot \frac{dS_{n,m}(t)}{dt}$$

$$\cdot (h(S_{k,l}(t)) - f(S_{n,m}(t)) + f(S_{n,m}(t)) \cdot I_{n,m})$$

$$= -\sum_{n,m} \frac{df(S_{n,m}(t))}{dS_{n,m}(t)} \cdot \left(\frac{dS_{n,m}(t)}{dt}\right)^2 \cdot C. \tag{9.39}$$

Since it is assumed that C is positive in physical meaning, the polarity of $E(t)/dt$ depends on $f'(S_{n,m})$. Assuming $f'(S_{n,m}) \geq 0$ leads to $E(t)/dt \leq 0$. From this theorem, the velocity-tuned filter is stable in a limited region where $f'(S_{n,m}) \geq 0$.

However, the investigated circuit is stable ultimately. If the initial condition of $f'(S_{n,m}) < 0$, $S_{n,m}$ enters the positive differential resistive region where $f'(S_{n,m}) \geq 0$ due to the external current source. The external current source gives a driving force to $S_{n,m}$ from negative differential region to positive differential region.

3) **Corollary 1**: The output system is stabilized when velocity-tuned filter is stabilized.

Proof:

Assume the Lyapunov function of the output system is $V(t)$. From Equation (9.15), the stability of $g(O_{n,m}(t))$ is affected only by $O_{n,m}$ since it is known that $S_{n,m}$ is a stable function. As a result, the stability of the output system can be examined by the same procedure as that of the velocity-tuned filter. This leads to the same condition of stability, that is, the output system is stable only when $f(S_{n,m}) \geq 0$.

4) **Corollary 2**: The velocity-tuned filter is asymptotically stable.

Proof:

From Theorems 1, 2, and Corollary 1, there exists any initial value of $S_{n,m}$ which satisfies the condition below

$$\|S_{0,0}\| < \delta \Rightarrow \|S_{n,m}(t, t_0, S_{0,0})\| < \varepsilon, \quad \forall t \geq t_0. \tag{9.40}$$

In addition, it is attractive so that there exists a number $\delta > 0$ such that, for all $t_0 \geq 0$ which can be described by

$$\|S_{0,0}\| < \delta \Rightarrow \|S_{n,m}(t, t_0, S_{0,0})\| \to 0, \quad t \to \infty. \tag{9.41}$$

However, the velocity-tuned filter does not satisfy the above condition assuming an infinite voltage source. Because the infinite voltage source can generate infinite current source, the Lyapunov function $E(t)$ is unbounded from Theorem 1.

L. O. Chua's work demonstrates the stability of general CNNs and shows that the system is globally stable [15]. In his work, the cell function is defined as a function which has a maximum and minimum and a positive slope region. However, the system is different from [15] using physical model of RTD which is used as a cell function. Also, it is claimed that the system is asymptotically stable, while Chua's system is stable at all conditions.

References

[1] M Egmont-Petersen, D de Ridder, and H Handels, "Image processing with neural networks – a review", *Pattern Recognition*, 2002, 2279–2301.

[2] P. Kinget and M. S. Steyaert, "A programmable analog cellular neural network CMOS chip for high speed image processing", *IEEE Journal of Solid-State Circuits*, vol. 30, 1995, 235–243.

[3] E. H. Adelson, *Layered representation for image coding.* Technical Report 181, Vision and Modeling Group, The MIT Media Lab, 1991.

[4] N. Franceschini, "Early processsig of color and motion in a mosaic visual system," *Neuro-science research*, vol. 2, 1985, 17–49.

[5] A. B. Torralba and J. Herault, "An efficient neuromorphic analog network for motion estimation" *IEEE Trans. on Circuits and Systems*, vol. 46, no. 2, 1999, 269–280.

[6] L. Esaki and R. Tsu, "Superlattice and negative differenctial conductivity in semiconductors," *IBM J. Res. Develop.*, vol. 14, no. 1, 1970, 61–65.

[7] L. L. Chang, L. Esaki, and R. Tsu, "Resonant tunneling in semiconductor double barriers," *Appl. Phys. Lett.*, vol. 24, no. 12, 1974, 593–595.

[8] P. Mazumder et al., "Digital circuit applications of resonant tunneling device," *Proc. IEEE*, vol. 86, no. 4, 1998, 664–686.

[9] Y. Su et al., "Novel AlInAsSb/InGaAs double-barrier resonant tunneling diode with high peak-to-valley current ratio at room temperature," *IEEE Electron Device Letters*, vol. 21, 2000, 146–148.

[10] J. N. Schulman, H. J. D. L. Santos, and D. H. Chow, "Physics based rtd current-voltage equation," *IEEE Electron Device Lett.*, vol. 17, 1996, 220–222.

[11] T. P. E. Broekaert et al., "A monolithic 4-bit 2-gsps resonant tunneling analog-to-digital converter," *IEEE J. Solid-State Circuits*, vol. 33, 1998, 1342–1349.

[12] J. P. Sun, G. I. Haddad, P. Mazumder, and J. N. Schulman, "Resonant tunneling diodes: models and properties," *Proc. IEEE*, vol. 86, 1998, 641–660.

[13] R. D. Kenney, M. J. Schulte, and M. A. Erle, "A high-frequency decimal multiplier," *IEEE International Conference on Computer Design*, 2004, 26–29.

[14] Yang et al., "Unequal packet loss resilience for mpeg-4 video over the internet," *ISCAS*, vol. 2, 2000, 832–835.

[15] L. O. Chua, and L.Yang, "Cellular neural networks: Theory," *IEEE Trans. Circuits and Systems*, vol. 35, October 1988, 1257–1272.

10

Image Processing by a Programmable Artificial Retina Comprising Quantum Dots and Variable Resistance Devices

Yalcin Yilmaz and Pinaki Mazumder

In this chapter, an analog programmable resistive grid-based architecture mimicking the cellular connections of a biological retina in the most basic level, which is capable of performing various real time image processing tasks such as edge and line detections, is presented. The unit cell structure employs 3-D confined resonant tunneling diodes called quantum dots for signal amplification and latching, and these dots are interconnected between neighboring cells through non-volatile continuously variable resistive elements. A method to program connections is introduced and verified through circuit simulations. Various diffusion characteristics, edge detection and line detection tasks have been demonstrated through simulations using a 2-D array of the proposed cell structure and analytical models have been provided.

10.1 Introduction

Feature extraction is a fundamental task in vision systems as extracted features provide bases for correlation. In digital general purpose processors, many image processing applications require an immense number of operations per second, albeit these applications do not require floating point accuracy [1]. Use of fast, simple and relatively accurate extraction systems in vision machines directly reduces the processing time and required iterations. The main processor element can thereby rely on the reduced dataset that provides quality information on the extracted features for decision making.

Inherent parallel processing capabilities of Cellular Nonlinear Network (CNN)-based architectures make them an efficient platform for various image processing tasks [2, 3]. Real-time operation provides fast processing times, and local connections provide simplicity, scalability and power efficiency for VLSI implementations [4]. Therefore, much effort has been put into developing novel methods and finding adequate CNN templates to perform detail extraction tasks in vision systems, such as edge detection [5–7] which benefit greatly from immense parallelism and computational efficiency.

Resistive grid-based architectures are shown to provide simple yet efficient ways to perform many image processing tasks and motion detection, and they are simple forms of CNNs [2]. Additional advantages including compact area, noise immunity and lower power consumption compared to digital computation structures, make them attractive for researchers. They are also relatively insensitive to mismatches in component values in VLSI chips [8]. However, most of the resistive grid-based architectures in literature are static application specific structures and do not have the functional flexibility of their digital counterparts. Therefore, novel methods and devices should be introduced in these architectures to achieve functional versatility.

Resonant tunneling diodes (RTDs) have been employed in many applications including various CNN architectures due to their negative differential resistance (NDR) and fast switching characteristics. RTDs have been introduced as variable resistors to introduce versatility and compactness to CNN unit cells. In [9], a CNN architecture employing RTDs is investigated for its operation and it is shown that RTDs support fast settling times for various image processing applications.

In this chapter, a variable resistance grid-based architecture that improves the velocity-tuned filter (VTF) architecture proposed by our group [10] is presented. It is demonstrated that when variable resistance connections are incorporated, various diffusion characteristics are obtained, and the developed architecture can be programmed for different image processing applications such as edge detection and line detection providing flexible analog processing environment that can perform various tasks. In addition, RTDs are utilized to provide high speed signal detection and amplification. A method to program resistive connections in four directions is also demonstrated.

10.2 CNN Architecture

10.2.1 Resonant Tunneling Diode Model and Biasing

RTDs have been employed in many circuit applications utilizing their fundamental characteristic of negative differential resistance (NDR). NDR implies that for certain range, the increase in applied voltage across an NDR device will result in decreased current through it, indicating increased resistance with increased voltage.

RTD conductance is determined by two mechanisms: the first mechanism is resonant tunneling, which provides the NDR characteristic, and the other mechanism is diode conduction.

The NDR property of the RTD I-V characteristics is shown in Figure 10.1 utilizing the physics-based model laid out in [11]. The RTD current $J_{RTD}(V)$ is given by:

$$J_1(V) = \frac{qm^*kT\Gamma}{4\pi^2\hbar^3} ln\left(\frac{1 + e^{(E_F - E_r + n_1 qV/2)/kT}}{1 + e^{(E_F - E_r - n_1 qV/2)/kT}}\right) *$$
$$\left(\frac{\pi}{2} + arctan\left(\frac{E_r - n_1 qV/2}{\Gamma/2}\right)\right) \tag{10.1a}$$

$$J_2(V) = H(en2qV/kT - 1) \tag{10.1b}$$

$$J_{RTD}(V) = J_1(V) + J_2(V) \tag{10.1c}$$

where $J_1(V)$ is the current due to resonant tunneling and $J_2(V)$ is the diode conduction current. E_F is the Fermi energy, E_r is the resonant level energy, Γ is the resonant width, n_1 and n_2 are empirical model parameters. q, m^*, k, T, $\hbar$ are electron charge, effective mass, Boltzmann constant, absolute temperature and reduced Planck constant, respectively. V is the voltage across the device.

The main advantage of the NDR characteristic becomes apparent when RTD is biased with a static current source. If current magnitude of the source is selected such that it intersects RTD I-V curve in three places as shown in Figure 10.1, two stable voltage points are obtained. This result indicates that for the same amount of current passing through RTD, the voltage across it can take two stable values which correspond to the lowest and highest voltage intersection points. RTD does not stabilize in the middle intersection point, since any small disturbance causes it to switch to one of the outer intersection points.

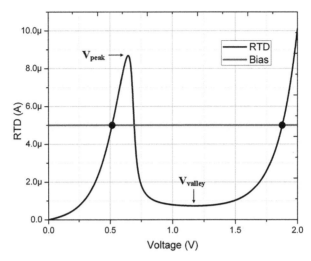

Figure 10.1 RTD I-V curve. The red line corresponds to the bias level. The intersections of the two lines represent the stable operating points.

The bistable characteristic of this structure can be utilized to build voltage-level detectors since any voltage below switching threshold results in stabilizing in the low state, and any voltage above threshold results in stabilizing in high state. RTD switching threshold can be approximated as:

$$V_{th_{RTD}} = \frac{V_{peak} + V_{valley}}{2} \tag{10.2}$$

where V_{peak} and V_{valley} are the peak and the valley voltages of the RTD, respectively. When used in the detection mode, as the system starts all the RTDs are biased to the low voltage state, and a controlled disturbance toward a higher voltage results in the RTDs stabilizing at the higher stable level, allowing the detection and locking of the signal state.

10.2.2 Unit Cell Structure

Figure 10.2a shows the proposed unit cell structure. It is composed of variable resistance devices to provide resistive connections to neighboring cells, diodes to introduce unidirectionality to these connections, and RTDs to detect and latch signal levels. The proposed cell has an input node denoted by $I_{n,m}$, a center node $C_{n,m}$ and an output node $O_{n,m}$. The input is driven by voltage signals that correspond to the pixel intensity level which

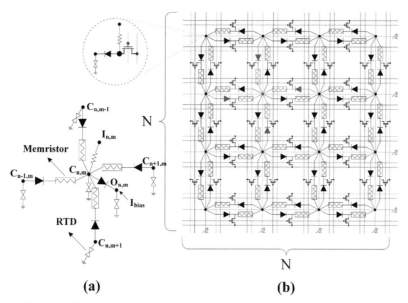

(a)   **(b)**

Figure 10.2   (a) Variable resistance unit cell, (b) Top view of the processing array. A unit cell is highlighted in red. Red and green lines denote programming connections to access transistors.

can be generated by a photo-detector. The connections employing variable resistance devices provide programmability for the realization of different characteristics including isotropic, anisotropic symmetrical and asymmetrical diffusion in grid architecture giving way to various spatiotemporal filter implementations.

    Four variable resistors are connected to the center nodes of the unit cell and its neighbors, making the center node voltage a function of the center node voltages of the neighboring cells. Resistances of connections determine how much neighbors' center voltages contribute to the center voltage of the cell. Series diodes allow current in one direction separating how outputs of the two neighboring cells affect each other. The output node is isolated from the center node by a diode providing a voltage barrier equal to the diode threshold. RTDs enable detection and latching of output signals. When biased with a current source, RTDs initially settle at the lower stable voltage. When the voltage level on the center node goes above detection threshold, RTDs settle at the higher stable voltage. Two stable states provide a binary output. The detection threshold is equal to the sum of the diode threshold and the switching threshold of the RTD.

Figure 10.2b shows unit cells connected in a 2D array fashion. A top view for a 4×4 sample processing array is provided to show the neighboring connections. The unit cell shown in Figure 10.2a is highlighted in red. In order to program certain functionalities in the array, the resistances need to be altered. The Green lines in Figure 10.2b indicate the programming connections to the cells. Each green connection denotes programming-enable signal and voltage driver connections. Access transistors are used to isolate the connections during normal array operations. Programming connections can also be made to share the same connections as cell inputs, thus reducing number of access transistors if the input resistances are designed to be small at the expense of increased programming time or increased programming voltages due to voltage drop across the input resistor. Connections shown in red and blue as well as access transistors are needed during cell erase to bypass reverse biased diodes. During erase operation voltage polarity across the variable resistance device is reversed.

10.3 Programming Variable Resistance Connections

To be able to implement different processing tasks in the same array, we need a procedure to program the resistances of the resistive connections.

Figure 10.3 shows the programming flow for an N × N array. Programming is performed in four directions (left-to-right, right-to-left, top-to-bottom and bottom-to-top) one direction at a time. Programming of the whole array is completed in four passes across the array in different directions to change the resistances in these directions. While a pass is being made in one direction, the resistances are set in a column by column fashion. Programming in this fashion reduces the total required time significantly compared to programming every resistive cell in the array individually. The duration and voltage amplitude of the write pulses determine the resistance to be stored in the connections.

Within one direction, same voltage amplitudes and pulse durations are used. However, pulse characteristics can be changed in different directions to program different resistances, hence to program different functionalities to the array.

A sample programming operation in the left to right direction is shown in Figure 10.4. All the connections in the array are initially at the low resistance state. The programming begins by setting the first column write voltage to high (indicated with a red line) and the remaining columns (indicated with a green line) to low (0V in our implementation).

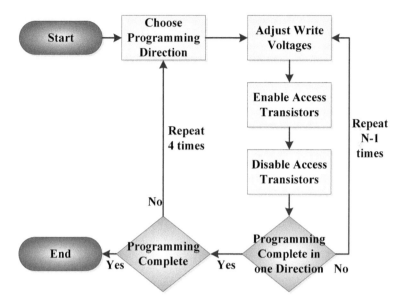

Figure 10.3 Array programming flow.

In this configuration the first column of connections observes a non-zero voltage difference across, whereas the remaining connections observe zero voltage difference. In the first column, only half of the connections are programmed due to the fact that half of the series diodes are forward-biased conducting high currents, and the other half are reverse-biased.

Once the resistances of the first column connections reach the desired level, the second column write voltage is set to high, making the voltage difference across these connections zero, thus stopping their programming. The rise of the voltage levels on the second column in turn causes the voltage difference across the next column connections to be non-zero. Once these connections reach the desired resistance, the next column's voltage is raised. This process is repeated until all the connections in the selected direction are programmed.

When programming in the selected direction is completed, another direction is selected, and the same process is repeated in this new direction. The use of different voltages or change of voltage raise-durations result in programming of different resistances in this direction.

In Figure 10.5, sample left-to-right direction programming voltages to the array are shown. As described earlier, write voltages are applied per column basis. Voltage levels are increased with same time intervals.

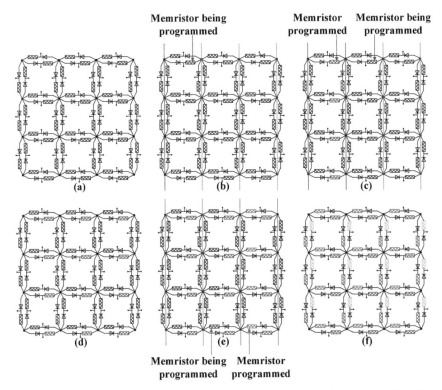

Figure 10.4 Programming in one direction. The green lines indicate the low voltage level (0 V); other colors indicate altered high voltage levels. The different colors of variable resistance devices indicate different final resistances: (a) Array in initial state, (b) Programming started in left to right direction, (c) Programming of first column completed in left to right direction, (d) Programming of all connections completed in left to right direction, (e) Programming of first column completed in right to left direction, (f) Whole array after all connections are programmed in all directions.

Figure 10.6 shows how the resistances of the connections change. The programming scheme succeeds in tuning all the connections in the same direction to the same resistive state.

For the proposed method to be feasible, two critical requirements must be met: The first requirement is that the variable resistive device should be able to be programmed even when there is a forward-biased diode connected in series. The second requirement is that the resistive state of the device should not change or should change negligibly when there is a reverse-biased diode connected in series.

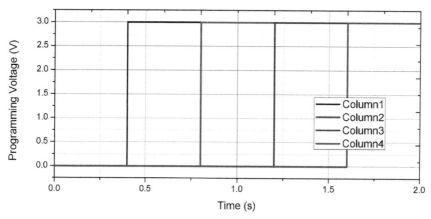

Figure 10.5   Programming in one direction in a 4 by 4 array: programming voltages.

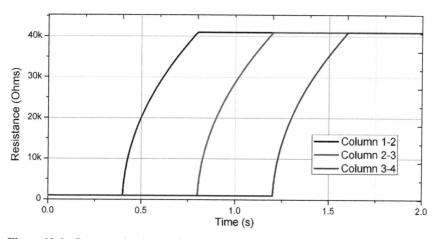

Figure 10.6   Programming in one direction in a 4 by 4 array: resistances in the same row.

Figure 10.7 shows effect of having a series diode with the variable resistance device while performing a programming operation. The results indicate that having a forward-biased series diode with the device causes the device to be programmed to a lower resistance than when programmed with no series diode. This reduction in resistance can be compensated for by increasing programming time or voltage amplitude. This result indicates that it is still possible to program connections with series diode.

A series reverse-biased diode causes no significant change in the resistance of the device during programming, effectively shielding it from

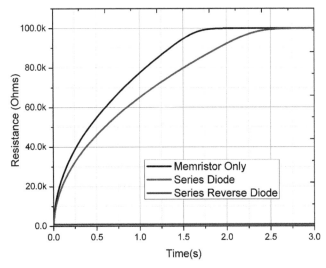

Figure 10.7 Connections under different bias conditions.

the high voltage bias. This property enables programming connections in opposite directions possible, which is crucial in our proposed scheme.

10.4 Analytical Modeling

10.4.1 Edge Detection

Edge detection provides physical information about object boundaries in processed images and is a fundamental feature extraction task in vision systems. An edge is located at the transition points between two different intensity levels.

The grid provides diffusion characteristics that can be adjusted by controlling the resistances. These characteristics combined together with bistable RTD biasing can be used to implement various image processing tasks including edge detection. When all the memristors are programmed to the same resistance, the grid shows symmetric diffusion properties that can be applied to detect edges or contours of an input image.

An edge exists whenever a low input is neighbored by a high input, since an edge is defined where the discontinuity between the input voltages occur.

A simplified analysis on one dimensional connection (Figure 10.8) is carried out to show that this structure can be used for edge detection. In the

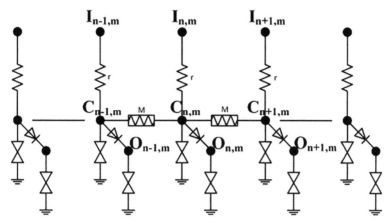

Figure 10.8   CNN circuitry in 1D case.

figure, $I_{n,m}$ is the input voltage level, $C_{n,m}$ is the center node voltage level and $O_{n,m}$ is the output voltage level for the nth node.

It is assumed that there is an edge between inputs $I_{n,m}$ and $I_{n+1,m}$. Thus, $I_{n,m}$ and inputs before it are high and $I_{n+1,m}$ and inputs after it are low (0V for this analysis purposes).

The analysis starts by applying Kirchoff's Current law to nodes $C_{n,m}$ and $C_{n+1,m}$ to obtain the node voltages. Initially the effects of the nearest neighborhood are ignored.

$$\frac{I_{n,m} - C_{n,m}}{r} = \frac{C_{n,m} - C_{n+1,m}}{M} + \frac{C_{n,m}}{R_{RTD_1}} \tag{10.3}$$

where r is the input resistance, M is the resistance of the device, R_{RTD1} is the effective resistance of center node RTD and $C_{n,m}/R_{RTD1}$ indicates current through RTD. The current branch through the diode to the output nodes is also ignored, since the diode current is order of magnitude less until the center node voltage reaches the switching threshold at the output node. When the RTD current equation is inserted, the equation becomes:

$$\frac{I_{n,m} - C_{n,m}}{r} = \frac{C_{n,m} - C_{n+1,m}}{M} + \frac{qm^*kT\Gamma}{4\pi^2\hbar^3}$$
$$ln\left(\frac{1 + e^{(E_F - E_r + n_1 q C_{n,m}/2)/kT}}{1 + e^{(E_F - E_r - n_1 q C_{n,m}/2)/kT}}\right) *$$

$$\left(\frac{\pi}{2} + arctan\left(\frac{E_r - n_1qC_{n,m}/2}{\Gamma/2}\right)\right)$$
$$+ H(e^{n_2qC_{n,m}/kT} - 1) \tag{10.4}$$

Similarly, at node $C_{n+1,m}$:

$$\frac{C_{n,m} - C_{n+1,m}}{M} = \frac{C_{n+1,m}}{r} + \frac{C_{n+1,m}}{R_{RTD2}} \tag{10.5}$$

and

$$\frac{C_{n,m} - C_{n+1,m}}{M} = \frac{C_{n+1,m}}{r} + \frac{qm^*kT\Gamma}{4\pi^2\hbar^3}$$
$$ln\left(\frac{1 + e^{(E_F - E_r + n_1qC_{n+1,m}/2)/kT}}{1 + e^{(E_F - E_r - n_1qC_{n+1,m}/2)/kT}}\right) *$$
$$\left(\frac{\pi}{2} + arctan\left(\frac{E_r - n_1qC_{n+1,m}/2}{\Gamma/2}\right)\right)$$
$$+ H(e^{n_2qC_{n+1,m}/kT} - 1) \tag{10.6}$$

R_{RTD2} is the effective resistance of output node RTD. Equations (10.4) and (10.6) can be evaluated numerically to obtain the intermediate node voltages $C_{n,m}$ and $C_{n+1,m}$. Designing what these voltages will be is essential to obtaining edge detection functionality. When there is an edge, the center node voltage $C_{n,m}$ should rise to disturb the detection node RTD.

Parameters should be picked such that:

$$C_{n,m} > V_{threshold} \quad \text{when } I_{n,m} = V_{high} \tag{10.7a}$$
$$C_{n+1,m} < V_{threshold} \quad \text{when } I_{n+1,m} = V_{low}(0V) \tag{10.7b}$$
$$V_{threshold} = V_d + V_{RTD} \tag{10.7c}$$

V_d is the diode threshold, and V_{RTD} is the switching threshold of the RTD. In this way, the output $O_{n,m}$ will switch to high stable point, indicating there is an edge and $O_{n+1,m}$ will remain at low stable point.

If the effects of the nearing neighbors are considered, one can see that since $I_{n+2,m}$ is also low, the actual voltage on the node $C_{n+1,m}$ will be lower than the above calculated value, hence not violating the condition $C_{n+1,m} < V_{threshold}$, but instead further helping to meet it. Similarly, $I_{n+1,m}$ helps node $C_{n,m}$ to be higher than $V_{threshold}$. In 2-D case, the state equation of the diffusion circuitry can be obtained as

$$\frac{I_{n,m} - C_{n,m}}{r} + \frac{C_{n-1,m} - C_{n,m}}{M} + \frac{C_{n,m-1} - C_{n,m}}{M} + \frac{C_{n+1,m} - C_{n,m}}{M}$$
$$+ \frac{C_{n,m+1} - C_{n,m}}{M} - I_{RTD} = c\frac{ds_{n,m}}{dt} \tag{10.8}$$

where
$$I_{RTD} = \frac{C_{n,m}}{R_{RTD}} \tag{10.9}$$

and c is parasitic capacitance of the RTD. Assuming RTD has a finite resistance, replacing (4.9) in (10.8):

$$I_{n,m} = C_{n,m}\left(1 + \frac{4r}{M} + \frac{r}{R_{RTD}}\right) - \frac{r}{M}\left(C_{n-1,m} + C_{n,m-1}\right.$$
$$+ C_{n+1,m} + C_{n,m+1}) + rc\frac{dC_{n,m}}{dt} \tag{10.10}$$

Taking Fourier transform, the transfer function is:

$$H(f_m, f_n, f_t) = \frac{S(f_m, f_n, f_t)}{E(f_m, f_n, f_t)}$$
$$= \frac{1}{\left(1 + \frac{4r}{M} + \frac{r}{R_{RTD}}\right) - \frac{2r}{M}\left(\cos(2\pi f_m) + \cos(2\pi f_n)\right) + rc2\pi j f_t} \tag{10.11}$$

As the RTD I-V curve indicates, it acts as a positive variable resistor, indicating that the real part of the denominator of the transfer function is always positive.

10.4.2 Line Detection

Introducing anisotropy in the vertical and horizontal directions in the resistive grid allows the implementation of line detection. In order to detect lines, the center node voltages should be made a weaker function of the neighboring cells' center node voltages in one direction and a stronger function in the other. For example, high resistance in the vertical and low resistance in the horizontal direction limits the effects of the neighboring cells in the vertical direction and enables diffusion in the horizontal direction, which means the detection of lines in the horizontal direction.

Low resistance in the vertical and high resistance in the horizontal direction limits the effects of the neighboring cells in the horizontal direction and enables diffusion in the vertical direction, which means the detection of vertical lines. In this case, the diffusion network state equation becomes:

$$I_{n,m} = C_{n,m} \left(1 + \frac{2r}{M_{high}} + \frac{2r}{M_{low}} + \frac{r}{R_{RTD}} \right) - \frac{r}{M_{high}}$$
$$(C_{n-1,m} + C_{n+1,m}) - \frac{r}{M_{low}} \left(C_{n,m-1} + C_{n,m+1} + rc\frac{dC_{n,m}}{dt} \right)$$

$$(10.12)$$

where M_{high} is the resistance of the variable resistance device when programmed to high, and M_{low} is the resistance when programmed to low. The state equation is symmetrical for vertical and horizontal line detection cases.

10.5 Simulation Results

Simulation results verifying various diffusion configurations and demonstrating edge detection and line detection operations are presented in this section. Simulations are carried out on a 64 × 64 array. RTDs based on device characteristics shown in Figure 10.1 as well as the variable resistance device model from [12] are used. Simulations are carried out with nominal parameters to provide proof of concept. However, studies carried out in [8] show that resistive grid-based architectures are variation tolerant and can operate with non-optimal values. Therefore, the proposed architecture is expected to be variation tolerant. For example, for the edge detection case, this tolerance depends on how much margin is left between the designed high/low voltages and the threshold.

The proposed memristive grid can support the different diffusion characteristics mentioned above. These characteristics are important in many image processing applications. Anisotropic diffusion can be used for edge extraction applications [13], anisotropic symmetrical diffusion characteristics can be used for line detection, and anisotropic asymmetrical diffusion characteristics can be used for motion detection [14].

Figure 10.9 shows diffusion characteristics that can be realized in the proposed architecture. When all the connections are programmed to the same resistance, isotropic diffusion (Figure 10.9b) is obtained.

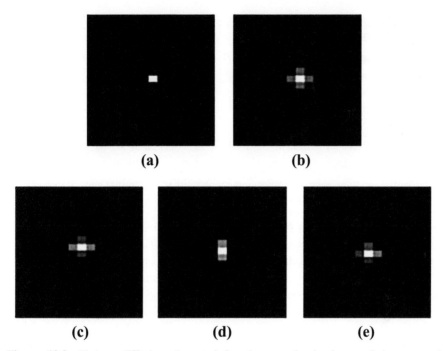

Figure 10.9 Various diffusion characteristics that can be implemented in proposed architecture. (a) Input, (b) Isotropic Diffusion, (c) Anisotropic symmetrical diffusion in horizontal direction, (d) Anisotropic symmetrical diffusion in vertical direction, (e) Anisotropic asymmetrical diffusion.

When horizontal connections are programmed to low and vertical ones are programmed to high resistance, anisotropic symmetrical diffusion in horizontal direction (Figure 10.9c) is obtained. When the resistances of the horizontal and vertical memristors are programmed reverse with respect to the previous case, anisotropic symmetrical diffusion in vertical direction (Figure 10.9d) is achieved. Finally, when all resistances are programmed to different resistances, anisotropic asymmetrical diffusion (Figure 10.9e) is achieved. These characteristics or combinations of them can be utilized to perform various vision tasks.

10.5.1 Edge Detection

Edge detection simulations are carried out on two types of input images: First, with irregular edges and intermediate pixel intensity values around the

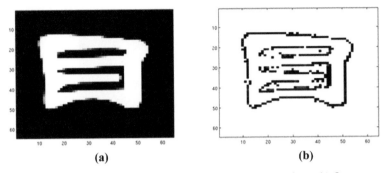

Figure 10.10   (a) Edge detection sample input with irregular edges, (b) Output result.

edges shown in Figure 10.10; second, with regular edges and maximum pixel intensity difference around the edges shown in Figure 10.11.

In the edge detection mode, the proposed architecture is initiated with high and low input values corresponding to black and white pixels with scaled voltages in between corresponding to shades of gray. When the first input type shown in Fig. 10.9a is applied to the array, the edge pattern shown in Figure 10.10b is observed at the $O_{n,m}$ nodes of the architecture. In Figure 10.10b, black lines correspond to high RTD voltage level and white lines correspond to low RTD voltage level on $O_{n,m}$ nodes.

The results of the above simulation suggest that edges are extracted with relative accuracy. In the regions with thicknesses of a few pixels or where the borders include shades of gray, discontinuities or jumps in the border lines are observed. However, the quality of the results can be improved by fine tuning grid resistance as well as RTD design parameters.

The second set of sample filtering is provided in Figure 10.11a–c. The second type of image contains a circle with large continuous areas of the same color pixels. In this image there is no region with several pixels thickness (except the outer line due to the finite number of pixels available to represent a circle). The architecture is able to clearly outline the edges without any discontinuities.

Although the speed of operation can be tuned by scaling the current capabilities of the active devices, the above simulation shows that the results obtained after 2.5 *ns* into the operation are almost the same as the results obtained after 100 *ns*.

Simulations on the variable resistance devices using the model provided in [12] indicated that a read pulse of 100 *ns* duration and 2*V* amplitude does not change the resistance detectably, and a write process usually takes in the

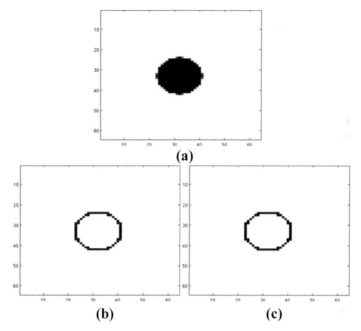

Figure 10.11 (a) Edge detection sample 2 with regular edges, (b) Output at 2.5 ns (c) Output at 100 ns.

range of a few seconds depending on the resistance to be encoded. Therefore, the architecture can perform the tuned operation repeatedly without detectably altering the tuning, thus minimizing (effectively eliminating) the need for a refresh operation.

The effect of component mismatch is more significant in input resistors (i.e vertical resistors) compared to grid resistors (i. e. horizontal resistors) [1]. A set of simulation results are listed in Figure 10.12a–f, showing how the variation in input resistance changes edge detection results. Simulation results indicate that edges are detected less accurately when the input resistance deviates from the optimum value obtained through simulations. The exact resistance values depend on various circuit and device parameters such as voltage levels used, RTD and diode current characteristics. Therefore, the resistance variation is presented in percentages. Edge detection quality directly depends on the variation of the input resistance mainly due to two factors. The first factor is that input resistance changes the spatial frequency tuning of the architecture making it less sensitive to edges. The second factor is that large input resistances cause input voltage drops, thus putting the architecture off the operating region.

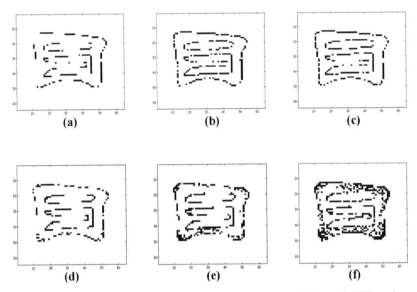

Figure 10.12   Edge detection results with input resistance variation: (a) 50% resistance, (b) 75% resistance, (c) 100% resistance (nominal case), (d) 500% resistance, (e) 1000% resistance, (f) 2000% resistance.

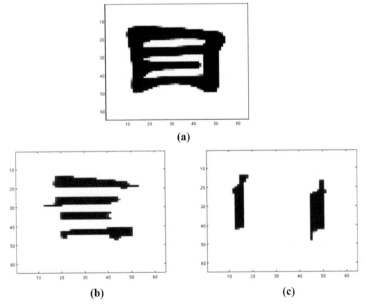

Figure 10.13   (a) Line detection sample input, (b) Horizontal line detection, (c) Vertical line detection.

10.5.2 Line Detection

Figure 10.13 shows line detection results. The plotted results are obtained 500 ns after system initialization. When the image in Figure 10.10a is considered, if we assume the black pixels represent the high voltages and white pixels represent the low voltages, the diffusion from high voltages to low voltages will result in the low voltages increasing and stabilizing at the higher voltage level. The results presented in Figure 10.13 are inverted to clearly show the detected lines.

10.6 Conclusion

An analog grid-based architecture incorporating variable resistance connections for programmability and RTDs for signal detection and latching was revealed in this chapter. The architecture can be programmed to perform various image processing tasks. A method to change the resistive state of the memristors in an array configuration was also provided and demonstrated through circuit simulations. Analytical models characterizing edge detection and line detection configurations were discussed and simulation results were provided to verify functionality. The simulation data establish that the presented architectural configuration incorporating programmable analog resistive elements can be reused to perform a wide gamut of image processing functions at extremely high speeds.

References

[1] T. Roska, "Analogic CNN Computing: Architectural, Implementation, and Algorithmic Advances – a Review," in *Cellular Neural Networks and Their Applications Proceedings, 1998 Fifth IEEE International Workshop on*, no. April, 1998, 3–10.

[2] B. E. Shi and L. O. Chua, "Resistive grid image filtering: Input/output analysis via the CNN framework," *IEEE Transactions on Circuits and Systems I: Fundamental Theory and Applications*, vol. 39, no. 7, 531–548, 1992.

[3] P. Kinget and M. S. J. Steyaert, "A programmable analog cellular neural network CMOS chip for high speed image processing," *IEEE Journal of Solid-State Circuits*, vol. 30, no. 3, 235–243, 1995.

[4] H. Li, X. Liao, C. Li, H. Huang, and C. Li, "Edge detection of noisy images based on cellular neural networks," *Communications in Nonlinear Science and Numerical Simulation*, vol. 16, no. 9, 3746–3759, 2011. [Online]. Available: http://dx.doi.org/10.1016/j.cnsns.2010.12.017

[5] J. Zhao, H. Wang, and D. Yu, "A new approach for edge detection of noisy image based on CNN," *International Journal of Circuit Theory and Applications*, vol. 31, no. 2, 119–131, 2003. [Online]. Available: http://doi.wiley.com/10.1002/cta.210

[6] T. Yoshida, J. Kawata, T. Tada, a. Ushida, and J. Morimoto, "Edge detection method with CNN," *SICE 2004 Annual Conference*, vol. 2, 1721–1724, 2004.

[7] I. N. Aizenberg, "Processing of noisy and small-detailed gray-scale images using cellular neural networks," *Journal of Electronic Imaging*, vol. 6, 272–285, July 1997.

[8] B. E. Shi, "The effect of mismatch in current- versus voltage-mode resistive grids," *International Journal of Circuit Theory and Applications*, vol. 37, 53–65, 2009.

[9] P. Mazumder, S. R. Li, and I. E. Ebong, "Tunneling-based cellular nonlinear network architectures for image processing," *IEEE Transactions on Very Large Scale Integration (VLSI) Systems*, vol. 17, no. 4, 487–495, 2009.

[10] W. H. Lee and P. Mazumder, "Motion detection by quantum-dots-based velocity-tuned filter," *IEEE Transactions on Nanotechnology*, vol. 7, no. 3, 355–362, 2008.

[11] J. Schulman, H. De Los Santos, and D. Chow, "Physicsbased RTD current-voltage equation," *IEEE Electron Device Letters*, vol. 17, no. 5, 220–222, May 1996. [Online]. Available: http://ieeexplore.ieee.org/lpdocs/epic03/wrapper.htm?arnumber=491835

[12] Z. Biolek, D. Biolek, and V. Biolkova, "SPICE model of memristor with nonlinear dopant drift," *Radioengineering*, vol. 18, no. 2, 210–214, 2009.

[13] P. Perona and J. Malik, "Scale-space and edge detection using anisotropic diffusion," 629–639, 1990.

[14] A. B. Torralba, "Analogue Architectures for Vision Cellular Neural Networks and Neuromorphic Circuits," Ph.D. dissertation, 1999.

11

Memristor-based Cellular Nonlinear/Neural Network: Design, Analysis and Applications

Xiaofang Hu, Shukai Duan, Wenbo Song,
Jiagui Wu and Pinaki Mazumder

In this chapter, a compact CNN model based on memristors is presented along with its performance analysis and applications. In the new CNN design, the memristor bridge circuit acts as the synaptic circuit element and substitutes the complex multiplication circuit used in traditional CNN architectures. Additionally, the negative differential resistance (NDR) and nonlinear I–V characteristics of the memristor have been leveraged to replace the linear resistor in conventional CNNs. The proposed CNN design has several merits, for example, high-density, non-volatility, and programmability of synaptic weights. The proposed memristor-based CNN design operations for implementing several image processing functions are illustrated through simulation and contrasted with conventional CNNs. Monte-Carol simulation has been used to demonstrate the behavior of the proposed CNN due to variations in memristor synaptic weights.

11.1 Introduction

Cellular nonlinear/neural network (CNN) was proposed in 1988 by Chua and Yang in their twin seminal papers [1, 2] by demonstrating how the regularity of cellular automata and local computation of neural networks can be melded together in order to accelerate numerous real-world computational tasks. Many artificial, physical, chemical, as well as biological systems have been represented using the CNN model [3]. Its continuous-time analog operation allows real-time signal processing at high precision, while its local interaction between constituent processing elements (PEs) obviate the need of global buses and long interconnects, leading to several digital and analog implementations of CNN architectures in the form of very large-scale integrated (VLSI) chips. In [4–9], several powerful applications of CNNs were reported including pattern and image analysis such as vertical line detection, noise reduction, edge detection, feature detection and character recognition.

However, in order to improve the resolution in static and dynamic image analysis, the size of PEs and the connectivity between the PEs describing the control (feed-forward) and feedback templates that are used as programming artifacts in the CNN model of computation [5, 6], must be significantly reduced. The current version of CNN arrays in a CMOS VLSI chip is typically limited to 16K processing elements. Emerging technologies such as resonant tunneling diode (RTD) [7] and quantum dots [8] have been attempted recently to implement PEs more compactly to improve the resolution of CNN computation. However, lack of programmability associated with fixed connecting elements between PEs as well as wide variability of tunneling currents in RTDs and quantum dots are major limitations of these emerging technologies.

The recent advent of memristors [10, 11] has opened the possibility of significantly enhancing the resolution of on-chip CNN model of computation. The memristor was introduced [12] as a fundamental circuit element and recently nanofabrication technology has shown its superior device properties such as nonvolatility, binary as well as multiple memory states, and nanometer geometries that can be shrunk to the ultimate physical dimensions [13–16]. These versatile features of memristors have been exploited in showing their applications in nonvolatile memory [17, 18], artificial neural networks [19–23], composite circuits [24, 25] and so on. Because its conductance can change in response to the applied voltage or current like a biological synapse, the memristor has been demonstrated an artificial synapse for biological signal processing. Recently, Kim et al. presented a compact memristor bridge synapse in which both weighting and weight programming can be performed at different time slots [21, 22].

In this study, a novel type of memristor-based CNN (M-CNN) is addressed. Specifically, the memristor bridge circuit is employed to realize the interactions between the neighboring cells. Different from the memristor circuit in [21, 22], the memristor used here is a more practical model based on experimental data. Moreover, since the memristor exhibits nonlinear I-V characteristic with locally negative differential resistance (NDR), the memristor is also considered to replace the original linear resistor in a traditional CNN cell. As a result, an M-CNN, equipped with nonvolatile and programmable synapse circuits, is more versatile and compact and saves the traditional complex output function realization circuits.

11.2 Memristor Basics

The memristor is a nonlinear passive device with variable resistance states. It is mathematically defined by its constitutive relationship of the charge q and the flux φ

$$\frac{d\varphi}{dt} = \frac{d\varphi(q)}{dq} \cdot \frac{dq}{dt}.$$ (11.1)

Based on the basic circuit law, (11.1) leads to

$$v(t) = \frac{d\varphi(q)}{dq} i(t) = M(q)i(t),$$ (11.2)

where $M(q)$ is defined as the resistance of a memristor called memristance and it is a function of the internal current i and the state variable x.

The Simmons tunnel barrier model is the most accurate physical model of TiO_2/TiO_{2-x} memristor, reported by the HP Lab [13]. As shown in Figure 11.1, the memristor is made up of two TiO_2 and TiO_{2-x} components sandwiched between two platinum (Pt) electrodes. The memristance magnitude is determined by the electron tunnel barrier, in series with a resistor, Rs. In this case, the state variable x is the width of the Simmons tunnel barrier and its dynamics are represented by [13].

$$\frac{dx(t)}{dt} = \begin{cases} c_{off} \sinh\left(\frac{i}{i_{off}}\right) \exp\left[-\exp\left(\frac{x-a_{off}}{w_c} - \frac{|i|}{b}\right) - \frac{x}{w_c}\right], & i > 0 \\ c_{on} \sinh\left(\frac{i}{i_{on}}\right) \exp\left[-\exp\left(-\frac{x-a_{on}}{w_c} - \frac{|i|}{b}\right) - \frac{x}{w_c}\right], & i < 0 \end{cases},$$ (11.3)

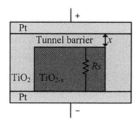

Figure 11.1   Physical model of Simmons tunnel barrier memristor [13].

where c_{off}, c_{on}, i_{off}, i_{on}, a_{off}, a_{on}, w_c, and b are fitting parameters, in which the parameters c_{off} and c_{on} influence the change in magnitude of x, the parameters i_{off} and i_{on} reflect the current thresholds, and a_{off} and a_{on} are the upper and the lower bound of x, respectively. Equation (11.3) is also interpreted as the velocity of the oxygen vacancy drift in TiO_{2-x} material. About this accurate model, two problems have been discussed [15]: (i) there is no explicit relationship between the current and the voltage of the memristive device, and (ii) it is too complicated to be used in numerical and mathematical analysis. Afterwards, an alternative model with simper expressions, which can reflect the same physical behavior, is proposed [15]. In this simplified model, the derivation of the state variable x is given by

$$\frac{dx(t)}{dt} = \begin{cases} k_{off} \left(\frac{i(t)}{i_{off}} - 1 \right)^{\alpha_{off}} \cdot f_{off}(x), & 0 < i_{off} < i \\ 0, & i_{on} < i < i_{off} , \\ k_{on} \left(\frac{i(t)}{i_{on}} - 1 \right)^{\alpha_{on}} \cdot f_{on}(x), & i < i_{on} < 0 \end{cases} \tag{11.4}$$

where k_{off} is a positive constant and k_{on} is a negative constant, a_{off} and a_{on} are fitting parameters, and i_{off} and i_{off} are current thresholds. Correspondingly, there are two window functions $f_{on}(x)$ amd $f_{off}(x)$ to represent the dependence of the derivation on the state variable x and guarantee the effective range of x, i.e. $x \in [x_{on}, x_{off}]$.

$$f_{off}(x) = \exp \left[-\exp \left(\frac{x - a_{off}}{w_c} \right) \right], \tag{11.5a}$$

$$f_{on}(x) = \exp \left[-\exp \left(\frac{x - a_{on}}{w_c} \right) \right]. \tag{11.5b}$$

The relationship between the current through and the voltage across the memristor can be written as

$$v(t) = \left[R_{on} + \frac{R_{off} - R_{on}}{x_{off} - x_{on}} (x - x_{on}) \right] \cdot i(t), \tag{11.6}$$

where $M(x) = R_{on} + \frac{R_{off} - R_{on}}{x_{off} - x_{on}} (x - x_{on})$ is the memristance, in which R_{on} and R_{off}, respectively, denote the low- and high- resistance of the memristor. Note that positive voltage or current can increase the undoped region width x, and thus increase the memristance, while negative excitation leads to memristance decrease.

A few fundamental simulations have been done to observe the behavior of the memristor subjected to an applied sine voltage $v = \sin(2\pi f)$ with results

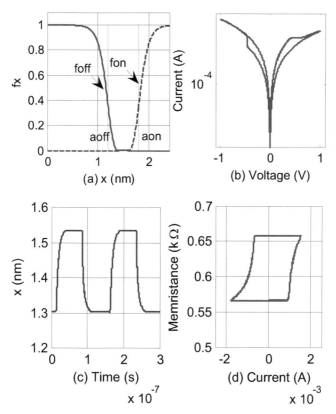

Figure 11.2 Characteristic curves of the memristor under $v = \sin(2\pi f)$: (a) Window functions $f_{off}(x)$ (red solid line) and $f_{on}(x)$ (blue dash line), (b) voltage and current relationship, (c) the change of the state variable x and (d) the relationship between memristance and the current through the device.

presented in Figure 11.2. Figure 11.2(a)–(d) shows the window functions $f_{on}(x)$ and $f_{off}(x)$, the current-voltage relationship with current presented in logarithm coordinate, the change of the state variable x, and the memristance change versus the current, respectively. In this simulation, the memristor model parameters are set as: $R_{off} = 1\ k\Omega$, $R_{on} = 50\ \Omega$, $i_{off} = 115\ \mu A$, $i_{off} = 8.9\ \mu A$, $a_{off} = 1.2$ nm, $a_{on} = 1.8$ nm and $w_c = 107$ pm.

11.3 Memristor-based Cellular Neural Network

11.3.1 Description of the M-CNN

A memristor-based cellular neural network (M-CNN) is constituted of an $M \times N$ rectangular array of cells $c(i, j)$ located at site (i, j),

$i = 1, 2, 3, \ldots, M; \; j = 1, 2, 3, \ldots, N$, as shown in Figure 11.3. A cell has $(2r + 1)^2$ neighboring cells, where $r \in [1, 2, 3, \ldots]$ is the radius of the neighborhood. It only communicates with its closest neighbors, which is the defined local interconnection of CNNs. In this chapter, we consider the most commonly used case $r = 1$ for our study, i.e. one cell only transmits information among its eight closest neighbors.

Like in a standard CNN [1], all cells have identical circuit structure and parameter values. Figure 11.4 shows a schematic implementation of the

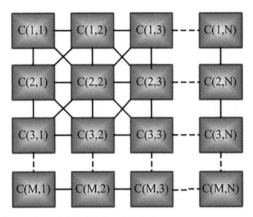

Figure 11.3 Topology of an M-CNN, in which the squares represent cells with identical structure.

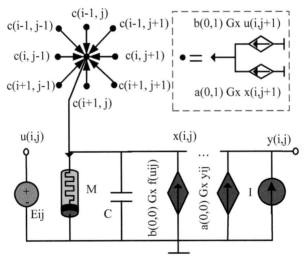

Figure 11.4 Schematic of a memristor-based CNN cell.

proposed M-CNN cell containing a capacitor, a memristor state resistor, and two variable-gain voltage-controlled current sources (VCCSs). Memristor bridge circuits were to used implement the VCCSs for weight setting as well as the weighting operation. Besides, a memristor (M) is employed to replace the linear state resistor. Therefore, the original state-output conversion part is removed.

In a CNN, each cell c(i, j) at ith row and jth column can receive eight input currents from a set of neighboring cells c (k', l') located at k'th row and l'th column of the array, where $(k', l') \in \{(i-1, j-1), (i-1, j), (i-1, j+1), (i, j-1), (i, j+1), (i+1, j-1), (i+1, j), (i+1, j+1)\}$. These inputs are indicated by solid dots in Figure 11.4. Specially, each solid dot represents an input current that is a summation of two VCCSs of the corresponding neighboring cells, as given in the dashed square. Likewise, each cell c(i, j) also provides eight output currents to its eight neighboring cells.

Note that the total input current of each cell contains not only the currents from its eight neighbors, but also from itself and an independent current source. Therefore, as a matter of simplicity, a cell itself is also considered as a neighbor cell included in its neighborhood in CNNs [1].

The dynamics of a cell c(i, j) usually depends on: a set of 18 weights called template elements, an independent current source denoted by I, an independent voltage source denoted by u_{ij}, as well as its own state x_{ij}. The template elements make up a 3×3 feedback template labeled by A, and a 3×3 control template labeled by B. They determine the gains of the interconnections between the neighboring cells. The independent current offers an offset current and the independent voltage source continuously provides the input for the network. In addition, because of the usage of the memristor bridge circuits and the memristor state resistor, the state x_{ij} can achieve stability itself and thus as the output directly, i.e., $y_{ij} = x_{ij}$ (detailed description is given later).

Based on the above descriptions and Kirchhoff's current law, the dynamics of each cell can be governed by

$$C\frac{dx_{ij}(t)}{dt} = -m(x_{ij}(t)) + \sum_{c(k,l)} (a_{ij,kl}x_{kl}(t) + b_{ij,kl}u_{kl}) + I, \quad (11.7)$$

where $c(k, l) \in N_r(i, j)$ denotes the r-neighborhood of $c(i, j)$ and here $r = 1$. As described above, for the cell c(i, j), its neighboring cells $c(k, l)$ contain all the neighbors (k', l') and itself.

The term $m(\cdot)$ is the current flowing through the memristor (M) in the form of

$$m(x_{ij}(t)) = \frac{v_M}{M(t)} = \frac{x_{ij}(t)}{M(t)}, \qquad (11.8)$$

where $M(t)$ is the memristance of the memristor state resistor. In the following section, the analog implementation scheme of the M-CNN based on the mathematical model (11.8) will be presented.

11.3.2 Implementation of Synaptic Connections with Memristor Bridge Circuits

The template elements (weights) play a crucial role in signal and image processing applications of the CNN. To execute different functions, the templates should be updated correspondingly. In traditional circuit implementation of CNNs, the weights and the weighting operation are achieved through a number of amplifiers and multipliers [7]. For an amplifier, the amplifying gain is fixed once the circuit is built, thereby not being easy to alter. Moreover, the multipliers are implemented with at least eight CMOS transistors that operate in a nonlinear way and hence consume significant power. Therefore, serious nonlinearity is unavoidable in the multiplication processing (weighting operation) [21].

Memristor bridge circuit was recently reported as a candidate of artificial synapse to replace the amplifiers and multipliers to implement weight programming and weighting operation [21–22]. Such a circuit, consisting of four identical memristors ($M_1, M_2, M_3,$ and M_4), is capable of performing zero, positive and negative synaptic weights, shown in Figure 11.5. When a pulse V_{in} (positive or negative) is applied at the input port, the memristance of each memristor changes correspondingly depending on its polarity. The output voltage between the positive and the negative terminals is governed by

$$V_W = V_+ - V_- = \left(\frac{M_2}{M_1 + M_2} - \frac{M_4}{M_3 + M_4} \right) V_{in}. \qquad (11.9)$$

In the form of the relationship between the synaptic weight ω and the synaptic input signal V_{in}, (9) can be rewritten as

$$V_W = \omega V_{in}, \qquad (11.10)$$

where the synaptic weight ω represents $M_2/(M_1 + M_2) - M_4/(M_3 + M_4)$ within the range of $[-0.9, 0.9]$. If $M_2/M_1 > M_4/M_3$, the synaptic weight

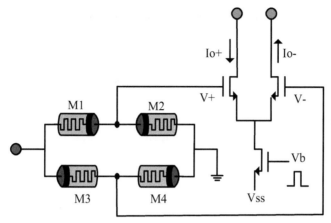

Figure 11.5 Memristor bridge circuit [21].

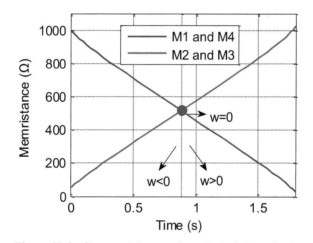

Figure 11.6 Change of the memristors in the bridge circuit.

is positive; if $M_2/M_1 < M_4/M_3$, it is negative; otherwise, it is zero and called balanced state. Figure 11.6 illustrates the change of memristors along with varying programming time in the bridge structure and indicates the three regions of the weights. In the simulation, M_1 and M_4 were initially set as R_{off}, while M_2 and M_3 were R_{on}. The amplitude of the programming voltage is 1 V.

In the bridge circuit, the input port is shared by the weight programming signal V_p and the synaptic input signal V_{in} for weighting operation at different time slots. The unintended change of the memristance during the weighting operation is negligible as long as the weighting signal amplitude is smaller than the thresholds [13–16] or the pulse is very narrow [21–22]. Therefore, the weighting factor ω is a constant and thus the relationship between the synaptic input and synaptic output is linear, which has been discussed in [21–22].

The differential amplifier on the right of Figure 11.5 performs the conversion of voltage to current with a transconductance parameter g_m. The currents at the positive and negative terminals are given by

$$\begin{cases} I_{o+} = -\frac{1}{2}g_m V_M = -\frac{1}{2}g_m \omega V_{in} \\ I_{o-} = \frac{1}{2}g_m V_M = \frac{1}{2}g_m \omega V_{in} \end{cases}. \tag{11.11}$$

Figure 11.7 illustrates the weighting operation of the bridge circuits. Four lines represent four levels of the synaptic weights as examples. Generally, the synaptic inputs are voltage, which eases the intra-chip distribution of multiplication factors (template coefficients) and the intra-chip distribution of the state variable to all the neuron synapses. But the synaptic output is usually in the form of current, which facilitates the summation at the state nodes [5]. The transconductance gain of the differential amplifier, g_m, can be set by the bias current and the transistor size of the amplifier according to the required template coefficients.

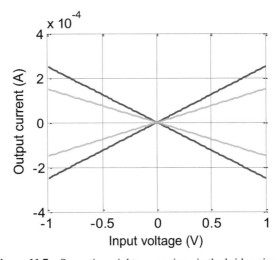

Figure 11.7   Synaptic weight processing via the bridge circuit.

11.3.3 Implementation of M-CNN cells

As described above, a cell $c(i, j)$ may receive 18 synaptic inputs from its neighboring cells including itself. These synaptic inputs are the weighted input $b_{ij,kl}u_{kl}$ and the weighted state $a_{ij,kl}x_{kl}$, and they are summed up and injected into the center cell $c(i, j)$ through the state node, namely, the memristor (M) shown in Figure 11.4.

Figure 11.8 illustrates an analog implementation schematic of the proposed M-CNN cell. The weighted signals ($I_{01}, \ldots, I_{0k}$) from k memristor bridge circuits input into the center cell from the ports at the lower right corner. The circuit located at the right bottom is a bias circuit that provides the direct current I. The synaptic currents and the bias current are aggregated by connecting all the output terminals together through a differential circuit at the left top. The total current is fed into the memristor and through an integrator at the right top, and then converted into the cell state variable x_{ij} also the cell output y_{ij} which is finally transmitted to neighboring cells.

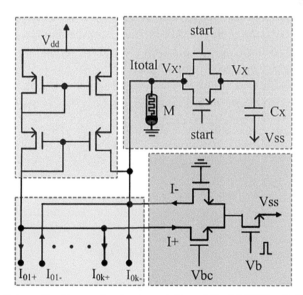

Figure 11.8 Schematic of an analog implementation of the M-CNN cell. The left bottom part indicates the synaptic inputs from the neighboring cells. The right bottom part is a bias circuit offering an offset current. These currents are summed up through the active load at the left top part. And the last part realizes the current-to-voltage conversion and integration.

The total current I_{total} is given by

$$I_{total} = \underbrace{I_{01} + I_{02} + I_{03} + \cdots + I_{0k}}_{I_0} + I, \qquad (11.12)$$

where $I_{0i} = I_{0i-} - I_{0i+}$ is the output current of the differential amplifier connected with ith memristor bridge circuit. As a result, I_0 that denotes the sum of the synaptic currents can be rewritten as

$$
\begin{aligned}
I_0 &= g_m(\omega_1 V_{in1} + \omega_2 V_{in2} + \omega_3 V_{in3} + \cdots \omega_k V_{in\,k}) \\
&= \sum_k g_m \omega_k V_{in\,k}. \qquad (11.13)
\end{aligned}
$$

Therefore the memristor bridge circuits can implement the accumulation terms in (11.7) in a more compact and simpler form. Note that, if both of the feedback and the control templates are non-zero, then $k = 18$; if one template is zero, then $k = 9$, which is a simple implementation method. Another choice is that if the memristor bridge synapses are time-multiplexed, then the 18 weighted values associated with the two templates are implemented by using only 9 of these synapse circuits. For instance, the control term $\sum b_{ij,kl} u_{kl}$ is first calculated in each computation cycle with the result stored in an analog current-mode memory. After that, the nine synapses are used to calculate the feedback term $\sum a_{ij,kl} x_{kl}$ and the result is added to the former result, finally realizing the two accumulation terms.

The total current flows through the state memristor and the output voltage corresponds to the derivation of the cell state, i.e., dx_{ij}/dt. Then, through an integrator, the cell state x can be obtained.

$$Vx = Cx \cdot M \left(\sum_k g_m \omega_k V_{in\,k} + I \right). \qquad (11.14)$$

Note that, besides performing the summation function, the active load part also plays the role of restricting the range of the state memristor voltage as [22]

$$Vx = \begin{cases} Cx \cdot (V_{dd} - 2V_{th}), & I_{total} \geq \frac{V_{dd} - 2V_{th}}{R_{out}} \\ Cx \cdot M \cdot I_{total}, & \frac{-V_{ss} + 2V_{th}}{R_{out}} \leq I_{total} \leq \frac{V_{dd} - 2V_{th}}{R_{out}} \\ Cx \cdot (-V_{ss} + 2V_{th}), & I_{total} \leq \frac{-V_{ss} + 2V_{th}}{R_{out}} \end{cases}. \qquad (11.15)$$

The original output function is no longer necessary, because the state can achieve stability (bounded) and support the binary output, which leads to the further reduction of the CNN circuit.

11.4 Mathematical Analysis

11.4.1 Stability

By definition, a CNN is stable if it has a steady output as time goes to infinity [9]. In circuit analysis, a continuous-time autonomous network is stable if its solution is bounded and its energy function is damped. Since the state variable (solution) of a CNN (a specific network) is limited by the circuit parameters, it is easy to guarantee its boundness. Then, generally, the energy degradation is considered as the key point to prove the stability of a CNN [9]. According to the Lyapunov stability theorem, the energy function of the M-CNN is chosen as

$$E(t) = \sum_{(i,j)} \int_0^{x_{ij}} m(s)ds - \frac{1}{2} \sum_{(i,j)} \sum_{(k,l)} a_{ij,kl} x_{kl}(t) x_{ij}(t)$$
$$- \sum_{(i,j)} \sum_{(k,l)} b_{ij,kl} u_{kl}(t) x_{ij}(t) - \sum_{(i,j)} I x_{ij}(t). \tag{11.16}$$

Theorem 1: The function $E(t)$ is bounded by

$$\max |E(t)| \le E_{\max}. \tag{11.17}$$

where

$$E_{\max} = \frac{1}{2} \sum_{(i,j)} \sum_{(k,l)} |a_{ij,kl} x_{kl}(t) x_{ij}(t)|$$
$$+ \sum_{(i,j)} \sum_{(k,l)} |b_{ij,kl} u_{kl}(t) x_{ij}(t)| + MN \left(|I| + \max_{i,j} \left| \int_0^{x_{ij}} m(s)ds \right| \right),$$

$$\tag{11.18}$$

for an $M{\times}N$ cellular neural network.

Proof: By the definition of $E(t)$, one can obtain

$$|E(t)| \le \frac{1}{2} \sum_{(i,j)} \sum_{(k,l)} |a_{ij,kl} x_{kl}(t) x_{ij}(t)|$$
$$+ \sum_{(i,j)} \sum_{(k,l)} |b_{ij,kl} u_{kl}(t) x_{ij}(t)| + \sum_{(i,j)} |I x_{ij}(t)| + \sum_{(i,j)} \left| \int_0^{x_{ij}} m(s)ds \right|$$

$$\leq \frac{1}{2}\sum_{(i,j)}\sum_{(k,l)}|a_{ij,kl}x_{kl}(t)x_{ij}(t)| + \sum_{(i,j)}\sum_{(k,l)}|b_{ij,kl}u_{kl}(t)x_{ij}(t)|$$

$$+ \sum_{(i,j)}|I|\cdot|x_{ij}(t)| + \sum_{(i,j)}\left|\int_0^{x_{ij}}m(s)ds\right|$$

$$\leq \frac{1}{2}\sum_{(i,j)}\sum_{(k,l)}|a_{ij,kl}x_{kl}(t)x_{ij}(t)| + \sum_{(i,j)}\sum_{(k,l)}|b_{ij,kl}u_{kl}(t)x_{ij}(t)|$$

$$+ MN(|I|) + MN\left(\max_{i,j}\left|\int_0^{x_{ij}}m(s)ds\right|\right)$$

$$\leq \frac{1}{2}\sum_{(i,j)}\sum_{(k,l)}|a_{ij,kl}x_{kl}(t)x_{ij}(t)| + \sum_{(i,j)}\sum_{(k,l)}|b_{ij,kl}u_{kl}(t)x_{ij}(t)|$$

$$+ MN\left(|I| + \max_{i,j}\left|\int_0^{x_{ij}}m(s)ds\right|\right) = E_{\max}. \tag{11.19}$$

It follows from (11.18) and (11.19) that $E(t)$ is bounded as claimed in (11.17). It is worth noting that the output voltage of a cell is constrained between two boundaries in circuit implementation (as described in (15)). Like in most of the CNN applications, the two limitations are used to represent its binary outputs, i.e., 1 and -1 in image processing, respectively. Thus, in the mathematical model of an M-CNN, the range of the state x_{ij} is guaranteed within $[-1, 1]$.

Theorem 2: The function $E(t)$ is monotonically decreasing, namely

$$\frac{dE(t)}{dt} \leq 0. \tag{11.20}$$

Proof: If the feedback template is symmetric, i.e., $a_{ij,kl} = a_{kl,ij}$, the derivative of the energy function with respect to time t is given by

$$\frac{dE(t)}{dt} = -\sum_{(i,j)}\sum_{(k,l)}a_{ij,kl}x_{kl}(t)\frac{dx_{ij}(t)}{dt} - \sum_{(i,j)}\sum_{(k,l)}b_{ij,kl}u_{kl}(t)\frac{dx_{ij}(t)}{dt}$$

$$- \sum_{(i,j)}I\frac{dx_{ij}(t)}{dt} + \sum_{(i,j)}\frac{dx_{ij}(t)}{dt}\frac{d}{dx_{ij}(t)}\int_0^{x_{ij}}m(s)ds. \tag{11.21}$$

According to the definition of the CNNs, one has

$$a_{ij,kl} = 0, b_{ij,kl} = 0 \text{ for } c(k,l) \notin N_r(i,j). \tag{11.22}$$

Afterwards

$$
\begin{aligned}
\frac{dE(t)}{dt} &= -\sum_{(i,j)} \frac{dx_{ij}(t)}{dt} \left\{ \sum_{(k,l)} a_{ij,kl} x_{kl}(t) + \sum_{(k,l)} b_{ij,kl} u_{kl}(t) \right. \\
&\quad \left. + I - \frac{d}{dx_{ij}(t)} \int_0^{x_{ij}} m(s) ds \right\} \\
&= -\sum_{(i,j)} \frac{dx_{ij}(t)}{dt} \left\{ \sum_{(k,l)} (a_{ij,kl} x_{kl}(t) + b_{ij,kl} u_{kl}(t)) \right. \\
&\quad \left. + I - m(x_{ij}(t)) \right\}.
\end{aligned}
\tag{11.23}
$$

Substituting (11.7) into (11.23), and recalling $C > 0$, one can obtain

$$
\frac{dE(t)}{dt} = -\sum_{(i,j)} C \left(\frac{dx_{ij}(t)}{dt} \right)^2 \leq 0,
\tag{11.24}
$$

which means the energy function is monotonic decreasing. Meanwhile, since the energy function is bounded under certain constraints (in Theorem 1), the state variable in the memristor-based CNN is bounded. As a result, such a network always generates a DC output, in other words, the memristor-based CNN is stable.

11.4.2 Fault Tolerance

In practice, some faulty cells may exist in a CNN because of device fault or other reasons. Generally, a CNN is expected to work normally even with the presence of faulty cells, namely, being capable of fault tolerance. Theoretically, a CNN is considered to possess fault tolerance if it is still stable when some cells do not work. In this study, the fault tolerance for the memristor-based CNN is verified.

Definition: The stuck-at-α fault is defined that state of faulty cell is not changeable along with the inputs and outputs of other cells, where α is a constant with $|\alpha| \leq 1$ [26].

In image processing, the normalized binary output of a CNN processor (cell) can be 1 and -1 (or 0) which denote the white pixel and the black pixel, respectively. A stuck-at-α faulty cell always keeps its state variable unchanged, and consequently, the corresponding pixel value is constant depending on the value of α.

Then the dynamics of a memristor-based CNN cell with a single stuck-at-α fault can be given by

$$C\frac{dx_{ij}(t)}{dt} = -m(x_{ij}(t)) + \sum_{c(k,l)} \left(\bar{a}_{ij,kl}x_{kl}(t) + b_{ij,kl}u_{kl}\right) + I + a_f\alpha,$$

(11.25)

where a_f and $\bar{a}$ denote the faulty and the rest template, respectively.

An energy function $V(t)$ for the memristor-based CNN with single stuck-at-α fault is chosen as

$$V(t) = \sum_{(i,j)} \int_0^{x_{ij}} m(s)ds - \frac{1}{2}\sum_{(i,j)}\sum_{(k,l)} \bar{a}_{ij,kl}x_{kl}(t)x_{ij}(t)$$
$$- \sum_{(i,j)}\sum_{(k,l)} b_{ij,kl}u_{kl}(t)x_{ij}(t) - \sum_{(i,j)} Ix_{ij}(t) - \sum_{(i,j)} a_f\alpha x_{ij}(t).$$

(11.26)

Theorem 3: The energy function $V(t)$ for an $M \times N$ memristor-based CNN is bounded by

$$\max |V(t)| \leq V_{\max},$$

(11.27)

where

$$V_{\max} = \frac{1}{2}\sum_{(i,j)}\sum_{(k,l)} |\bar{a}_{ij,kl}x_{kl}(t)x_{ij}(t)| + \sum_{(i,j)}\sum_{(k,l)} |b_{ij,kl}u_{kl}(t)x_{ij}(t)|$$
$$+ MN\left(|I| + \max_{i,j}\left|\int_0^{x_{ij}} m(s)ds\right| + |a_f\alpha|\right).$$

(11.28)

Proof: Based on the definition of $V(t)$ in (11.26), one has

$$|V(t)| \leq \frac{1}{2}\sum_{(i,j)}\sum_{(k,l)} |\bar{a}_{ij,kl}x_{kl}(t)x_{ij}(t)| + \sum_{(i,j)}\sum_{(k,l)} |b_{ij,kl}u_{kl}(t)x_{ij}(t)|$$
$$+ \sum_{(i,j)} |Ix_{ij}(t)| + \sum_{(i,j)}\left|\int_0^{x_{ij}} m(s)ds\right| + \sum_{(i,j)} |a_f\alpha \cdot x_{ij}(t)|$$
$$\leq \frac{1}{2}\sum_{(i,j)}\sum_{(k,l)} |\bar{a}_{ij,kl}x_{kl}(t)x_{ij}(t)| + \sum_{(i,j)}\sum_{(k,l)} |b_{ij,kl}u_{kl}(t)x_{ij}(t)|$$

$$+ \sum_{(i,j)} |I| \cdot |x_{ij}(t)| + \sum_{(i,j)} \left| \int_0^{x_{ij}} m(s)ds \right| + \sum_{(i,j)} |a_f \alpha| \cdot |x_{ij}(t)|$$

$$\leq \frac{1}{2} \sum_{(i,j)} \sum_{(k,l)} |\bar{a}_{ij,kl} x_{kl}(t) x_{ij}(t)| + \sum_{(i,j)} \sum_{(k,l)} |b_{ij,kl} u_{kl}(t) x_{ij}(t)|$$

$$+ MN(|I|) + MN \left(\max_{i,j} \left| \int_0^{x_{ij}} m(s)ds \right| \right) + MN(|a_f \alpha|)$$

$$\leq \frac{1}{2} \sum_{(i,j)} \sum_{(k,l)} |\bar{a}_{ij,kl} x_{kl}(t) x_{ij}(t)| + \sum_{(i,j)} \sum_{(k,l)} |b_{ij,kl} u_{kl}(t) x_{ij}(t)|$$

$$+ MN \left(|I| + \max_{i,j} \left| \int_0^{x_{ij}} m(s)ds \right| + |a_f \alpha| \right) = V_{\max}. \quad (11.29)$$

Therefore, *V(t)* is bounded as declared in (11.27).

Theorem 4: The energy function $V(t)$ is monotonic decreasing, namely

$$\frac{dV(t)}{dt} \leq 0, \quad (11.30)$$

Proof: With the same assumption for the feedback template as in Theorem 2, the derivative of $V(t)$ with respect to time t is given by

$$\frac{dV(t)}{dt} = -\sum_{(i,j)} \sum_{(k,l)} \bar{a}_{ij,kl} x_{kl}(t) \frac{dx_{ij}(t)}{dt}$$

$$- \sum_{(i,j)} \sum_{(k,l)} b_{ij,kl} u_{kl}(t) \frac{dx_{ij}(t)}{dt} - \sum_{(i,j)} I \frac{dx_{ij}(t)}{dt}$$

$$+ \sum_{(i,j)} \frac{dx_{ij}(t)}{dt} \frac{d}{dx_{ij}(t)} \int_0^{x_{ij}} m(s)ds - \sum_{(i,j)} a_f \alpha \frac{dx_{ij}(t)}{dt}. \quad (11.31)$$

Similar to the derivation of (11.23), one can get the following from (11.31) as

$$\frac{dV(t)}{dt} = -\sum_{(i,j)} \frac{dx_{ij}(t)}{dt} \left\{ \sum_{(k,l)} \bar{a}_{ij,kl} x_{kl}(t) \right.$$

$$\left. + \sum_{(k,l)} b_{ij,kl} u_{kl}(t) + I - \frac{d}{dx_{ij}(t)} \int_0^{x_{ij}} m(s)ds + a_f \alpha \right\}$$

$$= -\sum_{(i,j)} \frac{dx_{ij}(t)}{dt} \left\{ -m(x_{ij}(t)) \right.$$

$$\left. + \sum_{(k,l)} (\bar{a}_{ij,kl} x_{kl}(t) + b_{ij,kl} u_{kl}(t)) + I + \alpha_f \alpha \right\}. \quad (11.32)$$

Substituting (11.25) into (11.32), and recalling $C > 0$, one can get

$$\frac{dV(t)}{dt} = -\sum_{(i,j)} C \left(\frac{dx_{ij}(t)}{dt} \right)^2 \leq 0. \quad (11.33)$$

Then, it can be concluded that the memristor-based CNN with a single stuck-at-α fault is still stable, which indicates the fault tolerance property.

11.5 Computer Simulations

In this section, the M-CNN has been numerically analyzed on software Matlab, including its stability, fault tolerance property, and two typical applications in image processing.

11.5.1 Stability Analysis

The transient behavior of the cell $c(2,2)$ in a 4×4 memristor-based CNN subjected to six sets of initial conditions (shown in Figure 11.9) are presented in Figure 11.10. It can be observed that the output (state) of the M-CNN cells can reach stability and keep it ultimately. This is in accordance with the theoretical derivation in Section IV and essential for sequent applications. In this simulation, the templates and the offset current is chosen as

$$A = \begin{bmatrix} 0 & 1 & 0 \\ 1 & 2 & 1 \\ 0 & 1 & 0 \end{bmatrix}, \ B = 0, \ I = 0 \quad (11.34)$$

11.5.2 Analysis of Fault Tolerance Property

To verify the fault tolerance property, a 21×21 memristor-based CNN is considered to implement the image inversion function. The corresponding

0.8	0.7	1.0	-0.1
1.0	1.0	1.0	1.0
1.0	0.9	0.7	0.8
-0.1	1.0	0.8	1.0

(a)

0.8	1.0	1.0	0.6
1.0	1.0	1.0	1.0
-1.0	0.9	-1.0	-0.8
-0.9	-1.0	-0.7	-0.8

(b)

-0.8	1.0	-0.1	-0.6
1.0	1.0	1.0	-0.1
-1.0	0.9	-1.0	-0.8
-0.9	-1.0	-0.7	-0.8

(c)

-0.9	-1.0	1.0	1.0
-1.0	1.0	-1.0	1.0
1.0	-1.0	0.7	0.8
0.9	1.0	0.8	1.0

(d)

-0.9	-1.0	-0.9	-1.0
-1.0	1.0	-1.0	-1.0
1.0	-1.0	1.0	1.0
0.7	1.0	1.0	0.8

(e)

-0.8	-0.9	-1.0	-0.6
-1.0	1.0	-1.0	-1.0
-1.0	-0.8	-1.0	-0.8
-0.9	-1.0	-0.7	-0.8

(f)

Figure 11.9 (a)–(f) are six different sets of initial conditions in which the initial states of cell $c(2, 2)$.

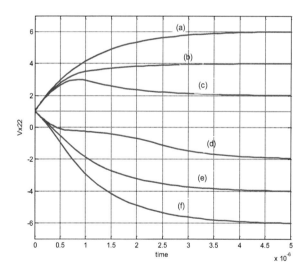

Figure 11.10 The transient behaviors of cell $c(2, 2)$ in the memristor-based CNN. Curves (a)–(f) denote the transient behaviors of the corresponding initial conditions in (a)–(f) in Figure 11.9, respectively.

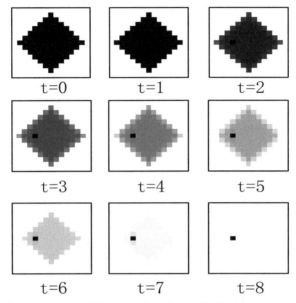

Figure 11.11   Image inversion of the memristor-based CNN with a single stuck-at-0 faulty cell at (8,5).

standard templates and the offset current (or threshold) are chosen as

$$A = \begin{bmatrix} 0 & 2 & 0 \\ 0 & 2 & 0 \\ 0 & 2 & 0 \end{bmatrix}, \quad B = \begin{bmatrix} 0 & 0 & 0 \\ 0 & 4 & 0 \\ 0 & 0 & 0 \end{bmatrix}, \quad I = 4 \qquad (11.35)$$

Figure 11.11 shows the time evolvement for image inversion of the memristor-based CNN with a single stuck-at-0 faulty cell $c(8,5)$. It can be observed that although there is a stuck-at-0 faulty cell, the memristor-based CNN can still accomplish the image inversion task satisfactorily, i.e., inversing the original black diamond into the white one successfully. It should be mentioned that the black-line boundary of each sub-figure is added to present the simulation results better, but it does not belong to the objective image.

11.5.3 Applications of the M-CNN

Two kinds of M-CNNs are used to perform several typical applications in image processing: (i) uncoupled M-CNN for horizontal line detection and edge extraction; (ii) coupled M-CNN for noise removal.

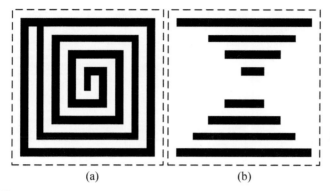

(a)   (b)

Figure 11.12 Horizontal line detection with uncoupled M-CNN: (a) Input image, (b) output image.

1) Uncoupled M-CNN for horizontal line detection and edge extraction

A CNN is considered as uncoupled if all of the elements of the feedback template A are zeroes or only with one non-zero element at center indicating self-feedback [9]. Horizontal line detection (HLD) is one of the functions an uncoupled CNN can complete. It means the uncoupled CNN will horizontally delete isolated points and the remaining horizontal lines should contain at least two points. For this example, the templates and the offset current are given by [21]

$$A = [0 \ 1 \ 0], \quad B = [1 \ 1], \quad I = -1. \tag{11.36}$$

Figures 11.12(a) and (b) shows the original image and the result respectively.

Another application is edge extraction, which is a common operation in image processing. Where, edge extraction is an important case of feature extraction, because the edges of an image contain most of the information regarding the shape of the image. Here, using the templates in (11.37), a practical example of a license plate (shown in Figure 11.13(a)) was respectively input into the M-CNN and the standard CNN to get its edges extracted. The processed results are presented in Figure 11.13(b) and (c), respectively.

$$A = \begin{bmatrix} 0 & 0 & 0 \\ 0 & 0 & 0 \\ 0 & 0 & 0 \end{bmatrix}, \quad B = \begin{bmatrix} -1 & -1 & -1 \\ -1 & 8 & -1 \\ -1 & -1 & -1 \end{bmatrix}, \quad I = -1. \tag{11.37}$$

As shown in Figure 11.13, (a) is the original image, (b) and (c) are the processed results obtained from the M-CNN and the standard CNN, respectively. Note that in the edge extraction the feedback template elements are

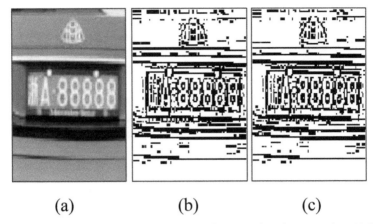

Figure 11.13 Illustration of image processing performance for edge extraction. (a) Original (input) images, (b) output of the M-CNN, and (c) output of a traditional CNN.

all zeroes, which means there is no feedback including self-feedback needed. The experiential results have also verified that the M-CNN model possesses similar processing performance in image processing with the standard CNN: the average pixel difference between Figure 11.13(b) and (c) is 1.47%.

2) Coupled M-CNN for noise removal

In CNNs, linear image processing with feedback and control templates is equivalent to spatial convolution with infinite impulse response (IIR) kernels [3]. For the image that is obtained from the real world by a camera or some other equipment, it is inevitable to be polluted by some noise. As a result, noise removal is a necessary and important operation. Now, we consider a 19×19 image (top left corner of Figure 11.14) which is polluted by Gaussian white noise with variance $\sigma = 0.02$, its mean value $m = 0$ (bottom left corner), as the input of a 19×19 M-CNN and a standard CNN, respectively. The processed results are illustrated on the first row and the second row correspondingly. The employed templates and threshold are given by (11.38). This simulation indicates that the M-CNN has satisfactory processing performance for noise removal as that of the standard CNN even with much simpler circuit structure.

$$A = \begin{bmatrix} 0 & 1 & 0 \\ 1 & 4 & 1 \\ 0 & 1 & 0 \end{bmatrix}, \quad B = \begin{bmatrix} 0 & 0 & 0 \\ 0 & 0 & 0 \\ 0 & 0 & 0 \end{bmatrix}, \quad I = 0 \quad (11.38)$$

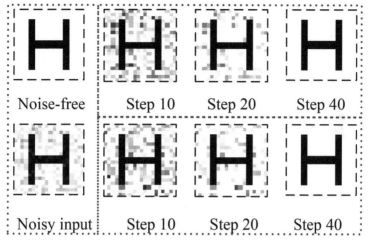

Figure 11.14 Image processing performance for noise removal is illustrated. The left two image are noise-free image and input image with 0.02 Gaussian white noise. On the right, the first row presents the process of the M-CNN, and the second row shows that of the standard CNN.

11.5.4 Impact of Variations of the Memristors on the M-CNN

The variations of the memristor synaptic weights may result from the process variation on memristors, the weight-setting inaccuracy, and weighting operation influence, etc. Taking the four ideal lines (corresponding to four levels of weights) for instance, assume that the overall variation impacts on the memristor weights follow a Gaussian distribution with variance $\sigma = 0.02$, 100 lines denoting the deviations of the ideal weights can be generated, as illustrated in Figure 11.15.

In this section, we take the HLD for example to evaluate the performance of the M-CNN under variations in the memristors. Since for the CNNs the image processing is a procedure of iteratively finding stable solution, we at first assume that during each iteration, the memristor synaptic weight (templates) varies following a Gaussian distribution with variance $\sigma = 0.02$ and then Monte-Carlo simulations are conducted 100 times for which the results are presented in Figure 11.16.

Furthermore, Figure 11.17 shows the average error (blue square) and the average running time (green star) of 100 Monte-Carlo simulations versus the increasing memristor weights variation. It can be observed that the variations of the memristor weights do not influence the processing result very much, because the small deviation is negligible compared to the binary quantization

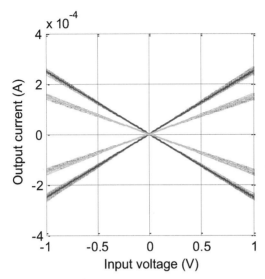

Figure 11.15   Synaptic weight processing via the memristor bridge circuit with memristor variations.

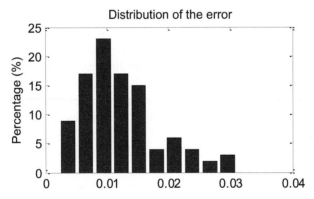

Figure 11.16   Histogram of the errors of 100 times of simulations for HLD with memristor weights variations, $\sigma = 0.02$.

of the output signals. For instance, even with variance of 1.0, the finial output is the same as the original one. However, since the image processing is a series of iterative operations, if some variations arise in the templates, the process will thus need more iterations to get the stable solution.

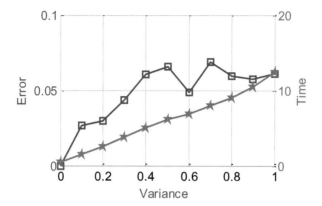

Figure 11.17   Average error (blue square) and average running time (green star) in seconds versus memristor weights variation's variance in 100 simulations for HLD.

11.6 Conclusions and Discussions

The memristor has been extensively studied in biological sciences and electrical engineering as a means to compactly implement synaptic function in neural networks. This chapter presents a novel implementation scheme for memristor-based cellular neural networks along with detailed mathematical analysis. In the proposed memristor-based CNN, memristors are employed not only to implement the synaptic weights (the templates) in programmable bridge structure, but also to replace the original linear state resistor. The weight programming and weighing process can both be performed in the memristor bridge circuit, and the original "sigmoid" output circuit can be eliminated to further reduce the size of the processing element as well as to improve the speed of computation. Therefore, the proposed architecture is more compact and versatile, as well as suitable for VLSI implementation.

Mathematical analysis including stability and fault tolerance has been presented using the Lyapunov theorem, thereby providing a theoretical foundation for memristor based cellular neural network (M-CNN) implementation. The simulation results indicate the performance of M-CNN in image processing for numerous common functions such as horizontal line detection, edge extraction, and noise removal. Finally, the effects of variations of memristors characteristics have been investigated by performing Monte-Carlo simulation. It was observed that variations of memristor characteristics influence the performance of an M-CNN somewhat in speed without impairing the quality of the final results.

References

[1] L. O. Chua and L. Yang, "Cellular neural networks: theory," *IEEE Trans. Circuits Syst.*, vol. CAS-35, no. 10, 1257–1272, 1988.

[2] L. O. Chua and L. Yang, "Cellular neural networks: applications," *IEEE Trans. Circuits Syst.*, vol. CAS-35, no. 10, 1273–1290, 1988.

[3] L. O. Chua and T. Roska, "Cellular Neural Networks and Visual Computing," Cambridge, England: *Cambridge University Press*. 2002, 272–275.

[4] T. Yang, "Cellular Neural Network and Image Processing," *Nova Science Publisher*, Inc. 2002, 1st Edition.

[5] R. Domínguez-Castro, S. Espejo, A. Rodríguez-Vázquez, R. A. Carmona, P. Földesy, Á. Zarándy, P. Szolgay, T. Szirányi, and T. Roska, "A 0.8-μm CMOS two-dimensional programmable mixed-signal focal-plane array processor with on-chip binary imaging and instructions storage," *IEEE J. Solid-State Circuits*, vol. 32, no.7, 1013–1026, 1997.

[6] J. M. Cruz and L. O. Chua, "A 16×16 cellular neural network universal chip: The first complete single-chip dynamic computer array with distributed memory and with gray-scale input-output," *Analog Integrated. Circuits Signal Process,* vol. 15, 227–237, 1998.

[7] P. Mazumder, S-R. Li, and I. E. Ebong, "Tunneling-based cellular nonlinear network architectures for image processing," *IEEE Trans. Very Large Scale Integr. (VLSI) Syst.*, vol. 17, no. 4, 487–495, 2009.

[8] W. H. Lee and P. Mazumder, "Motion detection by quantum dots based velocity-tuned filter", *IEEE Trans. Nanotechnol.*, vol. 7, No. 3, 357–362, 2008.

[9] S.-R. Li, P. Mazumder, and L. O. Chua, "Cellular neural/nonlinear networks using resonant tunneling diode," *The Fourth IEEE conference on nanotechnology*, 164–167, 2004.

[10] D. B. Strukov, G. S. Snider, D. R. Stewart, and R. S. Williams, "The missing memristor found," *Nature*, vol. 453, no. 7191, 80–83, 2008.

[11] R. S. Williams, "How we found the missing memristor," *IEEE spectr.*, vol. 45, no. 12, 28–35, 2008.

[12] L. O. Chua, "Memristor—the missing circuit element," *IEEE Trans. Circ. Theor.*, vol. 18, no. 5, 507–519, 1971.

[13] M. D. Pickett, D. B. Strukov, J. L. Borghetti, J. J. Yang, G. S. Snider, D. R. Stewart, and R. S. Williams, "Switching dynamics in titanium dioxide memristive devices," *J. Appl. Phys.*, vol. 106, no. 7, 074508, 2009.

[14] H. Abdalla and M. D. Pickett, "SPICE modeling of memristors," *In Circuits and Systems (ISCAS), 2011 IEEE International Symposium* on, 1832–1835, 2011.

[15] S. Kvatinsky, E. G. Friedman, A. Kolodny, and U. C. Weiser, "TEAM: ThrEshold adaptive memristor model," *IEEE Trans. Circuits Syst.,* vol. 60, no. 1, 211–221, 2013.

[16] F. Alibart, L. Gao, B. D. Hoskins, and D. B. Strukov, "High precision tuning of state for memristive devices by adaptable variation-tolerant algorithm," *Nanotechnology,* vol. 23, no. 7, 075201, 2012.

[17] X. Hu, S. Duan, L. Wang, and X. Liao, "Memristive crossbar array with applications in image processing," *Sci. China: Inf. Sci.,* vol. 55, no. 2, 461–472, 2012.

[18] S. Duan, X. Hu, L. Wang, C. Li, and P. Mazumder, "Memristor-Based RRAM with Applications," *Sci. China: Inf. Sci.,* vol. 55, no. 6, 1446–1460, 2012.

[19] S. H. Jo, T. Chang, I. Ebong, et al., "Nanoscale Memristor Device as Synapse in Neuromorphic Systems," *Nano Letters*, vol. 10, no. 4. 1297–1301, 2010.

[20] X. Hu, S. Duan, and L. Wang, "A Novel chaotic neural network using memristors with applications in associative memory," *Abstr. Appl. Anal.,* vol. 2012, 1–19, 2012.

[21] H. Kim, M. Sah, C. Yang, T. Roska, et al., "Memristor bridge synapses," *Proc. IEEE*, vol. 100, no. 6, 2061–2070, 2012.

[22] S.P. Adhikari, C. Yang, H. Kim, and L. O. Chua, "Memristor bridge synapse-based neural network and its learning," *IEEE trans. Neural Netw. Learn. Syst.*, vol. 23, no. 9, 1426–1435, 2012.

[23] F. Corinto, A. Ascoli, Y.-S. Kim, and K.-S. Min, "Cellular Nonlinear Networks with Memristor Synapses," In Memristor Networks, 267–291. *Springer International Publishing*, 2014.

[24] L. Wang, E. Drakakis, S. Duan, and P. He, "Memristor model and its application for chaos generation," *Int. J. Bifurcat. Chaos,* vol. 22, no. 8, 1250205, 2012.

[25] S. Liu, L.Wang, S. Duan, C. Li, and J. Wang, "Memristive device based filter and integration circuits with applications," *Adv. Sci. Letters*, vol. 8, no. 1, 194–199, 2012.

[26] L. Wang, X. Yang, and S. Duan, "Analysis of Fault Tolerance of Cellular Neural Networks and Applications to Image Processing," *in Third International Conference on Natural Computation, Washington, DC*, 252–256, 2007.

12

Dynamic Analysis of Memristor-based Neural Network and its Application

**Yongbin Yu, Lefei Men, Qingqing Hu, Shouming Zhong,
Nyima Tashi, Pinaki Mazumder, Idongesit Ebong,
Qishui Zhong and Xingwen Liu**

In this chapter, the memristive neural network is highlighted, in particular the memristor-based WTA neural network and the memristor-based recurrent neural network. Firstly, the design of two memristive neural networks is explained theoretically. Then, the dynamic analysis is presented to study the behaviors of these neural networks. Based on this theoretical analysis, the application of WTA neural network is realized which is employed to skin diseases classifier with an improving simulation results, and two examples are provided to verify this proposed memristive recurrent neural network.

12.1 Introduction

Memristor is postulated in 1971 as the fourth basic circuit element [1]. Since then, memristor remains as a theory until the first physical model is built by HP lab in 2008 [2, 3]. Recent years, more and more researchers are devoted to studying memristor and push forward to its potential applications such as non-volatile memory device with high density and speed as fast as DRAM [4, 5], neural networks [6, 7], and so on. Note that neural networks have acknowledged wide-ranging consideration owing to their advantage of parallel computation, learning ability, function approximation, and combinational optimization [8]. Numerous neural network models such as Hopfield neural network, cellular neural network, convolutional neural network, Winner-Take-All neural network, Cohen-Grossberg neural network, Lotka-Volterra neural network and adaptive neural network have been investigated [9–13].

Specially in artificial neural networks, memristor has been shown to be a promising candidate for application to synapses consisting of a memristor synapse-based multi-layer neural network [14, 15]. We know that the hopfield neural network model can be implemented in a circuit, where the self feedback connection weights and the connection weights are implemented by resistors [16, 17]. If we use memristors instead of resistors, then a new model can be builded where the connection weights change according to its state, i.e, that is, it is a state-dependent switching recurrent network which is called the memristor-based recurrent neural networks (MRNNs).

Dynamics analysis of MRNNs has attracted increasing attentions. In 2010, the differential inclusion theory was firstly used to study the dynamic analysis of MRNNs (see [18]). On one hand, the stability of MRNNs [19–22] and synchronization control of MRNNs [23–27] are investigated. On the other hand, the periodicity and chaotic characteristics [28] are studied. In addition, passivity of MRNNs studied by some theoretical results is presented in [29–33]. Passivity, and its generalization dissipativity, characterizes the energy consumption of a dynamical system. Dissipativity is an important theory used in sciences and control engineering, which provides strong connection between physics, system theory and control engineering. Therefore, research on the dissipativity of MRNNs is an interesting challenge. The global exponential dissipativity of MRNNs with time-varying delays is addressed [34], and the dissipativity of memristor-based complex-valued neural networks with time-varying delays is studied [35]), moreover strict (Q, S, R)-γ-dissipativity of MRNNs with leakage and time-varying delays is presented [36]. However, the dissipativity of drive-response system of MRNNs has not been studied or less in the literature. At the same time, leakage delay happens in this kind of MRNNs and leads to instability and poor performance. Inspired by related works in Table 12.1, therefore, considerable attention has been paid to the dynamics of MRNNs with the leakage terms.

In recent years, neural network has been developed to tackle skin classifier problem [37–40]. Neuroscientists find that winner-take-all (WTA) behavior always happens in many cognitive processing. Moreover, WTA is the underlying mechanism for competitive-learning, decision making and pattern recognition, which widely exists in society and nature. WTA neural

Table 12.1 Related works

Content Papers	Systems	Delays	Dynamic Analysis
Radhika [38]	Stochastic MRNN	Discrete and distributed	Dissipativity
Cai [5]	Switching jumps MRNN	Time-varying	Synchronization
Rakkiyappan [41]	Complex-valued Drive-response MRNN	Leakage and Time-varying	Dissipativity
Our published paper [57]	MRNN	Leakage and Time-varying	Dissipativity
Our paper	Drive-response MRNN	Leakage and Time-varying	Dissipativity
Comparison results	New model	Novel general delay	Relaxed criteria

network has the dual role of picking the maximun input to attend to as well as reinforcing that pattern after the input disappears, which means that only the maximally stimulated neuron responds and the others are inhibited [41–46]. Recently, considerable effort has been made to study WTA neural network and its application to image classification and feature extraction [47–52]. Not only WTA neural network, but also the other kind of neural network, have powerful capacities in medical field [37, 53–57]. In the current literature, most works focus on learning rule, architecture design, circuit implementation, and functional application. However, these works seldom provide memristor architecture with dynamic analysis and its application to skin classifier.

As for WTA neural network, four-neuron MOSFETs implemented neural network and its dynamic analysis [46], integrate-and fire neurons and RRAM synapses [47], stability analysis [58], and memristor networks with a WTA training algorithm [44] have been investigated in recent developments. In light of [46], a novel model with the consideration of memristor crossbar and its dynamic analysis are presented. Inspired by related works in Table 12.2, we will provide a novel point view of how WTA is used to solve real-life problems with a focus on how memristor arrays can be utilized to learn and make medical decision.

Table 12.2 Related works

Content Papers	Model	Dynamic Analysis	Application
Fang [11]	MOSFET+WTA	WTA behavior	No
Filimon [13]	BP Neural Network	No	Skin diseases classifier
Bavandpour [4]	STDP (Memristor Crossbar)	No	Image segmentation
Wu [55]	WTA	Stability analysis	No
Our published paper [10]	STDP+WTA (Memristor Crossbar)	No	Position detection
Our paper	BP-MWNN (Memristor Crossbar)	WTA behavior	Skin diseases classifier
Comparison results	New model	New sufficient conditions	New model, high percision

12.2 Some Definitions and Lemmas

For convenience, some preliminaries and terminologies are introduced: Throughout this section, solutions of all the systems considered in the following are intended in Filippov's sense [62]. $co[\Pi_1, \Pi_2]$ represents the closure of the convex hull generated by real numbers Π_1 and Π_2 or real matrices Π_1 and Π_2 [63]. $\mathbb{R}^+$ denotes the set of positive real number; $\mathbb{R}^n$ denotes the n-dimensional Euclidean space; $\mathbb{R}^{n \times m}$ is the set of all $n \times m$ real matrices. Let $C([-\tau, 0], \mathbb{R}^n)$ denotes the family of continuous functions $\phi(s)$ from $[-\tau, 0]$ to $\mathbb{R}^n$. The superscripts 'T' and '−1' stand for the transpose and inverse of a matrix, respectively; $A > 0 (A < 0)$ means that the matrix Λ is symmetric positive definite (negative definite), symmetric terms in a symmetric matrix are denoted by '$*$'; I is an appropriately dimensioned identity matrix. L_2^n is the space of square integrable functions on $\mathbb{R}^+$ with values in $\mathbb{R}^n$; L_{2e}^n is the extended L_2^n space defined by $L_{2e}^n = \{f : f$ is a measurable function on $\mathbb{R}^+$, $P_\tau f \in L_2^n, \forall \tau \in \mathbb{R}^+\}$, where $P_\tau f(t) = f(t)$ if $t \leq \tau$, and 0 if $t > \tau$; for any function $x(t), y(t) \in L_{2e}^n$, matrix M, we define $< x, My >_\tau = \int_0^\tau x^T(t) M y(t) \mathrm{d}t$.

We present some necessary Definitions and Lemmas.

Definition 1: ([64]) Let $E \subseteq \mathbb{R}^n$, $x \mapsto F(x)$ is called a set-valued map from E into R^n, if to each point x of a set $E \subseteq \mathbb{R}^n$, there corresponds a nonempty set $F(x) \subseteq \mathbb{R}^n$.

Definition 2: ([64]) A set-valued map F with nonempty values is said to be upper semi-continuous at $x_0 \in E \subseteq \mathbb{R}^n$ if, for any open set N containing $F(x_0)$, there exists a neighborhood M of x_0 such that $F(M) \subseteq N$. $F(x)$ is said to have a closed (convex, compact) image if for each $x \in E$, $F(x)$ is closed (convex, compact).

Definition 3: ([64]) For differential system $\frac{dx}{dt} = f(t, x)$, where $f(t, x)$ is discontinuous in x. The set-valued map of $f(t, x)$ is defined as

$$F(t, x) = \bigcap_{\varepsilon > 0} \bigcap_{\rho(N) = 0} co[f(t, B(x, \varepsilon) \backslash N)],$$

where $B(x, \varepsilon) = \{y \colon \|y - x\| \le \varepsilon\}$ is the ball of center x and radius ε; intersection is taken over all sets N of measure zero and over all $\varepsilon > 0$; and $\rho(N)$ is the Lebesgue measure of set N.

A Filippov solution of system $\frac{dx}{dt} = f(t, x)$ with initial condition $x(0) = x_0$ is absolutely continuous on any subinterval $t \in [t_2, t_2]$ of $[0, t_0]$, which satisfies $x(0) = x_0$ and the differential inclusion:

$$\frac{dx}{dt} \in F(t, x) \quad \text{for a.a. } t \in [0, t_0].$$

Lemma 1: ([65]) For any constant Hermitian matrix $W \in \mathbb{R}^{n \times n}$ and $W > 0$, a vector function $\omega(s) : [a, b]_S \to \mathbb{R}^n$ with scalars $a < b$ such that the integrations concerned are well defined, then

(1) $\left(\int_a^b \omega(s) ds \right)^{\mathrm{T}} W \left(\int_a^b \omega(s) ds \right) \le (b - a) \int_a^b \omega^{\mathrm{T}}(s) W \omega(s) ds,$

(2) $\left(\int_a^b \int_{t+\lambda}^t \omega(s) ds d\lambda \right)^{\mathrm{T}} W \left(\int_a^b \int_{t+\lambda}^t \omega(s) ds d\lambda \right) \le$

$\frac{(b-a)^2}{2} \int_a^b \int_{t+\lambda}^t \omega^{\mathrm{T}}(s) W \omega(s) ds d\lambda.$

Lemma 2: ([66]) Given constant matrices A, B, C, where $A = A^T, C = C^T$, then the linear matrix inequality

$$\begin{bmatrix} A & B^{\mathrm{T}} \\ B & -C \end{bmatrix} < 0$$

is equivalent to $C > 0$, $A + B^{\mathrm{T}} C^{-1} B < 0$.

12.3 Design of Memristor-based Neural Networks

Based on the survey of MRNN and MWNN in the Introduction and the relevant preliminaries including some definitions, lemmas in the previous section, MRNN and MWNN are designed in this section.

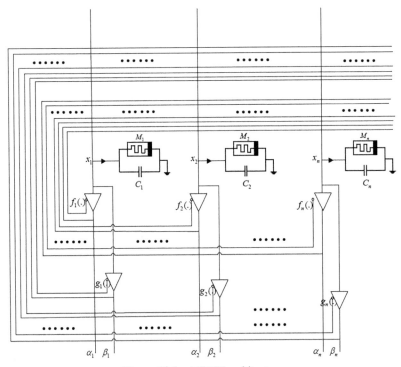

Figure 12.1   MRNN architecture.

12.3.1 Design of MRNN

Based on a class of MRNNs with leakage delay and time-varying delay, the drive-response system is obtained in this section.

We use the piecewise linear model proposed by Itoh (see [67]) to build a class of MRNNs as Figure 12.1, described as follows:

$$
\begin{cases}
\dot{x}_i(t) = & -c_i(x_i(t-\delta))x_i(t-\delta) \\
& + \sum_{j=1}^{n} a_{ij}(f_j(x_j(t)) - x_j(t))f_j(x_j(t)) \\
& + \sum_{j=1}^{n} b_{ij}(g_j(x_j(t-\tau_j(t)))) - x_j(t)) \\
& \times g_j(x_j(t-\tau_j(t))) + I_i(t), t \geq 0, i = 1, 2, \dots, n, \\
x_i(t) = & \psi_i(t), t \in [-\tau, 0], i = 1, 2, \dots, n,
\end{cases}
\tag{12.1}
$$

where

$$c_i(x_i(t - \delta)) = \begin{cases} c'_i, & x_i(t-\delta) \leq 0, \\ c''_i, & x_i(t-\delta) > 0, \end{cases}$$

$$a_{ij}(f_j(x_j(t)) - x_j(t)) = \begin{cases} a'_{ij}, & f_j(x_j(t))-x_j(t) \leq 0, \\ a''_{ij}, & f_j(x_j(t)) - x_j(t) > 0, \end{cases}$$

$$b_{ij}(g_j(x_j(t - \tau_j(t)))-x_j(t)) = \begin{cases} b'_{ij}, & g_j(x_j(t-\tau_j(t)))-x_j(t) \leq 0, \\ b''_{ij}, & g_j(x_j(t-\tau_j(t)))-x_j(t) > 0, \end{cases}$$

in which $c'_i > 0$, $c''_i > 0$, a'_{ij}, a''_{ij}, b'_{ij}, b''_{ij} $(i, j = 1, \ldots, n)$ are all constant numbers, f_i and g_i are bounded continuous functions, $I_i(t)$ is a continuous periodic external input function, c_i is the function of $x_i(t - \delta)$, a_{ij} is the function of $f_j(x_j(t)) - x_j(t)$, b_{ij} is the function of $g_j(x_j(t - \tau_j(t)))-x_j(t)$. And the activation function satisfy the assumption:

Assumption 1: For $j = 1, 2, \ldots, n$, $\forall a, b \in R$, $a \neq b$, the functions f_i, g_i are bounded and satisfy condition $l_i^- \leq \frac{f_j(a)-f_j(b)}{a-b} \leq l_i^+$, $k_i^- \leq \frac{g_j(a)-g_j(b)}{a-b} \leq k_j^+$, where l_i^-, l_i^+, k_i^-, k_j^+ are known real positive constants.

δ is leakage delay, $\tau_j(t)$ is time-varying delay, and the relations satisfy the following assumption:

Assumption 2: The leakage delay δ satisfies $\delta > 0$ and the time-varying delay $\tau_j(t)$ satisfies $0 \leq \tau_j(t) \leq \overline{\tau}$, $\dot{\tau}_j(t) \leq \mu$, where $\overline{\tau}$ and μ are constants.

Let $\tau = max\{\overline{\tau}, \delta\}$, $\overline{c}_i = max\{c'_i, c''_i\}$, $\underline{c}_i = min\{c'_i, c''_i\}$, $\overline{a}_{ij} = max \{a'_{ij}, a''_{ij}\}$, $\underline{a}_{ij} = min\{a'_{ij}, a''_{ij}\}$, $\overline{b}_{ij} = max\{b'_{ij}, b''_{ij}\}$, $\underline{b}_{ij} = min\{b'_{ij}, b''_{ij}\}$, and the relations satisfy the following assumption:

Assumption 3: For the differential inclusion, the following conditions hold

$$co[\underline{c}_i, \overline{c}_i]y_i(t) - co[\underline{c}_i, \overline{c}_i]x_i(t) \subseteq co[\underline{c}_i, \overline{c}_i](y_i(t) - x_i(t)),$$

$$co[\underline{a}_{ij}, \overline{a}_{ij}]f_j(y_j(t)) - co[\underline{a}_{ij}, \overline{a}_{ij}]f_j(x_j(t)) \subseteq co[\underline{a}_{ij}, \overline{a}_{ij}]$$
$$(f_j(y_j(t)) - f_j(x_j(t))),$$

$$co[\underline{b}_{ij}, \overline{b}_{ij}]g_j(y_j(t)) - co[\underline{b}_{ij}, \overline{b}_{ij}]g_j(x_j(t)) \subseteq co[\underline{b}_{ij}, \overline{b}_{ij}]$$
$$(g_j(y_j(t)) - g_j(x_j(t))).$$

It follows from (12.1) that

$$\dot{x}_i(t) \in - co[\underline{c}_i, \overline{c}_i] x_i(t - \delta) + \sum_{j=1}^{n} co[\underline{a}_{ij}, \overline{a}_{ij}] f_j(x_j(t))$$

$$+ \sum_{j=1}^{n} co[\underline{b}_{ij}, \overline{b}_{ij}] g_j(x_j(t - \tau_j(t))) + I_i(t), \tag{12.2}$$

or equivalently, for $i, j = 1, 2, \ldots, n$, there exist $c^* \in co[\underline{c}_i, \overline{c}_i]$, $a_{ij}^* \in co[\underline{a}_{ij}, \overline{a}_{ij}]$, $b_{ij}^* \in co[\underline{b}_{ij}, \overline{b}_{ij}]$, such that

$$\dot{x}_i(t) = -c_i^* x_i(t - \delta) + \sum_{j=1}^{n} (a_{ij}^* f_j(x_j(t))$$

$$+ b_{ij}^* g_j(x_j(t - \tau_j(t)))) + I_i(t). \tag{12.3}$$

To state conveniently, transform system (12.2) into the vector form as

$$\dot{x}(t) \in - co[\underline{C}, \overline{C}] x(t - \delta) + co[\underline{A}, \overline{A}] f(x(t))$$

$$+ co[\underline{B}, \overline{B}] g(x(t - \tau(t))) + I(t), \tag{12.4}$$

where $x(t) = [x_1(t), \ldots, x_n(t)]^{\mathrm{T}}$, $f(x(t)) = [f_1(x_1(t)), \ldots, f_n(x_n(t))]^{\mathrm{T}}$, $g(x(t - \tau(t))) = [g_1(x_1(t - \tau_1(t))), \ldots, g_n(x_n(t - \tau_n(t)))]^{\mathrm{T}}$, $\underline{C} = diag(\underline{c}_1, \ldots, \underline{c}_n)$, $\overline{C} = diag(\overline{c}_1, \ldots, \overline{c}_n)$, $\underline{A} = (\underline{a}_{ij})_{n \times n}$, $\overline{A} = (\overline{a}_{ij})_{n \times n}$, $\underline{B} = (\underline{b}_{ij})_{n \times n}$, $\overline{B} = (\overline{b}_{ij})_{n \times n}$, $I(t) = [I_1(t), \ldots, I_n(t)]^{\mathrm{T}}$.

Applying the theories of set-valued maps and differential inclusions, system (12.4) is equivalent to: there exist $C' \in co[\underline{C}, \overline{C}]$, $A' \in co[\underline{A}, \overline{A}]$, $B' \in co[\underline{B}, \overline{B}]$, such that

$$\dot{x}(t) = -C' x(t - \delta) + A' f(x(t)) + B' g(x(t - \tau(t))) + I(t). \tag{12.5}$$

In this section, consider system (1.2) or (1.3) as the drive system and the corresponding response system is as follows:

$$\dot{y}(t) \in - co[\underline{C}, \overline{C}] y(t - \delta) + co[\underline{A}, \overline{A}] f(y(t)) + co[\underline{B}, \overline{B}] g(y(t - \tau(t)))$$

$$+ I(t) + u(t), \tag{12.6}$$

or equivalently, there exist $C'' \in co[\underline{C}, \overline{C}]$, $A'' \in co[\underline{A}, \overline{A}]$, $B'' \in co[\underline{B}, \overline{B}]$, such that

$$\dot{y}(t) = -C'' y(t - \delta) + A'' f(y(t)) + B'' g(y(t - \tau(t))) + I(t) + u(t), \tag{12.7}$$

with initial conditions $y_i(t) = \omega_i(t)(t \in [-\tau, 0], i = 1, 2, \ldots, n)$, where $y(t) = [y_1(t), \ldots, y_n(t)]^{\mathrm{T}}$, $f(y(t)) = [f_1(y_1(t)), \ldots, f_n(y_n(t))]^{\mathrm{T}}$, $g(y(t - \tau(t))) = [g_1(y_1(t - \tau_1(t))), \ldots, g_n(y_n(t - \tau_n(t)))]^{\mathrm{T}}$, $u(t) = [u_1(t), \ldots, u_n(t)]^{\mathrm{T}}$ is the appropriate control input.

Applying the above theories and assumptions, let $e_i(t) = y_i(t) - x_i(t)$ be the error state, $\phi_i(e_i(t)) = f_i(y_i(t)) - f_i(x_i(t))$, $\varphi_i(e_i)) = g_i(y_i(t)) - g_i(x_i(t))$, we can obtain the error system as

$$\dot{e}(t) \in -co[\underline{C}, \overline{C}]e(t - \delta) + co[\underline{A}, \overline{A}]\phi(e(t)) + co[\underline{B}, \overline{B}]\varphi(e(t - \tau(t))) + u(t) \tag{12.8}$$

or equivalently, there exist $C \in co[\underline{C}, \overline{C}]$, $A \in co[\underline{A}, \overline{A}]$, $B \in co[\underline{B}, \overline{B}]$, such that

$$\begin{cases} \dot{e}(t) = -Ce(t - \delta) + A\phi(e(t)) + B\varphi(e(t - \tau(t))) + u(t) \\ \varpi(t) = \omega(t) - \psi(t), t \in [-\tau, 0]. \end{cases} \tag{12.9}$$

where $e(t) = [e_1(t), \ldots, e_n(t)]^{\mathrm{T}}$, $\phi(e(t)) = [\phi_1(e_1(t)), \ldots, \phi_n(e_n(t))]^{\mathrm{T}}$, $\varphi(e(t - \tau(t))) = [\varphi_1(e_1(t - \tau_1(t))), \ldots, \varphi_n(e_n(t - \tau_n(t)))]^{\mathrm{T}}$, with its initial condition $\varpi(t) = \omega(t) - \psi(t)(t \in [-\tau, 0])$, where $\varpi(t) = [\varpi_1(t), \ldots, \varpi_n(t)]^{\mathrm{T}}$, $\varpi_i(t) = \omega_i(t) - \psi_i(t)$.

Definition 4 ([68]) Given real symmetric matrix Q, R, real matrix S, the system (12.9) is called strictly (Q, S, R)-γ-dissipativity, if there exists a scalar $\gamma > 0$ such that the inequality

$$\langle \phi(e), \ Q\phi(e) \rangle + 2\langle \phi(e), \ Su \rangle + \langle u, Ru \rangle \ \geq \gamma \langle u, u \rangle \tag{12.10}$$

holds for all $u \in L_{2e}^n, t_f \geq 0$ and under the zero initial condition.

Remark 1: In dissipativity analysis, the system (12.9) will be regarded as an open-loop control system when $u(t) \in L_{2e}^n$. And, $u(t)$ and the error state $e(t)$ satisfy (12.10).

In next chapter, we will give the main results on the strict (Q, S, R)-γ-dissipativity criteria of the drive-response system (12.9) and some corollaries.

12.3.2 Design of MWNN

The memristor is a nonlinear passive electronic device defined mathematically by its constitutive relationship of the charge q and the flux φ

$$M(q) \triangleq \frac{\dot{\varphi}(q)}{\dot{q}}, \ W(q) \triangleq \frac{\dot{q}(\varphi)}{\dot{\varphi}}, \tag{12.11}$$

where $M(q)$ and $W(\varphi)$ are the memristance and memductance respectively, and overdot denotes derivative. Without loss of generality, we study the memductance, and assume that the memristor is characterized by "monotone-increasing" and "piecewise-linear" nonlinearity as follows.

$$q(\varphi) = d\varphi + 0.5(c - d)(|\varphi + 1| - |\varphi - 1|) \tag{12.12}$$

Consequently, the memductance $W(\varphi)$ is

$$W(\varphi) = \frac{\dot{q}(\varphi)}{\dot{\varphi}} = \frac{\dot{q}(\varphi)}{t}\frac{t}{\dot{\varphi}} = \frac{i}{v} = \begin{cases} c, & |\varphi| < 1, \\ d, & |\varphi| > 1, \end{cases} \tag{12.13}$$

In neural network, synapses can be modeled with memristors and network connectivity can be attained with the crossbars. The design of MWNN is inspired by the work [46], whose structure is implemented with the metal-oxide-semiconductor field effect transistors. On the basis of architecture design and circuit implementation, we introduce the memristor to design the MWNN. The neurons, realized on MOSFET, resistor, and capacitor, can connect to every other neuron through memristor synapses on the crossbar structure. A four-neuron WTA neural network implemented by MOSFETs and memristors is shown in Figure 12.2. The MOSFET function is given by

$$h(x, y) = \begin{cases} K[2(x + V_T)y - (x + V_T)^2], & y \geq 0, 0 \leq x + V_T \leq y \\ Ky^2, & y \geq 0, \ x + V_T > y \\ 0, & otherwise \end{cases} \tag{12.14}$$

where V_T is the voltage threshold, and K is the constant determined by the physical characteristics of the MOSFET. Firstly, we consider the first neuron and obtain the following state equation according to Kirchhoff's circuit law.

$$I_1 = C\dot{v}_1 + Gv_1 + h(v_1, v_2) + h(v_1, v_3) + h(v_1, v_4) + 3v_1W(\varphi_1), \tag{12.15}$$

then

$$C\dot{v}_1 = I_1 - Gv_1 - \sum_{j \neq 1} h(v_1, v_j) - 3v_1W(\varphi_1), \tag{12.16}$$

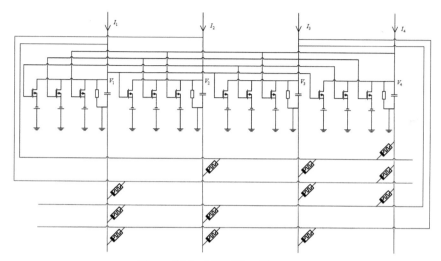

Figure 12.2 MWNN architecture.

where φ_1 denotes time integration of v_1. In a similar way, we present the state equation of the ith neuron,

$$C\dot{v}_i = I_i - Gv_i - \sum_{j\neq i} h(v_i, v_j) - 3v_i W(\varphi_i), \quad i = 1, 2, 3, 4. \quad (12.17)$$

Motivated by this model, but allowing $h(.,.)$ to be a general function, we have the

$$C\dot{v}_i = I_i - Gv_i - \sum_{j\neq i} h(v_i, v_j) - (N-1)v_i W(\varphi_i), i = 1, 2, \ldots, N. \quad (12.18)$$

where v_i is the activation of the ith neuron, and I_i is the external input of the ith neuron, and $W(\varphi_i)$ is the memductance of the ith memristor on the crossbar structure.

In the (12.18), the item of $\{-(N-1)v_i W(\varphi_i)\}$ is the key factor of MWNN. Obviously, the (12.18) without the item is changed into the model of [46].

In order to study the properties of the MWNN (12.18), we first introduce some notation, definitions, and lemmas [46]. Let $I_1, I_2, \ldots, I_N$ denote the external inputs, and I_{max}, v_{max} be the largest external inputs and the corresponding largest outputs respectively.

Remark 2: v is the WTA point if it is an equilibrium point of (12.18) such that $v_{max} > 0$ and $v_j \leq 0 (j \neq max)$.

Remark 3: According to the definition of $h(x, y)$, $h(x, y)$ is satisfied with Lipschitz condition. Therefore, there exits a constant $L > 0$ such that $|h(x_1, y) - h(x_2, y)| \leq L |x_1 - x_2|$ and $|h(x, y_1) - h(x, y_2)| \leq L |y_1 - y_2|$.

Remark 4: The function class F satisfying

$$F = \{ h(x, y) \geq 0 | h(x, y) \text{ is continuous, nondecreasing} \atop \text{in } x \text{ and } y, \ h(x, y) = 0 \text{ for } y \leq 0 \}. \tag{12.19}$$

Remark 5: The MWNN (12.18) is convergent, if the trajectory of (12.18) converges to an equilibrium point as time goes to infinity. Furthermore, (12.18) is exponentially convergent to x_0 if there exist an $M > 0$ and a $\sigma > 0$ such that $\| x(t) - x_0 \| \leq M e^{-\sigma t}$ for any $t \geq 0$ where $\| . \|$ is the Euclidean norm [69]. If system (12.18) is (exponentially, asymptotically) stable in the sense of Lyapunov [70], the MWNN (12.18) is (exponentially, asymptotically) stable.

Remark 6: The methods used to prove the theorem follow the idea in [46].

Lemma 3: Comparison Principle

1. Let $g(t, x)$ be a scalar continuous function, and let m(t) be the solution of the scalar differential equation: $\dot{m}(t) = g(t, m(t))$ with $m(t_0) = m_0$. Then, for any continuous function $x(t)$ such that the differential inequality: $\dot{x}(t) \leq g(t, x(t))$ with $x(t_0) = m_0$, we have $x(t) \leq m(t)$ for any $t \geq t_0$. Similarly, for any continuous function $x(t)$ satisfying the differential inequality: $\dot{x}(t) \geq g(t, x(t))$ with $x(t_0) = m_0$, we also have $x(t) \geq m(t)$ for any $t \geq t_0$.

2. If the continuous function $y(t) \geq 0$ such that $\dot{y}(t) \leq -\alpha y(t) + g(t)$ where $\alpha > 0$ and $g(t)$ exponentially converges to zero, i.e., there exist positive numbers $M > 0$ and $\beta > 0$ satisfying $|g(t)| \leq M e^{-\beta t}$, then $y(t)$ also exponentially converges to zero.

12.3.3 Conclusion

In this section, two novel neural networks including MRNN and MWNN are presented. First, the MRNN is designed and the preliminary is introduced to prepare for the dynamical analysis of the next section. Second, the MWNN is presented and some related knowledge is introduced.

12.4 Dynamical Analysis

According to the neural network models and the basic methods proposed in the previous section, the related dynamic analyses are carried out in this section. Firstly, the dynamic analysis of MRNN is presented. Then, propose sufficient conditions for WTA point to exist and the WTA behavior and convergence analysis are studied.

12.4.1 Dynamical Analysis of MRNN

In this section, we present the main results on the strict (Q, S, R)-γ-dissipativity criteria of the drive-response system (12.9) with time-varying delay and leakage delay.

Theorem 1: For given matrix S and symmetric matrix Q, R, the system (12.9) is strict (Q, S, R)-γ-dissipative if there exist symmetric matrices $Q_i = Q_i^T (i = 1, 2, 3)$, $R_i = R_i^T (i = 1, 2, 3)$, $S_i = S_i^T (i = 1, 2, 3)$, the positive definite matrices $P > 0$, $E > 0$, $F > 0$, $G > 0$, $H > 0$, $X > 0$, $Y > 0$, any matrices Z, $N_i(i = 1, 2, 3)$, the positive diagonal matrices $W_i = diag\{w_{i1}, \ldots, w_{in}\}(i = 1, 2, 3, 4)$, and a scalar $\gamma > 0$ such that the following matrix inequalities hold:

$$\Phi_1 = \begin{bmatrix} \vartheta_{11} & \vartheta_{12} & \vartheta_{13} \\ * & \vartheta_{14} & 0 \\ * & * & \vartheta_{15} \end{bmatrix} > 0, \tag{12.20}$$

$$\Phi_2 = \begin{bmatrix} E & -E & 0 & 0 \\ * & \vartheta_{21} & \vartheta_{22} & \vartheta_{23} \\ * & * & \vartheta_{24} & 0 \\ * & * & * & \vartheta_{25} \end{bmatrix} > 0, \tag{12.21}$$

$$\Phi_3 = \begin{bmatrix} G & -G & 0 & 0 \\ * & \vartheta_{31} & \vartheta_{32} & \vartheta_{33} \\ * & * & \vartheta_{34} & 0 \\ * & * & * & \vartheta_{35} \end{bmatrix} > 0, \tag{12.22}$$

$$\begin{bmatrix} E & Z \\ * & E \end{bmatrix} > 0, \tag{12.23}$$

$$\begin{bmatrix} \Pi & \mathcal{A}^T \ell \\ * & -\ell \end{bmatrix} < 0, \tag{12.24}$$

where

$$
\Pi =
\begin{bmatrix}
\Pi_{1,1} & \Pi_{1,2} & Z & G & \Pi_{1,5} & 0 & 0 & 0 & \Pi_{1,9} & PB & 0 & 0 & \bar{\tau}X & \Pi_{1,14} & P & 0 \\
* & \Pi_{2,2} & \Pi_{2,3} & 0 & 0 & 0 & 0 & 0 & 0 & 0 & 0 & 0 & 0 & 0 & 0 & 0 \\
* & * & \Pi_{3,3} & 0 & 0 & 0 & 0 & 0 & 0 & 0 & 0 & 0 & 0 & 0 & 0 & 0 \\
* & * & * & \Pi_{4,4} & 0 & 0 & 0 & 0 & 0 & 0 & 0 & 0 & 0 & 0 & 0 & 0 \\
* & * & * & * & \Pi_{5,5} & 0 & 0 & 0 & 0 & 0 & 0 & 0 & 0 & \Pi_{5,14} & -S & 0 \\
* & * & * & * & * & \Pi_{6,6} & 0 & 0 & 0 & \Pi_{6,10} & 0 & 0 & 0 & 0 & 0 & \Pi_{6,16} \\
* & * & * & * & * & * & -R_2 & 0 & 0 & 0 & 0 & 0 & 0 & 0 & 0 & 0 \\
* & * & * & * & * & * & * & -S_2 & 0 & 0 & 0 & 0 & 0 & 0 & 0 & 0 \\
* & * & * & * & * & * & * & * & \Pi_{9,9} & 0 & 0 & 0 & 0 & 0 & 0 & 0 \\
* & * & * & * & * & * & * & * & * & \Pi_{10,10} & 0 & 0 & 0 & \Pi_{10,14} & 0 & \Pi_{10,16} \\
* & * & * & * & * & * & * & * & * & * & -R_3 & 0 & 0 & 0 & 0 & 0 \\
* & * & * & * & * & * & * & * & * & * & * & -S_3 & 0 & 0 & 0 & 0 \\
* & * & * & * & * & * & * & * & * & * & * & * & \Pi_{13,13} & 0 & 0 & 0 \\
* & * & * & * & * & * & * & * & * & * & * & * & * & \Pi_{14,14} & -C^T P & 0 \\
* & * & * & * & * & * & * & * & * & * & * & * & * & * & \Pi_{15,15} & 0 \\
* & * & * & * & * & * & * & * & * & * & * & * & * & * & * & \Pi_{16,16}
\end{bmatrix}
$$

and

$$\vartheta_{11} = Q_1 + L_1 W_1 L_2 + K_1 W_2 K_2,$$

$$\vartheta_{12} = -\tfrac{1}{2}(L_1 + L_2)W_1, \vartheta_{13} = -\tfrac{1}{2}(K_1 + K_2)W_2,$$

$$\vartheta_{14} = Q_2 + W_1, \vartheta_{15} = Q_3 + W_2,$$

$$\vartheta_{21} = E + R_1 + L_1 W_1 L_2 + K_1 W_2 K_2,$$

$$\vartheta_{22} = -\tfrac{1}{2}(L_1 + L_2)W_1, \vartheta_{23} = -\tfrac{1}{2}(K_1 + K_2)W_2,$$

$$\vartheta_{24} = R_2 + W_1, \vartheta_{25} = R_3 + W_2,$$

$$\vartheta_{31} = G + S_1 + L_1 W_1 L_2 + K_1 W_2 K_2,$$

$$\vartheta_{32} = -\tfrac{1}{2}(L_1 + L_2)W_1, \vartheta_{33} = -\tfrac{1}{2}(K_1 + K_2)W_2,$$

$$\vartheta_{34} = S_2 + W_1, \vartheta_{35} = S_3 + W_2,$$

$$\Pi_{1,1} = Q_1 + R_1 + S_1 - PC - C^{\mathrm{T}}P - E - G + \bar{\tau}^2 F$$
$$+ \delta^2 H - \bar{\tau}^2 X - \delta^2 Y - L_1 W_1 L_2 - K_1 W_2 K_2,$$

$$\Pi_{1,2} = E - Z, \Pi_{1,5} = PA + \tfrac{1}{2}(L_1 + L_2)W_1 + L_2^{\mathrm{T}} W_3^{\mathrm{T}},$$

$$\Pi_{1,9} = \tfrac{1}{2}(K_1 + K_2)W_2 + K_2^{\mathrm{T}} W_4^{\mathrm{T}}, \Pi_{1,14} = C^{\mathrm{T}}PC + \delta Y,$$

$$\Pi_{2,2} = -(1-\mu)Q_1 - 2E + Z + Z^{\mathrm{T}}, \Pi_{2,3} = E - Z,$$

$$\Pi_{3,3} = -R_1 - E, \Pi_{4,4} = -S_1 - G,$$

$$\Pi_{5,5} = Q_2 + R_2 + S_2 - W_1 - W_3 - W_3^{\mathrm{T}} - Q,$$

$$\Pi_{5,14} = -A^{\mathrm{T}}PC, \Pi_{6,6} = -(1 - \mu)Q_2 + N_1 + N_1^{\mathrm{T}},$$

$$\Pi_{6,10} = N_1 + N_2^{\mathrm{T}}, \Pi_{6,16} = -N_1 + N_3^{\mathrm{T}},$$

$$\Pi_{9,9} = Q_3 + R_3 + S_3 - W_2 - W_4 - W_4^{\mathrm{T}},$$

$$\Pi_{10,10} = -(1-\mu)Q_3 + N_2 + N_2^{\mathrm{T}}, \Pi_{10,14} = -B^{\mathrm{T}}PC,$$

$$\Pi_{10,16} = -N_2 + N_3^{\mathrm{T}}, \Pi_{13,13} = -F - X,$$

$$\Pi_{14,14} = -H - Y, \ \Pi_{15,15} = \gamma I - R, \Pi_{16,16} = -N_3 - N_3^{\mathrm{T}},$$

$$K_1 = diag\{k_1^-, k_2^-, \ldots, k_n^-\}, K_2 = diag\{k_1^+, k_2^+, \ldots, k_n^+\},$$

$$L_1 = diag\{l_1^-, l_2^-, \ldots, l_n^-\}, L_2 = diag\{l_1^+, l_2^+, \ldots, l_n^+\},$$

$$\mathcal{A} = [0, 0, 0, -C, A, 0, 0, 0, 0, B, 0, 0, 0, 0, I, 0],$$

$$\ell = \bar{\tau}^2 E + \delta^2 G + \tfrac{\bar{\tau}^4}{4}X + \tfrac{\delta^4}{4}Y.$$

Proof 1: Consider the following Lyapunov–Krasovskii functions:

$$V(t) = V_1(t) + V_2(t) + V_3(t) + V_4(t),$$

where

$$V_1(t) = \left(e(t) - C \int_{t-\sigma}^{t} e(s)\mathrm{d}s \right)^{\mathrm{T}} P \left(e(t) - C \int_{t-\sigma}^{t} e(s)\mathrm{d}s \right),$$

$$V_2(t) = \int_{t-\tau(t)}^{t} e^{\mathrm{T}}(s)Q_1 e(s)\mathrm{d}s + \int_{t-\tau(t)}^{t} \phi^{\mathrm{T}}(s)Q_2\phi(s)\mathrm{d}s$$

$$+ \int_{t-\tau(t)}^{t} \varphi^{\mathrm{T}}(s)Q_3\varphi(s)\mathrm{d}s + \int_{t-\overline{\tau}}^{t} e^{\mathrm{T}}(s)R_1 e(s)\mathrm{d}s$$

$$+ \int_{t-\overline{\tau}}^{t} \phi^{\mathrm{T}}(s)R_2\phi(s)\mathrm{d}s + \int_{t-\overline{\tau}}^{t} \varphi^{\mathrm{T}}(s)R_3\varphi(s)\mathrm{d}s$$

$$+ \int_{t-\delta}^{t} e^{\mathrm{T}}(s)S_1 e(s)\mathrm{d}s + \int_{t-\delta}^{t} \phi^{\mathrm{T}}(s)S_2\phi(s)\mathrm{d}s$$

$$+ \int_{t-\delta}^{t} \varphi^{\mathrm{T}}(s)S_3\varphi(s)\mathrm{d}s,$$

$$V_3(t) = \overline{\tau} \int_{-\overline{\tau}}^{0} \int_{t+\theta}^{t} \dot{e}^{\mathrm{T}}(s)E\dot{e}(s)\mathrm{d}s\mathrm{d}\theta + \overline{\tau} \int_{-\overline{\tau}}^{0} \int_{t+\theta}^{t} e^{\mathrm{T}}(s)Fe(s)\mathrm{d}s\mathrm{d}\theta$$

$$+ \delta \int_{-\delta}^{0} \int_{t+\theta}^{t} \dot{e}^{\mathrm{T}}(s)G\dot{e}(s)\mathrm{d}s\mathrm{d}\theta + \delta \int_{-\delta}^{0} \int_{t+\theta}^{t} e^{\mathrm{T}}(s)He(s)\mathrm{d}s\mathrm{d}\theta,$$

$$V_4(t) = \frac{\overline{\tau}^2}{2} \int_{-\overline{\tau}}^{0} \int_{\theta}^{0} \int_{t+\beta}^{t} \dot{e}^{\mathrm{T}}(s)X\dot{e}(s)\mathrm{d}s\mathrm{d}\beta\mathrm{d}\theta$$

$$+ \frac{\delta^2}{2} \int_{-\delta}^{0} \int_{\theta}^{0} \int_{t+\beta}^{t} \dot{e}^{\mathrm{T}}(s)Y\dot{e}(s)\mathrm{d}s\mathrm{d}\beta\mathrm{d}\theta.$$

First of all, the function $V(t)$ is required to be proved to be positive definite. From Lemma 1, it can be obtained

$$\overline{\tau} \int_{-\overline{\tau}}^{0} \int_{t+\theta}^{t} \dot{e}^{\mathrm{T}}(s)E\dot{e}(s)\mathrm{d}s\mathrm{d}\theta$$

$$\geq \bar{\tau} \int_{-\bar{\tau}}^{0} \left(-\frac{1}{\theta} \left(\int_{t+\theta}^{t} \dot{e}(s) ds \right)^{\mathrm{T}} E \left(\int_{t+\theta}^{t} \dot{e}(s) ds \right) \right) d\theta. \qquad (12.25)$$

For any scalar $-\bar{\tau} \leq \theta < 0$, we have $-\frac{1}{\theta} \geq \frac{1}{\bar{\tau}}$, then it follows from (12.25) that

$$\bar{\tau} \int_{-\bar{\tau}}^{0} \int_{t+\theta}^{t} \dot{e}^{\mathrm{T}}(s) E \dot{e}(s) ds d\theta$$

$$\geq \int_{-\bar{\tau}}^{0} \left(\int_{t+\theta}^{t} \dot{e}(s) ds \right)^{\mathrm{T}} E \left(\int_{t+\theta}^{t} \dot{e}(s) ds \right) d\theta$$

$$= \int_{t-\bar{\tau}}^{t} \begin{bmatrix} e(t) \\ e(s) \end{bmatrix}^{\mathrm{T}} \begin{bmatrix} E & -E \\ -E & E \end{bmatrix} \begin{bmatrix} e(t) \\ e(s) \end{bmatrix} ds. \qquad (12.26)$$

Similarly,

$$\delta \int_{-\delta}^{0} \int_{t+\theta}^{t} \dot{e}^{\mathrm{T}}(s) G \dot{e}(s) ds d\theta$$

$$\geq \int_{t-\delta}^{t} \begin{bmatrix} e(t) \\ e(s) \end{bmatrix}^{\mathrm{T}} \begin{bmatrix} G & -G \\ -G & G \end{bmatrix} \begin{bmatrix} e(t) \\ e(s) \end{bmatrix} ds. \qquad (12.27)$$

In addition, the inequalities

$$\bar{\tau} \int_{-\bar{\tau}}^{0} \int_{t+\theta}^{t} e^{\mathrm{T}}(s) F e(s) ds d\theta > 0, \qquad (12.28)$$

$$\delta \int_{-\delta}^{0} \int_{t+\theta}^{t} e^{\mathrm{T}}(s) H e(s) ds d\theta > 0 \qquad (12.29)$$

hold due to the positive definite of F, H. And $V_1(t) > 0$, $V_4(t) > 0$ hold due to the positive definite of P, X and Y.

Assumption 1 guarantees that for $i = 1, 2, \ldots, n$,

$$\left(\phi_i(e_i(t)) - l_i^{-} e_i(t) \right) \left(l_i^{+} e_i(t) - \phi_i(e_i(t)) \right) \geq 0, \qquad (12.30)$$

and

$$\left(\varphi_i(e_i(t)) - k_i^{-} e_i(t) \right) \left(k_i^{+} e_i(t) - \varphi_i(e_i(t)) \right) \geq 0, . \qquad (12.31)$$

Then we can get that

$$m_1(t) = \phi^{\mathrm{T}}(e(t)) W_1 \phi(e(t)) - e^{\mathrm{T}}(t)(L_1 + L_2) W_1 \phi(e(t))$$
$$\quad + e^{\mathrm{T}}(t) L_1 W_1 L_2 e(t)$$

$$= \begin{bmatrix} e(t) \\ \phi(e(t)) \end{bmatrix}^{\mathrm{T}} \begin{bmatrix} L_1 W_1 L_2 & -\frac{1}{2}(L_1 + L_2) W_1 \\ * & W_1 \end{bmatrix} \begin{bmatrix} e(t) \\ \phi(e(t)) \end{bmatrix} \leq 0.$$
$$\qquad (12.32)$$

Similarly, we have

$$
\begin{aligned}
m_2(t) =& \varphi^{\mathrm{T}}(e(t))W_2\varphi(e(t)) - e^{\mathrm{T}}(t)(K_1 + K_2)W_2\varphi(e(t)) \\
& + e^{\mathrm{T}}(t)K_1W_2K_2e(t) \\
=& \left[\begin{array}{c} e(t) \\ \varphi(e(t)) \end{array}\right]^{\mathrm{T}} \left[\begin{array}{cc} K_1W_2K_2 & -\frac{1}{2}(K_1 + K_2)W_2 \\ * & W_2 \end{array}\right] \left[\begin{array}{c} e(t) \\ \varphi(e(t)) \end{array}\right] \le 0.
\end{aligned}
\tag{12.33}
$$

It is easy to obtain that

$$
h_1(t) = \int_{t-\tau(t)}^{t} m_1(s)\mathrm{d}s + \int_{t-\bar{\tau}}^{t} m_1(s)\mathrm{d}s + \int_{t-\delta}^{t} m_1(s)\mathrm{d}s \le 0, \tag{12.34}
$$

$$
h_2(t) = \int_{t-\tau(t)}^{t} m_2(s)\mathrm{d}s + \int_{t-\bar{\tau}}^{t} m_2(s)\mathrm{d}s + \int_{t-\delta}^{t} m_2(s)\mathrm{d}s \le 0. \tag{12.35}
$$

It follows from (12.26)–(12.35), if (12.20)–(12.22) hold,

$$
\begin{aligned}
V(t) \ge& V_1(t) + V_2(t) + V_3(t) + V_4(t) + h_1(t) + h_2(t) \\
\ge& V_2(t) + \int_{t-\bar{\tau}}^{t} \left[\begin{array}{c} e(t) \\ e(s) \end{array}\right]^{\mathrm{T}} \left[\begin{array}{cc} E & -E \\ -E & E \end{array}\right] \left[\begin{array}{c} e(t) \\ e(s) \end{array}\right] \mathrm{d}s \\
& + \int_{t-\delta}^{t} \left[\begin{array}{c} e(t) \\ e(s) \end{array}\right]^{\mathrm{T}} \left[\begin{array}{cc} G & -G \\ -G & G \end{array}\right] \left[\begin{array}{c} e(t) \\ e(s) \end{array}\right] \mathrm{d}s + h_1(t) + h_2(t) \\
=& \int_{t-\tau(t)}^{t} \eta_1^{\mathrm{T}}(s)\Phi_1\eta_1(s)\mathrm{d}s + \int_{t-\bar{\tau}}^{t} \eta_2^{\mathrm{T}}(t, s)\Phi_2\eta_2(t, s)\mathrm{d}s \\
& + \int_{t-\delta}^{t} \eta_3^{\mathrm{T}}(t, s)\Phi_3\eta_3(t, s)\mathrm{d}s > 0, \tag{12.36}
\end{aligned}
$$

where

$$
\eta_1^{\mathrm{T}}(s) = [e^{\mathrm{T}}(s), \phi^{\mathrm{T}}(e(s)), \varphi^{\mathrm{T}}(e(s))]^{\mathrm{T}},
$$

$$
\eta_2^{\mathrm{T}}(t, s) = [e^{\mathrm{T}}(t), e^{\mathrm{T}}(s), \phi^{\mathrm{T}}(e(s)), \varphi^{\mathrm{T}}(e(s))]^{\mathrm{T}},
$$

$$
\eta_3^{\mathrm{T}}(t, s) = [e^{\mathrm{T}}(t), e^{\mathrm{T}}(s), \phi^{\mathrm{T}}(e(s)), \varphi^{\mathrm{T}}(e(s))]^{\mathrm{T}}.
$$

So far, $V(t)$ has been proved to be positive definite. Next the time-derivative of $V(t)$ needs to be taken along the solution of (12.9).

$$\dot{V}_1(t) = 2\left(e(t) - C\int_{t-\sigma}^{t} e(s)\mathrm{d}s\right)^{\mathrm{T}} P\left(\dot{e}(t) - C(e(t) - e(t-\delta))\right).$$

(12.37)

$$
\begin{aligned}
\dot{V}_2(t) \leq\ & e^{\mathrm{T}}(t)Q_1 e(t) - (1-\mu)e^{\mathrm{T}}(t-\tau(t))Q_1 e(t-\tau(t)) \\
& + \phi^{\mathrm{T}}(e(t))Q_2\phi(e(t)) \\
& - (1-\mu)\phi^{\mathrm{T}}(e(t-\tau(t)))Q_2\phi(e(t-\tau(t))) \\
& + \varphi^{\mathrm{T}}(e(t))Q_3\varphi(e(t)) \\
& - (1-\mu)\varphi^{\mathrm{T}}(e(t-\tau(t)))Q_3\varphi(e(t-\tau(t))) \\
& + e^{\mathrm{T}}(t)R_1 e(t) - e^{\mathrm{T}}(t-\overline{\tau})R_1 e(t-\overline{\tau}) \\
& + \phi^{\mathrm{T}}(e(t))R_2\phi(e(t)) - \phi^{\mathrm{T}}(e(t-\overline{\tau}))R_2\phi(e(t-\overline{\tau})) \\
& + \varphi^{\mathrm{T}}(e(t))R_3\varphi(e(t)) - \varphi^{\mathrm{T}}(e(t-\overline{\tau}))R_3\varphi(e(t-\overline{\tau})) \\
& + e^{\mathrm{T}}(t)S_1 e(t) - e^{\mathrm{T}}(t-\delta)S_1 e(t-\delta) \\
& + \phi^{\mathrm{T}}(e(t))S_2\phi(e(t)) - \phi^{\mathrm{T}}(e(t-\delta))S_2\phi(e(t-\delta)) \\
& + \varphi^{\mathrm{T}}(e(t))S_3\varphi(e(t)) - \varphi^{\mathrm{T}}(e(t-\delta))S_3\varphi(e(t-\delta)). \quad (12.38)
\end{aligned}
$$

$$
\begin{aligned}
\dot{V}_3(t) =\ & \overline{\tau}^2\dot{e}^{\mathrm{T}}(t)E\dot{e}(t) - \overline{\tau}\int_{t-\overline{\tau}}^{t}\dot{e}^{\mathrm{T}}(s)E\dot{e}(s)\mathrm{d}s \\
& + \overline{\tau}^2 e^{\mathrm{T}}(t)Fe(t) - \overline{\tau}\int_{t-\overline{\tau}}^{t} e^{\mathrm{T}}(s)Fe(s)\mathrm{d}s \\
& + \delta^2\dot{e}^{\mathrm{T}}(t)G\dot{e}(t) - \delta\int_{t-\delta}^{t}\dot{e}^{\mathrm{T}}(s)G\dot{e}(s)\mathrm{d}s \\
& + \delta^2 e^{\mathrm{T}}(t)He(t) - \delta\int_{t-\delta}^{t} e^{\mathrm{T}}(s)He(s)\mathrm{d}s. \quad (12.39)
\end{aligned}
$$

Let $\eta_4(t) = \int_{t-\tau(t)}^{t} \dot{e}(s)\mathrm{d}s$, $\eta_5(t) = \int_{t-\bar{\tau}}^{t-\tau(t)} \dot{e}(s)\mathrm{d}s$, according to Lemma 2.1, partition integral interval $[t-\bar{\tau}, t]$ into $[t-\bar{\tau}, t-\tau(t)]$ and $[t-\tau(t), t]$, it follows that

$$\bar{\tau} \int_{t-\bar{\tau}}^{t} \dot{e}^{\mathrm{T}}(s)E\dot{e}(s)\mathrm{d}s \geq \frac{\bar{\tau}}{\tau(t)}\eta_4^{\mathrm{T}}(t)E\eta_4(t) + \frac{\bar{\tau}}{\bar{\tau}-\tau(t)}\eta_5^{\mathrm{T}}(t)E\eta_5(t)$$

$$=\eta_4^{\mathrm{T}}(t)E\eta_4(t) + \frac{\bar{\tau}-\tau(t)}{\tau(t)}\eta_4^{\mathrm{T}}(t)E\eta_4(t) + \eta_5^{\mathrm{T}}(t)E\eta_5(t)$$

$$+ \frac{\tau(t)}{\bar{\tau}-\tau(t)}\eta_5^{\mathrm{T}}(t)E\eta_5(t). \tag{12.40}$$

It is clear that,

$$\left[\begin{array}{c} \sqrt{\frac{\bar{\tau}-\tau(t)}{\tau(t)}}\eta_4^{\mathrm{T}}(t) \\ -\sqrt{\frac{\tau(t)}{\bar{\tau}-\tau(t)}}\eta_5^{\mathrm{T}}(t) \end{array} \right]^{\mathrm{T}} \left[\begin{array}{cc} E & Z \\ & E \end{array} \right] \left[\begin{array}{c} \sqrt{\frac{\bar{\tau}-\tau(t)}{\tau(t)}}\eta_4^{\mathrm{T}}(t) \\ -\sqrt{\frac{\tau(t)}{\bar{\tau}-\tau(t)}}\eta_5^{\mathrm{T}}(t) \end{array} \right] \geq 0, \tag{12.41}$$

which implies that

$$\frac{\bar{\tau}-\tau(t)}{\tau(t)}\eta_4^{\mathrm{T}}(t)E\eta_4(t) + \frac{\tau(t)}{\bar{\tau}-\tau(t)}\eta_5^{\mathrm{T}}(t)E\eta_5(t)$$

$$\geq \eta_4^{\mathrm{T}}(t)Z\eta_5(t) + \eta_5^{\mathrm{T}}(t)Z^T\eta_4(t). \tag{12.42}$$

From (12.40)–(12.42), if (12.23) holds, we have

$$\bar{\tau} \int_{t-\bar{\tau}}^{t} \dot{e}^{\mathrm{T}}(s)E\dot{e}(s)\mathrm{d}s$$

$$\geq \eta_4^{\mathrm{T}}(t)E\eta_4(t) + \eta_5^{\mathrm{T}}(t)E\eta_5(t) + \eta_4^{\mathrm{T}}(t)Z\eta_5(t) + \eta_5^{\mathrm{T}}(t)Z\eta_4(t)$$

$$= \left[\begin{array}{c} \eta_4^{\mathrm{T}}(t) \\ \eta_5^{\mathrm{T}}(t) \end{array} \right]^{\mathrm{T}} \left[\begin{array}{cc} E & Z \\ & E \end{array} \right] \left[\begin{array}{c} \eta_4^{\mathrm{T}}(t) \\ \eta_5^{\mathrm{T}}(t) \end{array} \right] \geq 0. \tag{12.43}$$

It is clear that (12.43) implies

$$-\bar{\tau} \int_{t-\bar{\tau}}^{t} \dot{e}^{\mathrm{T}}(s)E\dot{e}(s)\mathrm{d}s \leq \omega^{\mathrm{T}}(t)\Upsilon\omega(t), \tag{12.44}$$

where $\omega^{\mathrm{T}}(t) = [e^{\mathrm{T}}(t), e^{\mathrm{T}}(t-\tau(t)), e^{\mathrm{T}}(t-\bar{\tau})]$,

$$\Upsilon = \begin{bmatrix} -E & E-Z & Z \\ & -2E+Z^T+Z & E-Z \\ * & & -E \end{bmatrix}.$$

In addition, from Lemma 1, it is easy to get

$$-\delta \int_{t-\delta}^{t} \dot{e}^{\mathrm{T}}(s)G\dot{e}(s)\mathrm{d}s$$

$$\leq -(e(t)-e(t-\sigma))^{\mathrm{T}}G\,(e(t)-e(t-\sigma))$$

$$= \begin{bmatrix} e(t) \\ e(t-\sigma) \end{bmatrix}^{\mathrm{T}} \begin{bmatrix} -G & G \\ & -G \end{bmatrix} \begin{bmatrix} e(t) \\ e(t-\sigma) \end{bmatrix}, \qquad (12.45)$$

$$-\bar{\tau}\int_{t-\bar{\tau}}^{t} e^{\mathrm{T}}(s)Fe(s)\mathrm{d}s \leq -\int_{t-\bar{\tau}}^{t} e^{\mathrm{T}}(s)\mathrm{d}s F \int_{t-\bar{\tau}}^{t} e(s)\mathrm{d}s, \qquad (12.46)$$

$$-\delta \int_{t-\delta}^{t} e^{\mathrm{T}}(s)He(s)\mathrm{d}s \leq -\int_{t-\delta}^{t} e^{\mathrm{T}}(s)\mathrm{d}s H \int_{t-\delta}^{t} e(s)\mathrm{d}s. \qquad (12.47)$$

Using (12.2) in Lemma 1, we have

$$\dot{V}_4(t) = \frac{\bar{\tau}^4}{4}\dot{e}^{\mathrm{T}}(t)X\dot{e}(t) - \frac{\bar{\tau}^2}{2}\int_{-\bar{\tau}}^{0}\int_{t+\theta}^{t}\dot{e}^{\mathrm{T}}(s)X\dot{e}(s)\mathrm{d}s\mathrm{d}\theta$$

$$+ \frac{\delta^4}{4}\dot{e}^{\mathrm{T}}(t)Y\dot{e}(t) - \frac{\delta^2}{2}\int_{-\delta}^{0}\int_{t+\theta}^{t}\dot{e}^{\mathrm{T}}(s)Y\dot{e}(s)\mathrm{d}s\mathrm{d}\theta$$

$$\leq \frac{\bar{\tau}^4}{4}\dot{e}^{\mathrm{T}}(t)X\dot{e}(t) - \bar{\tau}^2 e^{\mathrm{T}}(t)Xc(t) + \bar{\tau}e^{\mathrm{T}}(t)X\int_{t-\bar{\tau}}^{t} e(s)\mathrm{d}s$$

$$+ \bar{\tau}\int_{t-\bar{\tau}}^{t} e^{\mathrm{T}}(s)\mathrm{d}s Xe(t) - \int_{t-\bar{\tau}}^{t} e^{\mathrm{T}}(s)\mathrm{d}s X\int_{t-\bar{\tau}}^{t} e(s)\mathrm{d}s$$

$$+ \frac{\delta^4}{4}\dot{e}^{\mathrm{T}}(t)Y\dot{e}(t) - \delta^2 e^{\mathrm{T}}(t)Ye(t) + \delta e^{\mathrm{T}}(t)Y\int_{t-\delta}^{t} e(s)\mathrm{d}s$$

$$+ \delta\int_{t-\delta}^{t} e^{\mathrm{T}}(s)\mathrm{d}s Ye(t) - \int_{t-\delta}^{t} e^{\mathrm{T}}(s)\mathrm{d}s Y\int_{t-\delta}^{t} e(s)\mathrm{d}s. \qquad (12.48)$$

Let $m(t) = \phi(e(t - \tau(t))) + \varphi(e(t - \tau(t)))$, we have

$$2[\phi^{\mathrm{T}}(e(t - \tau(t)))N_1 + \varphi^{\mathrm{T}}(e(t - \tau(t)))N_2 + m^{\mathrm{T}}(t)N_3]$$
$$\times [\phi(e(t - \tau(t))) + \varphi(e(t - \tau(t))) - m(t)] = 0. \tag{12.49}$$

Moreover, it can be verified under Assumption 1 that

$$m_3(t) = 2\phi^{\mathrm{T}}(e(t))W_3(L_2 e(t) - \phi(e(t))) \geq 0, \tag{12.50}$$

$$m_4(t) = 2\varphi^{\mathrm{T}}(e(t))W_4(K_2 e(t) - \varphi(e(t))) \geq 0. \tag{12.51}$$

From (12.32)–(12.33) and (12.37)–(12.51), we have

$$\dot{V}(t) - \phi^{\mathrm{T}}(e(t))Q\phi(e(t)) - 2\phi^{\mathrm{T}}(e(t))Su(t)$$
$$- u^{\mathrm{T}}(t)(R - \gamma I)u(t)$$
$$\leq \dot{V}(t) - \phi^{\mathrm{T}}(e(t))Q\phi(e(t)) - 2\phi^{\mathrm{T}}(e(t))Su(t)$$
$$- u^{\mathrm{T}}(t)(R - \gamma I)u(t) - m_1(t) - m_2(t) + m_3(t) + m_4(t)$$
$$+ 2(\phi^{\mathrm{T}}(e(t - \tau(t)))N_1 + \varphi^{\mathrm{T}}(e(t - \tau(t)))N_2 + m^{\mathrm{T}}(t)N_3)$$
$$\times (\phi(e(t - \tau(t))) + \varphi(e(t - \tau(t))) - m(t))$$
$$\leq \xi^{\mathrm{T}}(t)\Pi\xi(t) + \dot{e}^{\mathrm{T}}(t)\ell\dot{e}(t)$$
$$= \xi^{\mathrm{T}}(t)(\Pi + \mathcal{A}^{T}\ell\mathcal{A})\xi(t), \tag{12.52}$$

where

$$\xi^{\mathrm{T}}(t) = [e^{\mathrm{T}}(t), \ e^{\mathrm{T}}(t - \tau(t)), \ e^{\mathrm{T}}(t - \bar{\tau}), \ e^{\mathrm{T}}(t - \delta),$$
$$\phi^{\mathrm{T}}(e(t)), \ \phi^{\mathrm{T}}(e(t - \tau(t))), \ \phi^{\mathrm{T}}(e(t - \bar{\tau})),$$
$$\phi^{\mathrm{T}}(e(t - \delta)), \ \varphi^{\mathrm{T}}(e(t)), \ \varphi^{\mathrm{T}}(e(t - \tau(t))),$$
$$\varphi^{\mathrm{T}}(e(t - \bar{\tau})), \ \varphi^{\mathrm{T}}(e(t - \delta)), \ \int_{t - \bar{\tau}}^{t} e^{\mathrm{T}}(s)\mathrm{d}s,$$
$$\int_{t - \delta}^{t} e^{\mathrm{T}}(s)\mathrm{d}s, \ u^{\mathrm{T}}(t), \ m^{\mathrm{T}}(t)].$$

Applying Lemma 2 to (12.24) yields $\Pi + \mathcal{A}^{T} \ell \mathcal{A} < 0$. Then we can get that

$$
\dot{V}(t) - \phi^{T}(e(t))Q\phi(e(t)) - 2\phi^{T}(e(t))Su(t)
$$

$$
- u^{T}(t)(R - \gamma I)u(t) \leq 0. \tag{12.53}
$$

For any $\tau \geq 0$, integrating (12.53) from 0 to τ gives

$$
\int_{0}^{\tau} \left(\dot{V}(t) - \phi^{T}(e(t))Q\phi(e(t)) - 2\phi^{T}(e(t))Su(t) \right) dt
$$

$$
- \int_{0}^{\tau} \left(u^{T}(t)(R - \gamma I)u(t) \right) dt \leq 0, \tag{12.54}
$$

which implies that, under zero initial conditions,

$$
\int_{0}^{\tau} \left(-\phi^{T}(e(t))Q\phi(e(t)) - 2\phi^{T}(e(t))Su(t) \right) dt
$$

$$
- \int_{0}^{\tau} \left(u^{T}(t)(R - \gamma I)u(t) \right) dt \leq -V(t) \leq 0. \tag{12.55}
$$

This is equivalent to the strict (Q, S, R)-γ-dissipativity in (12.10). Therefore, if LMI (12.20)–(12.24) hold, the system (12.9) is strictly (Q, S, R)-γ-dissipative. The proof is completed.

Remark 7: Compared to [57], there are more two triple integral terms $\frac{\bar{\tau}^{2}}{2} \int_{-\bar{\tau}}^{0} \int_{\theta}^{0} \int_{t+\beta}^{t} \dot{e}^{T}(s)X\dot{e}(s)dsd\beta d\theta$ and $\frac{\delta^{2}}{2} \int_{-\delta}^{0} \int_{\theta}^{0} \int_{t+\beta}^{t} \dot{e}^{T}(s)Y\dot{e}(s)dsd\beta d\theta$ in the considered LKF, whose function is to further reduce the conservativeness.

Remark 8: In this paper, time-varying delay and leakage delay are considered in neural networks, which can be extended to the more general circumstances.

Corollary 1: Let $u = 0$, the condition of Theorem 1 is changed as follows.
(1) $\Pi_{55} = Q_2 + R_2 + S_2 - W_1 - W_3 - W_3^{T}$,
(2) deleting the column 15 and row 15 from matrix Π,
(3) changing the subscript 16 to subscript 15.

Then the system (8) or (9) is asymptotically stable, that is, the drive system (6) is synchronized with the response system (4).

Compared to general memristor-based recurrent neural networks, system (12.1) includes not only time-varying delay but also leakage delay, which is more general and applicable. Specially, when there is no leakage delay, the system (12.1) turns into the following:

$$
\begin{cases}
\dot{x}_i(t) = -c_i(x_i)x_i(t) \\
\quad + \sum_{j=1}^{n} a_{ij}(f_j(x_j(t)) - x_j(t))f_j(x_j(t)) \\
\quad + \sum_{j=1}^{n} b_{ij}(g_j(x_j(t - \tau_j(t)))) - x_j(t)) \\
\quad \times g_j(x_j(t - \tau_j(t))) + I_i(t), t \geq 0, i = 1, 2, \ldots, n, \\
x_i(t) = \psi_i(t), t \in [-\tau, 0], i = 1, 2, \ldots, \text{n},
\end{cases}
\tag{12.56}
$$

Similarly as before, we can get the corresponding drive-response system in vector form as following:

$$
\begin{cases}
\dot{e}(t) = -Ce(t) + A\phi(e(t)) + B\varphi(e(t - \tau(t))) \\
\quad + u(t), t \geq 0, \\
\varpi(t) = \omega(t) - \psi(t), t \in [-\tau, 0].
\end{cases}
\tag{12.57}
$$

Following the similar procedures discussed in Theorem 2, the following corollaries can be immediately obtained:

Corollary 2: For given matrix S and symmetric matrix Q, R, the system (12.57) is strict $(Q, S, R) - \gamma$-dissipative if there exist symmetric matrices $Q_i = Q_i^T (i = 1, 2, 3)$, $R_i = R_i^T (i = 1, 2, 3)$, the positive matrices $P > 0$, $E > 0$, $F > 0$, $X > 0$, any matrices Z, $N_i(i = 1, 2, 3)$, the positive diagonal matrices $W_i = diag\{w_{i1}, \ldots, w_{in}\}(i = 1, 2, 3, 4)$, and a scalar $\gamma > 0$ such that the following matrix inequalities hold:

$$
\Phi_1 = \begin{bmatrix} \vartheta_{11}^1 & \vartheta_{12}^1 & \vartheta_{13}^1 \\ * & \vartheta_{14}^1 & 0 \\ * & * & \vartheta_{15}^1 \end{bmatrix} > 0, \Phi_2 = \begin{bmatrix} E & -E & 0 & 0 \\ * & \vartheta_{21}^1 & \vartheta_{22}^1 & \vartheta_{23}^1 \\ * & * & \vartheta_{24}^1 & 0 \\ * & * & * & \vartheta_{25}^1 \end{bmatrix} > 0,
$$

$$
\begin{bmatrix} E & Z \\ * & E \end{bmatrix} > 0, \begin{bmatrix} \Pi & \mathcal{A}^T \ell \\ * & -\ell \end{bmatrix} < 0,
$$

where

$$
\Pi = \begin{bmatrix}
\Pi_{1,1} & \Pi_{1,2} & Z & \Pi_{1,4} & 0 & 0 & \Pi_{1,7} & PB & 0 & \overline{\tau}X & P & 0 \\
* & \Pi_{2,1} & \Pi_{2,3} & 0 & 0 & 0 & 0 & 0 & 0 & 0 & 0 & 0 \\
* & * & \Pi_{33} & 0 & 0 & 0 & 0 & 0 & 0 & 0 & 0 & 0 \\
* & * & * & \Pi_{44} & 0 & 0 & 0 & 0 & 0 & 0 & -S & 0 \\
* & * & * & * & \Pi_{55} & 0 & \Pi_{5,7} & 0 & 0 & 0 & 0 & \Pi_{5,12} \\
* & * & * & * & * & -R_2 & 0 & 0 & 0 & 0 & 0 & 0 \\
* & * & * & * & * & * & \Pi_{77} & 0 & 0 & 0 & 0 & 0 \\
* & * & * & * & * & * & * & \Pi_{88} & 0 & 0 & 0 & \Pi_{8,12} \\
* & * & * & * & * & * & * & * & -R_3 & 0 & 0 & 0 \\
* & * & * & * & * & * & * & * & * & \Pi_{10,10} & 0 & 0 \\
* & * & * & * & * & * & * & * & * & * & \Pi_{11,11} & 0 \\
* & * & * & * & * & * & * & * & * & * & * & \Pi_{12,12}
\end{bmatrix}
$$

$$
\begin{aligned}
\vartheta_{11}^1 &= Q_1 + L_1 W_1 L_2 + K_1 W_2 K_2, \\
\vartheta_{12}^1 &= -\tfrac{1}{2}(L_1 + L_2)W_1, \vartheta_{13}^1 = -\tfrac{1}{2}(K_1 + K_2)W_2, \\
\vartheta_{14}^1 &= Q_2 + W_1, \vartheta_{15}^1 = Q_3 + W_2, \\
\vartheta_{21}^1 &= E + R_1 + L_1 W_1 L_2 + K_1 W_2 K_2, \\
\vartheta_{22}^1 &= -\tfrac{1}{2}(L_1 + L_2)W_1, \vartheta_{23}^1 = -\tfrac{1}{2}(K_1 + K_2)W_2, \\
\vartheta_{24}^1 &= R_2 + W_1, \vartheta_{25}^1 = R_3 + W_2, \\
\Pi_{1,1} &= Q_1 + R_1 + S_1 - PC - C^{\mathrm{T}}P - E + \overline{\tau}^2 F - \overline{\tau}^2 X \\
&\quad - L_1 W_1 L_2 - K_1 W_2 K_2, \\
\Pi_{1,2} &= E - Z, \Pi_{1,4} = PA + \tfrac{1}{2}(L_1 + L_2)W_1 + L_2^{\mathrm{T}}W_3^{\mathrm{T}}, \\
\Pi_{1,7} &= \tfrac{1}{2}(K_1 + K_2)W_2 + K_2^{\mathrm{T}}W_4^{\mathrm{T}}, \\
\Pi_{2,2} &= -(1-\mu)Q_1 - 2E + Z + Z^{\mathrm{T}}, \Pi_{2,3} = E - Z, \\
\Pi_{3,3} &= -R_1 - E, \Pi_{4,4} = Q_2 + R_2 - W_1 - W_3 - W_3^{\mathrm{T}} - Q, \\
\Pi_{5,5} &= -(1-\mu)Q_2 + N_1 + N_1^{\mathrm{T}}, \Pi_{5,7} = N_1 + N_2^{\mathrm{T}}, \\
\Pi_{5,12} &= -N_1 + N_3^{\mathrm{T}}, \Pi_{6,16} = -N_1 + N_3^{\mathrm{T}}, \\
\Pi_{7,7} &= Q_3 + R_3 - W_2 - W_4 - W_4^{\mathrm{T}}, \\
\Pi_{8,8} &= -(1-\mu)Q_3 + N_2 + N_2^{\mathrm{T}}, \Pi_{8,12} = -N_2 + N_3^{\mathrm{T}},
\end{aligned}
$$

$$\Pi_{10,10} = -F - X, \Pi_{11,11} = \gamma I - R, \Pi_{12,12} = -N_3 - N_3^{\mathrm{T}},$$
$$K_1 \quad = diag\{k_1^-, k_2^-, \ldots, k_n^-\}, K_2 = diag\{k_1^+, k_2^+, \ldots, k_n^+\},$$

$$L_1 = diag\{l_1^-, l_2^-, \ldots, l_n^-\}, L_2 = diag\{l_1^+, l_2^+, \ldots, l_n^+\},$$
$$\mathcal{A} = [-C, 0, 0, A, 0, 0, 0, B, 0, 0, I, 0], \quad \ell = \bar{\tau}^2 E + \frac{\bar{\tau}^4}{4} X.$$

Next, we consider another general situation. When $f_i = g_i$, we get another general system:

$$\begin{cases} \dot{x}_i(t) = -c_i(x_i(t - \delta))x_i(t - \delta) \\ \qquad + \sum_{j=1}^n a_{ij}(f_j(x_j(t)) - x_j(t))f_j(x_j(t)) \\ \qquad + \sum_{j=1}^n b_{ij}(f_j(x_j(t - \tau_j(t))) - x_j(t)) \\ \qquad \times f_j(x_j(t - \tau_j(t))) + I_i(t), t \geq 0, i = 1, 2, \ldots, n, \\ x_i(t) = \psi_i(t), t \in [-\tau, 0], i = 1, 2, \ldots, n, \end{cases} \qquad (12.58)$$

Then, the corresponding drive-response system in vector form is as following:

$$\begin{cases} \dot{e}(t) = -Ce(t - \delta) + A\phi(e(t)) + B\phi(e(t - \tau(t))) \\ \qquad + u(t), t \geq 0, \\ \varpi(t) = \omega(t) - \psi(t), t \in [-\tau, 0]. \end{cases} \qquad (12.59)$$

Following the similar procedures discussed in Theorem 2, the following corollaries can be immediately obtained:

Corollary 3: For given matrix S and symmetric matrix Q, R, the system (12.59) is strict $(Q, S, R) - \gamma$-dissipative if there exist symmetric matrices $Q_i = Q_i^T (i = 1, 2)$, $R_i = R_i^T (i = 1, 2)$, $S_i = S_i^T (i = 1, 2)$, the positive matrices $P > 0$, $E > 0$, $F > 0$, $G > 0$, $H > 0$, $X > 0$, $Y > 0$, any matrices Z, the positive diagonal matrices $W_i = diag\{w_{i1}, \ldots, w_{in}\}(i = 1, 2)$, and a scalar $\gamma > 0$ such that the following matrix inequalities hold:

$$\Phi_1 = \begin{bmatrix} \vartheta_{11}^2 & \vartheta_{12}^2 \\ * & \vartheta_{13}^2 \end{bmatrix} > 0, \Phi_2 = \begin{bmatrix} E & -E & 0 \\ * & \vartheta_{21}^2 & \vartheta_{22}^2 \\ * & * & \vartheta_{23}^2 \end{bmatrix} > 0,$$

$$\Phi_3 = \begin{bmatrix} G & -G & 0 \\ * & \vartheta_{31}^2 & \vartheta_{32}^2 \\ * & * & \vartheta_{33}^2 \end{bmatrix} > 0,$$

$$\begin{bmatrix} E & Z \\ * & E \end{bmatrix} > 0, \begin{bmatrix} \Pi & \mathcal{A}^{\mathrm{T}}\ell \\ * & -\ell \end{bmatrix} < 0,$$

Where

$$\Pi = \begin{bmatrix} \Pi_{1,1} & \Pi_{1,2} & Z & G & \Pi_{1,5} & PB & 0 & 0 & \bar{\tau}X & \Pi_{1,10} & P \\ * & \Pi_{22} & \Pi_{2,3} & 0 & 0 & 0 & 0 & 0 & 0 & 0 & 0 \\ * & * & \Pi_{33} & 0 & 0 & 0 & 0 & 0 & 0 & 0 & 0 \\ * & * & * & \Pi_{44} & 0 & 0 & 0 & 0 & 0 & 0 & 0 \\ * & * & * & * & \Pi_{55} & 0 & 0 & 0 & 0 & \Pi_{5,10} & -S \\ * & * & * & * & * & \Pi_{66} & 0 & 0 & 0 & 0 & 0 \\ * & * & * & * & * & * & -R_2 & 0 & 0 & 0 & 0 \\ * & * & * & * & * & * & * & -S_2 & 0 & 0 & 0 \\ * & * & * & * & * & * & * & * & \Pi_{1,1} & 0 & 0 \\ * & * & * & * & * & * & * & * & * & \Pi_{10,10} & -C^T P \\ * & * & * & * & * & * & * & * & * & * & \Pi_{11,11} \end{bmatrix}$$

$$\vartheta_{11}^2 = Q_1 + L_1 W_1 L_2, \vartheta_{12}^2 = -\tfrac{1}{2}(L_1 + L_2)W_1,$$

$$\vartheta_{13}^2 = Q_2 + W_1, \vartheta_{21}^2 = E + R_1 + L_1 W_1 L_2,$$

$$\vartheta_{22}^2 = -\tfrac{1}{2}(L_1 + L_2)W_1, \vartheta_{23}^2 = R_2 + W_1,$$

$$\vartheta_{31}^2 = G + S_1 + L_1 W_1 L_2, \vartheta_{32}^2 = -\tfrac{1}{2}(L_1 + L_2)W_1,$$

$$\vartheta_{33} = S_2 + W_1,$$

$$\Pi_{1,1} = Q_1 + R_1 + S_1 - PC - C^{\mathrm{T}}P - E - G + \bar{\tau}^2 F + \delta^2 H$$
$$\quad -\bar{\tau}^2 X - \delta^2 Y - L_1 W_1 L_2,$$

$$\Pi_{1,2} = E - Z, \Pi_{1,5} = PA + \tfrac{1}{2}(L_1 + L_2)W_1 + L_2^{\mathrm{T}}W_2^{\mathrm{T}},$$

$$\Pi_{1,10} = C^{\mathrm{T}}PC + \delta Y, \Pi_{2,2} = -(1-\mu)Q_1 - 2E + Z + Z^{\mathrm{T}},$$

$$\Pi_{2,3} = E - Z, \Pi_{3,3} = -R_1 - E, \Pi_{4,4} = -S_1 - G,$$

$$\Pi_{5,5} = Q_2 + R_2 + S_2 - W_1 - W_2 - W_2^{\mathrm{T}} - Q,$$

$$\Pi_{5,10} = -A^{\mathrm{T}}PC, \Pi_{6,6} = -(1-\mu)Q_2, \Pi_{6,10} = -B^{\mathrm{T}}PC,$$

$$\Pi_{9,9} = -F - X, \Pi_{10,10} = -H - Y, \Pi_{11,11} = \gamma I - R,$$

$$L_1 = diag\{l_1^-, l_2^-, \ldots, l_n^-\}, L_2 = diag\{l_1^+, l_2^+, \ldots, l_n^+\},$$

$$\mathcal{A} = [0,0,0,-C,A,B,0,0,0,0,I],$$

$$\ell = \bar{\tau}^2 E + \delta^2 G + \tfrac{\bar{\tau}^4}{4}X + \tfrac{\delta^4}{4}Y.$$

More general, when there is no leakage delay and $f_i = g_i$, we obtain a system:

$$\begin{cases} \dot{x}_i(t) = -c_i(x_i(t))x_i(t) \\ \qquad + \sum_{j=1}^n a_{ij}(f_j(x_j(t))-x_j(t))f_j(x_j(t)) \\ \qquad + \sum_{j=1}^n b_{ij}(f_j(x_j(t - \tau_j(t))) - x_j(t)) \\ \qquad \times f_j(x_j(t - \tau_j(t))) + I_i(t), t \geq 0, i = 1, 2, \ldots, n, \\ x_i(t) = \psi_i(t), t \in [-\tau, 0], \quad i = 1, 2, \ldots, n, \end{cases} \qquad (12.60)$$

Its corresponding drive-response system in vector form is as following:

$$\begin{cases} \dot{e}(t) = -Ce(t)+A\phi(e(t))+B\phi(e(t - \tau(t))) \\ \qquad +u(t), t \geq 0, \\ \phi(t) = \omega(t)-\psi(t), t \in [-\tau, 0]. \end{cases} \qquad (12.61)$$

Following the similar procedures discussed in Theorem 2, the following corollaries can be immediately obtained:

Corollary 4: For given matrix S and symmetric matrix Q, R, the system (12.61) is strict $(Q, S, R) - \gamma$-dissipative if there exist symmetric matrices $Q_i = Q_i^T (i = 1, 2)$, $R_i = R_i^T (i = 1, 2)$, the positive matrices $P > 0$, $E > 0$, $X > 0$, any matrices Z, the positive diagonal matrices $W_i = diag\{w_{i1}, \ldots, w_{in}\}(i = 1, 2)$, and a scalar $\gamma > 0$ such that the following matrix inequalities hold:

$$\Phi_1 = \begin{bmatrix} \vartheta_{11}^3 & \vartheta_{12}^3 \\ * & \vartheta_{13}^3 \end{bmatrix} > 0, \Phi_2 = \begin{bmatrix} E & -E & 0 \\ * & \vartheta_{21}^3 & \vartheta_{22}^3 \\ * & * & \vartheta_{23}^3 \end{bmatrix} > 0,$$

$$\begin{bmatrix} E & Z \\ * & E \end{bmatrix} > 0, \begin{bmatrix} \Pi & \mathcal{A}^T\ell \\ * & -\ell \end{bmatrix} < 0,$$

where

$$
\Pi = \begin{bmatrix}
\Pi_{1,1} & \Pi_{1,2} & Z & \Pi_{1,4} & PB & 0 & \bar{\tau}X & P \\
* & \Pi_{2,2} & \Pi_{2,3} & 0 & 0 & 0 & 0 & 0 \\
* & * & \Pi_{3,3} & 0 & 0 & 0 & 0 & 0 \\
* & * & * & \Pi_{4,4} & 0 & 0 & 0 & -S \\
* & * & * & * & \Pi_{5,5} & 0 & 0 & 0 \\
* & * & * & * & * & -R_2 & 0 & 0 \\
* & * & * & * & * & * & \Pi_{7,7} & 0 \\
* & * & * & * & * & * & * & \Pi_{8,8}
\end{bmatrix},
$$

$$
\begin{aligned}
\vartheta_{11}^3 &= Q_1 + L_1 W_1 L_2,\ \vartheta_{12}^3 = -\tfrac{1}{2}(L_1 + L_2)W_1, \\
\vartheta_{13}^3 &= Q_2 + W_1,\ \vartheta_{21}^3 = E + R_1 + L_1 W_1 L_2, \\
\vartheta_{22}^3 &= -\tfrac{1}{2}(L_1 + L_2)W_1,\ \vartheta_{23}^3 = R_2 + W_1, \\
\Pi_{1,1} &= Q_1 + R_1 - PC - C^{\mathrm{T}}P - E + \bar{\tau}^2 F - \bar{\tau}^2 X - L_1 W_1 L_2, \\
\Pi_{1,2} &= E - Z,\ \Pi_{1,4} = PA + \tfrac{1}{2}(L_1 + L_2)W_1 + L_2^{\mathrm{T}}W_2^{\mathrm{T}}, \\
\Pi_{2,2} &= -(1-\mu)Q_1 - 2E + Z + Z^{\mathrm{T}},\ \Pi_{2,3} = E - Z, \\
\Pi_{3,3} &= -R_1 - E,\ \Pi_{4,4} = Q_2 + R_2 - W_1 - W_2 - W_2^{\mathrm{T}} - Q, \\
\Pi_{5,5} &= -(1-\mu)Q_2,\ \Pi_{7,7} = -F - X,\ \Pi_{8,8} = \gamma I - R, \\
L_1 &= diag\{l_1^-, l_2^-, \ldots, l_n^-\},\ L_2 = diag\{l_1^+, l_2^+, \ldots, l_n^+\}, \\
\mathcal{A} &= [0, 0, 0, -C, A, B, 0, I],\ \ell = \bar{\tau}^2 E + \tfrac{\bar{\tau}^4}{4}X.
\end{aligned}
$$

12.4.2 Dynamical Analysis of MWNN

In order to make the system be a WTA network, we must guarantee that there exists the WTA point for system (12.18). Furthermore, we still need to demonstrate that the MWNN (12.18) will eventually settle down to the WTA point. In this section, we make dynamic analysis of the system (12.18). First, some sufficient conditions for the existence of the WTA point are presented. Then, WTA behavior and convergence analysis are studied.

12.4.3 Sufficient Conditions for WTA Point to Exist

Theorem 2: Suppose that the function $h(x, y)$ belongs to the class F. Then, the system (12.18) has a WTA point if and only if

$$I_j \leq h_{max}\left(0, \frac{I_{max}}{(G+(N-1)k)}\right) = h_{max}(0, v_{max}), \quad j \neq max, \quad j = 1, 2, \ldots, N,$$
(12.62)

where I_{max} is the largest input, and $k = min\{c, d\}$ is memductance of the memristor.

Proof 2: Without loss of generality, we assume that $I_1 > I_2 \ldots > I_N > 0$. Here, $I_{max} = I_1, h_1 = h_{max}$.

Sufficiency: Suppose that (12.62) is true, we want to show that system (12.18) has an equilibrium point v such that $v_1 > 0$ and $v_j \leq 0$ ($j \neq 1$). We only need to show that there is a WTA point. In fact, because $h(x, y)$ belongs to the class F, then we have $-Gv_1 + I_1 - (N-1)v_1k = 0$, i.e., $v_1 = I_1/(G+(N-1)k)$, and $-Gv_j + I_j - (N-1)v_jk - h_1(v_j, v_1) = 0$; next, we prove that for any $j \neq 1$, the equation $-Gv_j + I_j - (N-1)v_jk - h_1(v_j, v_1)$ has nonpositive solution. Let $g(x) = -Gx + I_j - (N-1)xk - h_1(x, v_1)$. $g(x)$ is continuous, $g(0) = I_j - h_1(0, v_1) \leq 0$, let $T > 0$, $g(-T) = GT + I_j - h_1(-T, v_1) + T(N-1)$ $k \geq GT - h_1(0, v_1) > 0$ for sufficiently large T. According to intermediate theorem for continuous functions, we obtain that there exists a points $v_j \in [-T, 0]$ such that $g(v_j) = 0$. This proves the existence of a WTA point.

Necessity: We show that if the system (12.18) has a WTA point, then (12.62) must be true. In fact, suppose that (12.62) is not true, then there is $j \neq 1$ satisfying $I_j > h_1(0, v_1)$. Let v be the WTA point, i.e., $v_1 > 0$, and $v_j \leq 0 (j \neq 1)$. We know that $v_1 = I_1/(G + (N-1)k)) > 0$ and $-Gv_j + I_j - h_1(v_j, v_1) - (N-1)v_jk = 0$. So, $0 > Gv_j + (N-1)v_jk = I_j - h_1(v_j, v_1)$, and $I_j < h_1(v_j, v_1)$. Because $h(x, y)$ is monotonic nondecreasing in x, we have $I_j < h_1(v_j, v_1) < h_1(0, v_1)$, which contradicts the supposition that $I_j > h_1(0, v_1)$. Therefore, (12.62) must be true.

Theorem 2 is for the fixed set of inputs. When we do not know which input will be the largest, we can obtain the following result.

Corollary 5: Suppose that the function $h_i(x, y)$ belongs to the class F, then system (12.18) has a WTA point if

$$I_j \leq \min_{1 \leq i \leq N} h_i(0, v_{max}), \quad j \neq max, j = 1, 2, \ldots, N.$$
(12.63)

Proof 3: This is straightforward from Theorem 3.1. So, the proof is omitted.

Specially, when $h_i(x, y) = h(x, y), i \in \{1, 2, \ldots, N\}$, we can derive the following result.

Corollary 6: If the function $h(x, y)$ belongs to class F, then the MWNN (12.18) has a WTA point if and only if

$$I_j \leq h(0, v_{max}), \quad j \neq max, j = 1, 2, \ldots, N.$$

Proof 4: This proof is similar to Theorem 3.1.

As for the MWNN implemented with MOSFET and memristor, we have the following result.

Corollary 7: The MWNN (12.18) with MOSFET function $h_i(x, y)$ has a WTA point if and only if either

$$G \leq min \left\{ \frac{I_{max}}{V_T} - (N-1)k, \frac{2KV_T I_{max}}{I_j + KV_T} - (N-1)k \right\}$$

or

$$\frac{I_{max}}{V_T} - (N-1)k \leq G \leq \frac{\sqrt{K}I_{max}}{\sqrt{I_j}} - (N-1)k$$

for all $j \neq max$.

12.4.4 WTA Behavior and Convergence Analysis

On the basis of sufficient conditions for WTA point to exist, we make the convergence analysis so that the MWNN (12.18) has the WTA behavior.

Theorem 3: If the functions $h_i(x, y)(i = 1, 2, \ldots, N)$ belong to the class F, the trajectory of the MWNN (12.18) is bounded.

Proof 5: We only need to show that the trajectory of system (8) will eventually stay in a bounded set. If $v_i > \frac{I_i}{G+(N-1)k}$, then according to the nonnegativity of $h_i(x, y)$, we have

$$\begin{aligned} C\dot{v}_i &= I_i - Gv_i - \sum_{j \neq i} h(v_i, v_j) - (N-1)v_i k \\ &\leq I_i - Gv_i - (N-1)v_i k < 0. \end{aligned} \quad (12.64)$$

Hence v_i will decrease until $v_i < \frac{I_i}{G+(N-1)k}$. So there exists a positive $U > 0$ such that $v_i \leq U(i = 1, 2, \ldots, N)$. If $v_i < 0$, from the fact that $h_i(x, y)$ is nondecreasing in both x and y, we have

$$
\begin{aligned}
C\dot{v}_i &= I_i - Gv_i - \sum_{j \neq i} h(v_i, v_j) - (N-1)v_i k \\
&\geq I_i - Gv_i - \sum_{j \neq i} h(0,U) - (N-1)v_i k.
\end{aligned}
\tag{12.65}
$$

Therefore, if $v_i < \frac{I_i - \sum_{j \neq i} h(0,U)}{G + (N-1)k}$, then, $\dot{v}_i > 0$ and v_i will increase, so there is a constant B satisfying $v_i \geq B(i = 1, 2, \dots, N)$. Here, we get an upper bound U and a lower bound B. So the trajectory of the MWNN (12.18) is bounded. This completes the proof.

The trajectory of the MWNN (12.18) has the following order-preserving property.

Theorem 4: If the functions $h_i(x, y)(i = 1, 2, \dots, N)$ are continuous functions such that $h_1(x, x) = h_2(x, x) = \dots = h_N(x, x)$ for any x, then MWNN (12.18) is order preserving.

Proof 6: We need to demonstrate that: for any $t > t_0$, $v_i(t) > v_j(t)$ when $I_i > I_j$ and $v_i(t_0) > v_j(t_0)$.

Let $\Delta v(t) = v_i(t) - v_j(t)$, so $\Delta v(t)$ is continuous and $\Delta v(t_0) > 0$. Suppose that the theorem is not true, then there exists a $t^* > t_0$ satisfying $\Delta v(t^*) = 0$ and $\Delta v(t) > 0$ for $t_0 < t^* < t$. At t^*, we have $v_i(t^*) = v_j(t^*)$. Subtracting the jth equation from the ith equation in (12.8), we obtain

$$
\begin{aligned}
C\dot{\Delta v}(t^*) \\
&= - G\Delta v(t^*) + (I_i - I_j) - [h_j(v_i(t^*), v_j(t^*)) - h_i(v_j(t^*), v_i(t^*))] \\
&\quad - \sum_{k \neq i, j} [h_k(v_i(t^*), v_k(t^*)) - h_k(v_j(t^*), v_k(t^*))] - (N-1)k\Delta v(t^*) \\
&= I_i - I_j > 0.
\end{aligned}
\tag{12.66}
$$

Since $\dot{\Delta v}(t)$ is also continuous, from the above inequality, there exists a small $\delta > 0(\delta < t^* - t_0)$ satisfying $\dot{\Delta v}(t) > 0$ in the interval $[t^* - \delta, t^* + \delta]$, i.e., $\Delta v(t)$ is strictly increasing in this interval, so $\Delta v(t^*) > \Delta v(t^* - \delta) > 0$, this contradicts the choice of t^*. So our supposition is not true. This completes the proof.

In fact, if the functions $h_i(x, y)(i = 1, 2, \dots, N)$ are smooth, we can show that the WTA point is exponentially stable.

Theorem 5: Suppose that functions $h_i(x, y)(i = 1, 2, \dots, N)$ belong to the class F and satisfy $h_1(x, x) = h_2(x, x) = \dots = h_N(x, x)$ for any x.

Then, whenever the trajectory of (12.18) starting from the origin enters the WTA region C^+, it will stay there forever. Moreover, if functions $h_i(x, y)(i = 1, 2, \ldots, N)$ satisfy the Lipschitz condition, it will converge exponentially to the WTA point with convergence rate at least γ for any $\gamma < G + (N - 1)k$.

The proof is given in the Appendix.

12.4.5 Conclusion

In this section, the dynamical analysis of MRNN and memristor-based WTA neural network are studied. Moreover, a good theoretical basis for the simulation and application of the next section are provided.

12.5 Application and Simulation

On the basis of dynamic analysis highlighted in the previous section, there exists the exponentially stable WTA point for the MWNN, which can be used in skin classifier, and to demonstrate the main results of proposed MRNN, two examples with simulation results will be adopted.

12.5.1 Simulation of MRNN

In this section, two examples are presented to verify the main results of MRNN.

Example 1: Let us consider two-dimensional MRNN with leakage delay and time-varying delays model as:

$$\dot{x}(t) = -Cx(t - \delta) + Af(x(t)) + Bg(x(t - \tau(t))) + I(t), \quad (12.67)$$

with the following parameter values

$$c_1(x_1(t - \delta)) = \begin{cases} 4.02, & x_1(t-\delta) \leq 0, \\ 3.98, & x_1(t-\delta) > 0, \end{cases}$$

$$c_2(x_2(t - \delta)) = \begin{cases} 4.02, & x_2(t-\delta) \leq 0, \\ 3.98, & x_2(t-\delta) > 0, \end{cases}$$

$$a_{11}(f_1(x_1(t)) - x_1(t)) = \begin{cases} -0.5, & f_1(x_1(t)) - x_1(t) \leq 0, \\ -0.6, & f_1(x_1(t)) - x_1(t) > 0, \end{cases}$$

$$a_{12}(f_2(x_2(t)) - x_2(t)) = \begin{cases} 1.2, & f_2(x_2(t)) - x_2(t) \le 0, \\ 1.1, & f_2(x_2(t)) - x_2(t) > 0, \end{cases}$$

$$a_{21}(f_1(x_1(t)) - x_1(t)) = \begin{cases} 2.4, & f_1(x_1(t)) - x_1(t) \le 0, \\ 2.5, & f_1(x_1(t)) - x_1(t) > 0, \end{cases}$$

$$a_{22}(f_2(x_2(t)) - x_2(t)) = \begin{cases} 2.5, & f_2(x_2(t)) - x_2(t) \le 0, \\ 2.4, & f_2(x_2(t)) - x_2(t) > 0, \end{cases}$$

$$b_{11}(g_1(x_1(t - \tau_1(t))) - x_1(t)) = \begin{cases} 1.4, & g_1(x_1(t - \tau_1(t))) - x_1(t) \le 0, \\ 1.3, & g_1(x_1(t - \tau_1(t))) - x_1(t) > 0, \end{cases}$$

$$b_{12}(g_2(x_2(t - \tau_2(t))) - x_2(t)) = \begin{cases} -1, & g_2(x_2(t - \tau_2(t))) - x_2(t) \le 0, \\ -1.1, & g_2(x_2(t - \tau_2(t))) - x_2(t) > 0, \end{cases}$$

$$b_{21}(g_1(x_1(t - \tau_1(t))) - x_1(t)) = \begin{cases} 2.5, & g_1(x_1(t - \tau_1(t))) - x_1(t) \le 0, \\ 2.4, & g_1(x_1(t - \tau_1(t))) - x_1(t) > 0, \end{cases}$$

$$b_{22}(g_2(x_2(t - \tau_2(t))) - x_2(t)) = \begin{cases} -0.1, & g_2(x_2(t - \tau_2(t))) - x_2(t) \le 0, \\ -0.2, & g_2(x_2(t - \tau_2(t))) - x_2(t) > 0, \end{cases}$$

Then, we study the corresponding drive-response system based on (12.67). We choose

$$C = \begin{bmatrix} 4 & 0 \\ 0 & 4 \end{bmatrix}, \quad A = \begin{bmatrix} -0.5 & 1.2 \\ 2.5 & 2.5 \end{bmatrix}, \quad B = \begin{bmatrix} 1.4 & -1 \\ 2.5 & -0.1 \end{bmatrix},$$

$$u = \begin{bmatrix} 0.01 \\ -0.04 \end{bmatrix}, \quad Q = \begin{bmatrix} 5 & 2.2 \\ 2.2 & 3 \end{bmatrix}, \quad R = \begin{bmatrix} 4.5 & -0.5 \\ -0.5 & 2.5 \end{bmatrix},$$

$$S = \begin{bmatrix} 0.3 & -0.6 \\ 0.4 & 0.5 \end{bmatrix}.$$

Thus, it is easy to check that Assumptions 1, 2, 3 are satisfied. To prove the results are very relaxed, we choose $\delta = 0.2$, $\bar{\tau} = 2$, $\mu = 0.2$, the activation function $f(x) = g(x) = \tanh x$. Now we find the solution to the LMI in the Theorem 2. The feasible solutions are

$$Q_1 = \begin{bmatrix} 0.0253 & -0.0120 \\ -0.0120 & 0.0136 \end{bmatrix}, \quad Q_2 = \begin{bmatrix} 0.03503 & 0.1175 \\ 0.1175 & 0.1429 \end{bmatrix},$$

$$Q_3 = \begin{bmatrix} 0.8663 & -0.2380 \\ -0.2380 & 0.3327 \end{bmatrix}, \quad S_1 = \begin{bmatrix} 0.2242 & -0.0871 \\ -0.0871 & 0.0996 \end{bmatrix},$$

$$S_2 = \begin{bmatrix} 0.2895 & 0.0856 \\ 0.0856 & 0.1176 \end{bmatrix}, \qquad S_3 = \begin{bmatrix} 0.0065 & 0.0000 \\ 0.0000 & 0.0041 \end{bmatrix},$$

$$R_1 = \begin{bmatrix} 0.0351 & -0.0117 \\ -0.0117 & 0.0183 \end{bmatrix}, \qquad R_2 = \begin{bmatrix} 0.3635 & 0.1248 \\ 0.1248 & 0.1481 \end{bmatrix},$$

$$R_3 = \begin{bmatrix} 0.0082 & 0.0012 \\ 0.0012 & 0.0053 \end{bmatrix}, \qquad P = \begin{bmatrix} 0.9150 & -0.2607 \\ -0.2607 & 0.3815 \end{bmatrix},$$

$$E = \begin{bmatrix} 0.0004 & -0.0002 \\ -0.0002 & 0.0002 \end{bmatrix}, \qquad F = \begin{bmatrix} 0.0056 & -0.0030 \\ -0.0030 & 0.0030 \end{bmatrix},$$

$$G = \begin{bmatrix} 0.1257 & -0.0498 \\ -0.0498 & 0.0610 \end{bmatrix}, \qquad H = \begin{bmatrix} 67.7842 & -18.7619 \\ -18.7619 & 25.3050 \end{bmatrix},$$

$$X = \begin{bmatrix} 0.0001 & -0.0001 \\ -0.0001 & 0.0001 \end{bmatrix}, \qquad Y = \begin{bmatrix} 15.2297 & -4.9328 \\ -4.9328 & 6.1341 \end{bmatrix},$$

$$Z = \begin{bmatrix} -0.000006 & 0.000004 \\ 0.000004 & -0.000004 \end{bmatrix}, \qquad N_1 = \begin{bmatrix} 347910 & -1780 \\ 1780 & -347910 \end{bmatrix},$$

$$N_2 = \begin{bmatrix} 347910 & -1780 \\ 1780 & -347910 \end{bmatrix}, \qquad N_3 = \begin{bmatrix} 347910 & -1780 \\ 1780 & -347910 \end{bmatrix},$$

$$W_1 = \begin{bmatrix} 0.0790 & 0 \\ 0 & 0.0379 \end{bmatrix}, \qquad W_2 = \begin{bmatrix} 0.0610 & 0 \\ 0 & 0.0312 \end{bmatrix},$$

$$W_3 = \begin{bmatrix} 0.2506 & 0 \\ 0 & 0.0986 \end{bmatrix}, \qquad W_4 = \begin{bmatrix} 0.8575 & 0 \\ 0 & 0.3313 \end{bmatrix},$$

and $\gamma = 0.3700$. Therefore, the corresponding drive-response system based on (12.67) is $(Q, S, R) - \gamma$-dissipative. Figure 12.3 represents the state response of all the trajectories of this error system when $\tau = 0.5$ with the initial conditions $e_1 = 0.5$, $e_2 = -0.15$. Even $\bar{\tau} > 1$, we still obtain the feasible solutions. Figure 12.3 represents the state response of all the trajectories when $\tau = 1.5$.

Example 2: Let us consider another two-dimensional MENN with leakage delay and time-varying delays model as:

$$\dot{x}(t) = -Cx(t - \delta) + Af(x(t)) + Bg(x(t - \tau(t))) + I(t), \qquad (12.68)$$

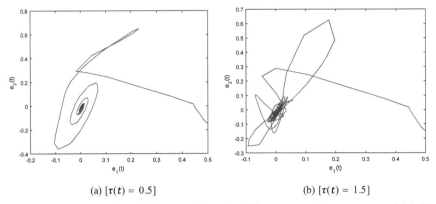

(a) $[\tau(t) = 0.5]$ (b) $[\tau(t) = 1.5]$

Figure 12.3   The behavior of all state trajectories of the correspond error system of (12.67).

with the following parameter values

$$c_1(x_1(t - \delta)) = \begin{cases} 5.02, & x_1(t-\delta) \le 0, \\ 4.98, & x_1(t-\delta) > 0, \end{cases}$$

$$c_2(x_2(t - \delta)) = \begin{cases} 6.02, & x_2(t-\delta) \le 0, \\ 5.98, & x_2(t-\delta) > 0, \end{cases}$$

$$a_{11}(f_1(x_1(t)) - x_1(t)) = \begin{cases} 4.8, & f_1(x_1(t)) - x_1(t) \le 0, \\ 4.7, & f_1(x_1(t)) - x_1(t) > 0, \end{cases}$$

$$a_{12}(f_2(x_2(t)) - x_2(t)) = \begin{cases} 0.4, & f_2(x_2(t)) - x_2(t) \le 0, \\ 0.5, & f_2(x_2(t)) - x_2(t) > 0, \end{cases}$$

$$a_{21}(f_1(x_1(t)) - x_1(t)) = \begin{cases} -4.1, & f_1(x_1(t)) - x_1(t) \le 0, \\ -4, & f_1(x_1(t)) - x_1(t) > 0, \end{cases}$$

$$a_{22}(f_2(x_2(t)) - x_2(t)) = \begin{cases} 3.9, & f_2(x_2(t)) - x_2(t) \le 0, \\ 3.8, & f_2(x_2(t)) - x_2(t) > 0, \end{cases}$$

$$b_{11}(g_1(x_1(t - \tau_1(t))) - x_1(t)) = \begin{cases} -2.6, & g_1(x_1(t-\tau_1(t))) - x_1(t) \le 0, \\ -2.5, & g_1(x_1(t-\tau_1(t))) - x_1(t) > 0, \end{cases}$$

$$b_{12}(g_2(x_2(t - \tau_2(t))) - x_2(t)) = \begin{cases} -1.2, & g_2(x_2(t-\tau_2(t))) - x_2(t) \le 0, \\ -1.3, & g_2(x_2(t-\tau_2(t))) - x_2(t) > 0, \end{cases}$$

$$b_{21}(g_1(x_1(t - \tau_1(t))) - x_1(t)) = \begin{cases} -1.2, & g_1(x_1(t-\tau_1(t))) - x_1(t) \le 0, \\ -1.3, & g_1(x_1(t-\tau_1(t))) - x_1(t) > 0, \end{cases}$$

$$b_{22}(g_2(x_2(t - \tau_2(t))) - x_2(t)) = \begin{cases} -2.9, & g_2(x_2(t-\tau_2(t))) - x_2(t) \le 0, \\ -2.8, & g_2(x_2(t-\tau_2(t))) - x_2(t) > 0, \end{cases}$$

Then, we study the corresponding drive-response system based on (12.68). We choose

$$C = \begin{bmatrix} 5 & 0 \\ 0 & 6 \end{bmatrix}, \quad A = \begin{bmatrix} 4.8 & 0.5 \\ -4 & 3.9 \end{bmatrix}, \quad B = \begin{bmatrix} -2.5 & -1.2 \\ -1.2 & -2.8 \end{bmatrix},$$

$$u = \begin{bmatrix} -0.03 \\ 0.05 \end{bmatrix}, \quad Q = \begin{bmatrix} 5 & 2.2 \\ 2.2 & 3 \end{bmatrix}, \quad R = \begin{bmatrix} 4.5 & -0.5 \\ -0.5 & 2.5 \end{bmatrix},$$

$$S = \begin{bmatrix} 0.3 & -0.6 \\ 0.4 & 0.5 \end{bmatrix}.$$

Thus, it is easy to check that Assumptions 1, 3, 2 are satisfied. To prove the results are very relaxed, we choose $\delta = 0.1$, $\overline{\tau} = 2$, $\mu = 0.2$, and the activation function $f(x) = g(x) = \tanh x$. Now we find the solution to the LMI in the Theorem 1. The feasible solutions are

$$Q_1 = \begin{bmatrix} 0.0605 & -0.0129 \\ -0.0129 & 0.0209 \end{bmatrix}, \quad Q_2 = \begin{bmatrix} 0.3651 & 0.2498 \\ 0.2498 & 0.2244 \end{bmatrix},$$

$$Q_3 = \begin{bmatrix} 0.8356 & 0.2186 \\ 0.2186 & 0.5467 \end{bmatrix}, \quad S_1 = \begin{bmatrix} 0.2113 & -0.0025 \\ -0.0025 & 0.1553 \end{bmatrix},$$

$$S_2 = \begin{bmatrix} 0.3277 & 0.2250 \\ 0.2250 & 0.2009 \end{bmatrix}, \quad S_3 = \begin{bmatrix} 0.0391 & -0.0147 \\ -0.0147 & 0.0136 \end{bmatrix},$$

$$R_1 = \begin{bmatrix} 0.0862 & -0.0091 \\ -0.0091 & 0.0332 \end{bmatrix}, \quad R_2 = \begin{bmatrix} 0.3759 & 0.2580 \\ 0.2580 & 0.2315 \end{bmatrix},$$

$$R_3 = \begin{bmatrix} 0.0470 & -0.0182 \\ -0.0182 & 0.0166 \end{bmatrix}, \quad P = \begin{bmatrix} 0.4421 & 0.0272 \\ 0.0272 & 0.2257 \end{bmatrix},$$

$$E = \begin{bmatrix} 0.00086 & -0.00009 \\ -0.00009 & 0.00017 \end{bmatrix}, \quad F = \begin{bmatrix} 0.0102 & -0.0017 \\ -0.0017 & 0.0038 \end{bmatrix},$$

$$G = \begin{bmatrix} 1.1829 & 0.0160 \\ 0.0160 & 0.3051 \end{bmatrix}, \quad H = \begin{bmatrix} 31.1825 & 2.2005 \\ 2.2005 & 39.8134 \end{bmatrix},$$

$$X = \begin{bmatrix} 0.00026 & -0.00002 \\ -0.00002 & 0.00008 \end{bmatrix}, \quad Y = \begin{bmatrix} 266.6554 & 13.0960 \\ 13.0960 & 98.9058 \end{bmatrix},$$

$$Z = \begin{bmatrix} -0.00001 & 0.000001 \\ 0.000001 & -0.000001 \end{bmatrix}, \quad N_1 = \begin{bmatrix} -1661.3 & 671.5 \\ -671.4 & -1661.3 \end{bmatrix},$$

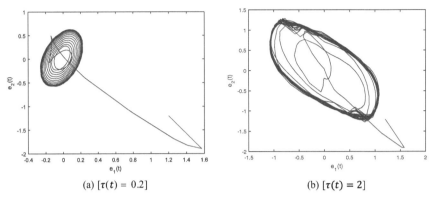

(a) [$\tau(t) = 0.2$]   (b) [$\tau(t) = 2$]

Figure 12.4   The behavior of all state trajectories of the correspond error system of (12.68).

$$N_2 = \begin{bmatrix} -1661.4 & 671.5 \\ -671.5 & -1661.4 \end{bmatrix}, \quad N_3 = \begin{bmatrix} 1661.4 & -671.5 \\ 671.5 & 1661.4 \end{bmatrix},$$

$$W_1 = \begin{bmatrix} 0.1574 & 0 \\ 0 & 0.0641 \end{bmatrix}, \quad W_2 = \begin{bmatrix} 0.1920 & 0 \\ 0 & 0.0817 \end{bmatrix},$$

$$W_3 = \begin{bmatrix} 0.1320 & 0 \\ 0 & 0.0468 \end{bmatrix}, \quad W_4 = \begin{bmatrix} 1.0405 & 0 \\ 0 & 0.6108 \end{bmatrix},$$

and $\gamma = 0.4154$. Therefore, the corresponding drive-response system based on (12.68) is $(Q, S, R) - \gamma$-dissipative. Figure 12.4 represents the state response of all the trajectories of this error system when $\tau = 0.2$ with the initial conditions $e_1 = 1.2$, $e_2 = -1.2$. Even $\overline{\tau} > 1$, we still obtain the feasible solutions. Figure 12.4 represents the state response of all the trajectories when $\tau = 2$.

Remark 9: Let $\Pi + \mathcal{A}^T \ell \mathcal{A} = 0$, and we can figure out the upper bound of τ.

12.5.2 Illustrative Examples of the BP-MWNN Classifier System

In this section, illustrative examples are presented to demonstrate our results obtained and further confirm the consistency between the theoretical results and the numerical results. We use Matlab to develop the BP-MWNN classifier system. As for the erythemato-squamous disease, 33 clinical and histopatho-logical features are input into the system. In the input vector, the family history has the value 1 if an erythemato-squamous disease has been observed in the family (otherwise, it is 0), and all the other features have an intensity degree between 0 (it is not present) and 3 (it is acute), moreover the values 1 and 2 are intermediate intensities. Six outputs obtained from the BP neural network are input into the MWNN.

The outputs with WTA behavior and classification result are presented in Table 12.3. The confusion matrixes for training, test and validation are shown in Figure 12.5. Observed from these matrixes, there are large number of correct answers and small number of wrong answers which means the outputs of the networks are accurate. Compared to [37], the precisions of the training, test and validation respectively increase 5.8%, 3.7% and 1.8%, and the system precision is 98.6% with increasement rate of 4.9%, which verifies the feasibility and efficiency of our classifier system.

Table 12.3 BP-MWNN classifier

	BP Neural Network		MWNN	
Record	Input Vector	Output Vector	Output with WTA Behavior	Classification Result
1	{2,3,1,2,1,0, 0,0,0,0,0,0, 0,2,0,0,1,0, 0,2,1,2,2,0, 0,0,0,0,0,0, 0,2,0}	{1.03082150, −0.06557043 0.00863432 0.01264964 0.06634284 −0.03107488}		psoriasis
2	{3,2,1,3,0,0, 0,0,0,0,1,0, 1,2,0,3,2,0, 1,0,1,0,0,0, 0,0,0,3,0,0, 0,1,0}	{−0.05508221, 0.91115232, 0.01084680, 0.10613479 0.00340466 0.09781484}		seborrheic dermatitis

(*Continued*)

Table 12.3 Continued

	BP Neural Network		MWNN	
Record	Input Vector	Output Vector	Output with WTA Behavior	Classification Result
3	{1,1,1,2,2,2, 0,2,0,0,0,2, 0,0,0,2,2,0, 1,0,0,0,0,0, 2,0,2,0,2,0, 0,3,3}	{0.04603663, −0.06130094, 0.98628654, 0.01201888, 0.05594145, −0.00411883}		lichen planus
4	{2,2,2,1,0, 0,0,0,0,0, 0,0,1,0,0, 0,2,0,0,0, 0,0,0,0,0, 1,0,2,0,0, 0,2,0}	{−0.13204039, 0.29050715, −0.06002621, 0.65121912, 0.08087675, 0.14979199}		pityriasis rosea
5	{3,2,2,0,0,0, 0,0,0,0,0,0, 0,0,1,0,2,0, 0,0,1,0,0,0, 0,0,0,0,0,0, 0,2,0,}	{−0.02883893, −0.15524329, −0.04431461, 0.16526339, 0.95013090, 0.13961385 }		chronic dermatitis
6	{2,2,1,0,0,0, 2,0,2,0,0,0, 0,0,0,3,2,0, 1,0,0,0,0,0, 0,0,0,3,0,2, 2,2,0}	{−0.09570358, 0.18649829, 0.00099481, −0.04136459, 0.03260639, 1.04644850 }		pityriasis rubra pilaris

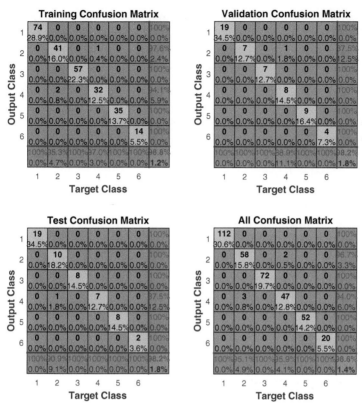

Figure 12.5 The confusion matrixes for training, test and validation.

12.5.3 Conclusion

In this section, the application and simulation results are presented to demonstrate the feasibility and applicability of memritive WTA neural network and recurrent neural network which are two typical types of neural network based on memristor.

References

[1] Shyam Prasad Adhikari, Hyongsuk Kim, Ram Kaji Budhathoki, Changju Yang, and Leon O Chua. A circuit-based learning architecture for multilayer neural networks with memristor bridge synapses. *IEEE Transactions on Circuits and Systems I: Regular Papers*, 62(1):215–223, 2015.

[2] SM Shafiul Alam and Mohammed Imamul Hassan Bhuiyan. Detection of seizure and epilepsy using higher order statistics in the emd domain. *IEEE journal of biomedical and health informatics*, 17(2):312–318, 2013.

[3] Filippo Amato et al., Artificial neural networks in medical diagnosis. *Journal of Applied Biomedicine*, 11(2):47–58, 2013.

[4] Mohammad Bavandpour, Saeed Bagheri-Shouraki, Hamid Soleimani, Arash Ahmadi, and Bernabé Linares-Barranco. Spiking neuro-fuzzy clustering system and its memristor crossbar based implementation. *Microelectronics Journal*, 45(11):1450–1462, 2014.

[5] Zuowei Cai, Lihong Huang, and Lingling Zhang. New conditions on synchronization of memristor-based neural networks via differential inclusions. *Neurocomputing*, 186:235–250, 2016.

[6] A Chandrasekar, R Rakkiyappan, Jinde Cao, and Shanmugam Lakshmanan. Synchronization of memristor-based recurrent neural networks with two delay components based on second-order reciprocally convex approach. *Neural Networks*, 57:79–93, 2014.

[7] A Chandrasekar and R Rakkiyappan. Impulsive controller design for exponential synchronization of delayed stochastic memristor-based recurrent neural networks. *Neurocomputing*, 173:1348–1355, 2016.

[8] Leon Chua. Memristor-the missing circuit element. *IEEE Transactions on circuit theory*, 18(5):507–519, 1971.

[9] Shukai Duan, Xiaofang Hu, Zhekang Dong, Lidan Wang, and Pinaki Mazumder. Memristor-based cellular nonlinear/neural network: design, analysis, and applications. *IEEE transactions on neural networks and learning systems*, 26(6):1202–1213, 2015.

[10] Idongesit E, Ebong and Pinaki Mazumder. Cmos and memristor-based neural network design for position detection. *Proceedings of the IEEE*, 100(6):2050–2060, 2012.

[11] Yuguang Fang, Michael A Cohen, and Thomas G Kincaid. Dynamics of a winner-take-all neural network. *Neural Networks*, 9(7):1141–1154, 1996.

[12] Jerome A Feldman and Dana H Ballard. Connectionist models and their properties. *Cognitive science*, 6(3):205–254, 1982.

[13] Delia-Maria Filimon and Adriana Albu. Skin diseases diagnosis using artificial neural networks. In *SACI*, pages 189–194, 2014.

[14] Aleksei Fedorovich Filippov. *Differential equations with discontinuous righthand sides: control systems*, volume 18. Springer Science & Business Media, 2013.

[15] Zhenyuan Guo, Jun Wang, and Zheng Yan. Global exponential dissipativity and stabilization of memristor-based recurrent neural networks with time-varying delays. *Neural Networks*, 48:158–172, 2013.

[16] Zhenyuan Guo, Jun Wang, and Zheng Yan. Passivity and passification of memristor-based recurrent neural networks with time-varying delays. *Neural Networks and Learning Systems, IEEE Transactions on*, 25(11):2099–2109, 2014.

[17] Zhishan Guo and Jun Wang. Information retrieval from large data sets via multiple-winners-take-all. In *2011 IEEE International Symposium of Circuits and Systems (ISCAS)*, 2669–2672, 2011.

[18] Morris W Hirsch. Convergent activation dynamics in continuous time networks. *Neural Networks*, 2(5):331–349, 1989.

[19] Nan Hou, Hongli Dong, Zidong Wang, Weijian Ren, and Fuad E Alsaadi. Non-fragile state estimation for discrete markovian jumping neural networks. *Neurocomputing*, 179:238–245, 2016.

[20] Jin Hu and Jun Wang. Global uniform asymptotic stability of memristor-based recurrent neural networks with time delays. In *Proceedings of 2010 International Joint Conference on Neural Networks (IJCNN)*, 1–8, 2010.

[21] Jin Hu and Jun Wang. Global uniform asymptotic stability of memristor-based recurrent neural networks with time delays. In *Proceedings of The 2010 International Joint Conference on Neural Networks (IJCNN)*, 1–8. IEEE, 2010.

[22] Xiaofang Hu, Gang Feng, Shukai Duan, and Lu Liu. Multilayer rtd-memristor-based cellular neural networks for color image processing. *Neurocomputing*, 162:150–162, 2015.

[23] Makoto Itoh and Leon O Chua. Memristor oscillators. *International Journal of Bifurcation and Chaos*, 18(11):3183–3206, 2008.

[24] L Lenhardt, I Zekovic′, T Dramic′anin, and MD Dramic′anin. Artificial neural networks for processing fluorescence spectroscopy data in skin cancer diagnostics. *Physica Scripta*, 2013(T157):014057, 2013.

[25] Richard Lippmann. An introduction to computing with neural nets. *IEEE Assp magazine*, 4(2):4–22, 1987.

[26] Hongjian Liu, Zidong Wang, Bo Shen, and Xiu Kan. Synchronization for discrete-time memristive recurrent neural networks with time-delays. In *Proceedings of 2015 Chinese Control Conference (CCC) on IEEE*, 3478–3483, 2015.

[27] Yurong Liu, Zidong Wang, and Xiaohui Liu. On global exponential stability of generalized stochastic neural networks with mixed time-delays. *Neurocomputing*, 70(1):314–326, 2006.

[28] Ning Li and Jinde Cao. New synchronization criteria for memristor-based networks: Adaptive control and feedback control schemes. *Neural Networks*, 61:1–9, 2015.

[29] Shuai Li, Yangming Li, and Zheng Wang. A class of finite-time dual neural networks for solving quadratic programming problems and its k-winners-take-all application. *Neural Networks*, 39:27–39, 2013.

[30] Xiaodi Li, R Rakkiyappan, and G Velmurugan. Dissipativity analysis of memristor-based complex-valued neural networks with time-varying delays. *Information Sciences*, 294:645–665, 2015.

[31] J. Lu and D. W. C. Ho. Globally exponential synchronization and synchronizability for general dynamical networks. *IEEE Transactions on Systems, Man, and Cybernetics, Part B (Cybernetics)*, 40(2):350–361, April 2010.

[32] K Mala, V Sadasivam, and S Alagappan. Neural network based texture analysis of ct images for fatty and cirrhosis liver classification. *Applied Soft Computing*, 32:80–86, 2015.

[33] Pinaki Mazumder, Sung-Mo Kang, and Rainer Waser. Memristors: devices, models, and applications. *Proceedings of the IEEE*, 100(6):1911–1919, 2012.

[34] Victor Emil Neagoe and Armand Dragos Ropot. Concurrent self-organizing maps for pattern classification. In *IEEE International Conference on Cognitive Informatics, 2002. Proceedings*, 304–312, 2002.

[35] Ben Niu, Hamid Reza Karimi, Huanqing Wang, and Yanli Liu. Adaptive output-feedback controller design for switched nonlinear stochastic systems with a modified average dwell-time method. *IEEE Transactions on Systems, Man, and Cybernetics: Systems*, 2017.

[36] Ben Niu and Lu Li. Adaptive backstepping-based neural tracking control for mimo nonlinear switched systems subject to input delays. *IEEE Transactions on Neural Networks and Learning Systems*, 2017.

[37] Ben Niu, Xudong Zhao, Lixian Zhang, and Hongyi Li. p-times differentiable unbounded functions for robust control of uncertain switched nonlinear systems with tracking constraints. *International Journal of Robust and Nonlinear Control*, 25(16):2965–2983, 2015.

[38] Thirunavukkarasu Radhika and Gnaneswaran Nagamani. Dissipativity analysis of stochastic memristor-based recurrent neural networks with

discrete and distributed time-varying delays. *Network: Computation in Neural Systems*, 27(4):237–267, 2016.

[39] Rajan Rakkiyappan, Arunachalam Chandrasekar, and Jinde Cao. Passivity and passification of memristor-based recurrent neural networks with additive time-varying delays. *Neural Networks and Learning Systems, IEEE Transactions on*, 26(9):2043–2057, 2015.

[40] R Rakkiyappan, A Chandrasekar, S Laksmanan, and Ju H Park. State estimation of memristor-based recurrent neural networks with time-varying delays based on passivity theory. *Complexity*, 19(4):32–43, 2014.

[41] R Rakkiyappan, G Velmurugan, Xiaodi Li, and Donal O'Regan. Global dissipativity of memristor-based complex-valued neural networks with time-varying delays. *Neural Computing and Applications*, 27(3): 629–649, 2016.

[42] Davide Sacchetto, Giovanni De Micheli, and Yusuf Leblebici. Multiterminal memristive nanowire devices for logic and memory applications: a review. *Proceedings of the IEEE*, 100(6):2008–2020, 2012.

[43] Li Shang, De-Shuang Huang, Ji-Xiang Du, and Zhi-Kai Huang. Palmprint recognition using ICA based on winner-take-all network and radial basis probabilistic neural network. In *International Symposium on Neural Networks*, pages 216–221. Springer, 2006.

[44] Patrick M Sheridan, Chao Du, and Wei D Lu. Feature extraction using memristor networks. *IEEE transactions on neural networks and learning systems*, 27(11):2327–2336, 2016.

[45] Qiankun Song, Jinling Liang, and Zidong Wang. Passivity analysis of discrete-time stochastic neural networks with time-varying delays. *Neurocomputing*, 72(7):1782–1788, 2009.

[46] Dmitri B Strukov, Gregory S Snider, Duncan R Stewart, and R Stanley Williams. The missing memristor found. *Nature*, 453(7191):80–83, 2008.

[47] M Vemis, G Economou, S Fotopoulos, and A Khodyrev. The use of boolean functions and local operations for edge detection in images. *Signal Processing*, 45(2):161–172, 1995.

[48] Mathukumalli Vidyasagar. *Nonlinear systems analysis*, volume 42. Siam, 2002.

[49] N Abdel Wahab, M Abdel Wahed, and Abdallah SA Mohamed. Texture features neural classifier of some skin diseases. In *Circuits and Systems, 2003 IEEE 46th Midwest Symposium on*, volume 1, 380–382, 2003.

[50] Shiping Wen, Gang Bao, Zhigang Zeng, Yiran Chen, and Tingwen Huang. Global exponential synchronization of memristor-based recurrent neural networks with time-varying delays. *Neural Networks*, 48:195–203, 2013.

[51] Shiping Wen, Zhigang Zeng, Tingwen Huang, and Yiran Chen. Passivity analysis of memristor-based recurrent neural networks with time-varying delays. *Journal of the Franklin Institute*, 350(8):2354–2370, 2013.

[52] Shiping Wen, Zhigang Zeng, and Tingwen Huang. Exponential stability analysis of memristor-based recurrent neural networks with time-varying delays. *Neurocomputing*, 97:233–240, 2012.

[53] R Stanley Williams. How we found the missing memristor. *IEEE spectrum*, 45(9):28–35, 2008.

[54] Ailong Wu, Shiping Wen, and Zhigang Zeng. Synchronization control of a class of memristor-based recurrent neural networks. *Information Sciences*, 183(1):106–116, 2012.

[55] Ailong Wu, Zhigang Zeng, and Jiejie Chen. Analysis and design of winner-take-all behavior based on a novel memristive neural network. *Neural Computing and Applications*, 24(7–8):1595–1600, 2014.

[56] Ailong Wu and Zhigang Zeng. Exponential stabilization of memristive neural networks with time delays. *Neural Networks and Learning Systems, IEEE Transactions on*, 23(9):1919–1929, 2012.

[57] Jianying Xiao, Shouming Zhong, and Yongtao Li. Relaxed dissipativity criteria for memristive neural networks with leakage and time-varying delays. *Neurocomputing*, 171:708–718, 2016.

[58] Jar-Ferr Yang, Chi-Ming Chen, Wen-Chung Wang, and Jau-Yien Lee. A general mean-based iterative winner-take-all neural network. *IEEE Transactions on Neural Networks*, 6(1):14–24, 1995.

[59] Yu Chao Yang, Feng Pan, Qi Liu, Ming Liu, and Fei Zeng. Fully room-temperature-fabricated nonvolatile resistive memory for ultrafast and high-density memory application. *Nano letters*, 9(4):1636–1643, 2009.

[60] Jui-Cheng Yen, Fu-Juay Chang, and Shyang Chang. A new winners-take-all architecture in artificial neural networks. *IEEE Transactions on Neural Networks*, 5(5):838–843, 1994.

[61] AL Yuille and D Geiger. Winner-take-all networks. *The handbook of brain theory and neural networks*, 1228–1231, 2003.

[62] Shimeng Yu. Orientation classification by a winner-take-all network with oxide rram based synaptic devices. In *2014 IEEE International Symposium on Circuits and Systems (ISCAS)*, 1058–1061, 2014.

[63] Guodong Zhang, Yi Shen, and Junwei Sun. Global exponential stability of a class of memristor-based recurrent neural networks with time-varying delays. *Neurocomputing*, 97:149–154, 2012.

[64] Guodong Zhang, Yi Shen, Quan Yin, and Junwei Sun. Global exponential periodicity and stability of a class of memristor-based recurrent neural networks with multiple delays. *Information Sciences*, 232:386–396, 2013.

[65] Guodong Zhang, Yi Shen, Quan Yin, and Junwei Sun. Passivity analysis for memristor-based recurrent neural networks with discrete and distributed delays. *Neural Networks*, 61:49–58, 2015.

[66] Cheng-De Zheng and Yongjin Xian. On synchronization for chaotic memristor-based neural networks with time-varying delays. *Neurocomputing*, 216:570–586, 2016.

Appendix

Idongesit Ebong and Pinaki Mazumder

Memristors

Chua's Modeling of Memristors and Memristive Systems

The name memristor is a portmanteau created by joining the words "memory" and "resistor". The element was introduced in Chua's seminal paper [1] due to the absence of an element that embodied a relationship between flux and charge. Chua noticed the four circuit variables—charge (q), voltage (V), flux (φ), and current (i)—constituted relationships between three circuit elements and hypothesized a fourth element, which he named memristor. Figure A.1 is a reconstruction of a diagram that relates all four circuit variables with the four circuit elements.

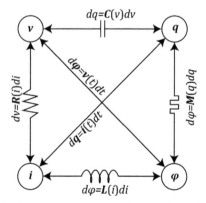

Figure A.1 Relationship between all four circuit variables and their constitutive circuit elements.

351

The four fundamental circuit elements (capacitor, inductor, resistor, and memristor) are shown in Figure A.1, and are thus named because they cannot be defined as a network of other circuit elements. The relationships are summarized in (A.1a) to (A.1d), where (A.1d) is the constitutive relationship of the memristor relating charge and flux.

$$dq = C(v)dv \tag{A.1a}$$
$$dv = R(i)di \tag{A.1b}$$
$$d\phi = L(i)di \tag{A.1c}$$
$$d\phi = M(q)dq \tag{A.1d}$$

From (A.1d), the term "memristance" (M) determines the relationship between q and ϕ. The relationship defined in (A.1d) when divided by dt yields

$$v(t) = M(q(t))i(t) \tag{A.2}$$

This equation is similar to the definition of a resistor in accordance to Ohm's Law, but instead of R we have an M. The linear resistor is a special case of (A.2) when M is a constant term. But when M is not a constant, then M behaves like a variable resistor that remembers its previous state based upon the amount of charge that has flowed through the device.

Description of (A.2) presents a charge-controlled memristor, but the converse view, a flux-controlled memristor, may be adopted as shown in (A.3).

$$dq = W(\phi(t))d\phi \tag{A.3}$$

From the perspective of (A.1d), M is dubbed the *incremental memristance* while from the perspective of (A.3), W is dubbed *incremental menduc-tance*. Along with these constitutive relationships comes some properties of memristors associated with circuit theory. In circuit theory, the fundamental elements are all passive elements, so in order to ensure memristor passivity, the passivity criterion states: "a memristor characterized by a differentiable charge-controlled curve is passive if and only if its incremental memristance is non-negative."

Although thorough, the inchoate charge-flux memristor is only a special case of a general class of dynamical systems. Chua and Kang's [2] memristor idea culminated in a general class of systems called memristive systems defined in state-space representation form:

$$\begin{cases} \dot{x} = f(x, u, t) \\ y = g(x, u, t)u \end{cases}$$

This state space representation of the system defines x as the state of the system, and u and y as inputs and outputs of the system, respectively. The function f is a continuous n-dimensional vector function, and g is a continuous scalar function. The special nature of the structure of y to u distinguishes memristive systems from other dynamic systems because whenever u is 0, y is 0, regardless of the value of x. By incorporating a memristor into this form, the result is (A.4).

$$\begin{cases} \dot{w} = f(w, i, t) \\ v = R(w, i, t)i \end{cases} \tag{A.4}$$

In (A.4), w, v, and i denote an n-dimensional state variable, port voltage, and current, respectively. This representation of the memristor has current as an input and voltage as the output, hence is a current-controlled memristor. The voltage controlled counterpart is provided in (A.5).

$$\begin{cases} \dot{w} = f(w, v, t) \\ i = G(w, v, t)v \end{cases} \tag{A.5}$$

In [2], memristive systems were used to model three disparate systems. Firstly, memristive systems were shown capable of modeling thermistors, specifically showing that using the characterized thermistor equation, the thermistor is not a memoryless temperature-dependent linear resistor but a first-order time-invariant currentcontrolled memristor. Secondly, memristive system analysis was applied to the Hodgkin Huxley model; the results identified the potassium channel as a first order timeinvariant voltage controlled memristor and the sodium channel as a second-order time-invariant voltage controlled memristor. Thirdly, the discharge tube was analyzed, showing it can be modeled as a first-order time-invariant current-controlled memristor. The generalization of memristive systems allowed for the proper classification of different models of dynamic systems. The definition of memristive systems allowed for generic properties of such systems with some of the properties listed here.

Memristive systems were defined in the context of circuits, even though the aforementioned examples of memristors need not be specific to circuits. From a circuit standpoint, memristive systems can be made passive if $R(x, i, t) = 0$ in (A.4). If the system is passive then there is no energy discharge from the device, i.e., the instantaneous power entering the device is always non-negative. A current controlled memristor under periodic operation will always form a $v - i$ Lissajous figure whose voltage v can be at most a double valued function of i. In the time invariant case, if $R(x, i) = R(x, -i)$, then the $v - i$ Lissajous figure will possess an odd symmetry with respect to the origin. For more memristive system properties and proofs of the listed properties, refer to [2].

Experimental Realizations of Memristors

Chua's work laid the groundwork for the theoretical concept of the memristor, but the device was not recognized until 2008 in HP Labs [3]. Although different devices and materials were investigated for resistive RAM [4–7], HP was the first to associate the properties of their device as a direct connection to memristive systems. This section will deal with the different types of memristors and their associated transport mechanisms. The information summarized in this section can be found in [3, 8–11].

The HP Labs memristor is in the MIM configuration whereby platinum electrodes sandwich a TiO_2 thin film as shown in Figure A.2. The TiO_2 thin film is composed of two parts: a stoichiometric highly resistive layer (TiO_2) and an oxygen deficient highly conductive layer (TiO_{2-x}).

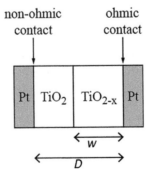

Figure A.2 HP memristor device showing appropriate layers and identifying the ohmic and non-ohmic contacts.

In the low resistive state, the memristor behavior is dominated by carrier tunneling through the metal-oxide layer while in the high resistive state, its behavior is rectifying [12]. The change in resistance of the memristor happens due to manipulation of oxygen vacancies in response to applied bias. This manipulation is in fact modeled through a modulation of an effective width w in Figure A.2. w can then serve as the state variable in order to determine the transport characteristics of the device in accordance with the definition of memristive system espoused in (A.5) and (A.4).

From device measurements [12], the fabricated memristor exhibits a rectifying behavior, thereby suggesting that the non-ohmic contact at the Pt/TiO_2 interface influences electrical transport in the device. The oxygen vacancies in the TiO_{2-x} side make the TiO_{2-x}/Pt contact an ohmic contact, allowing a model whereby a series resistance can be attributed to this side of the memristor and a more complex rectifying behavior is attributed to the other side of the memristor. Since tunneling through the TiO_2 barrier determines the current through the memristor, the non-ohmic interface is said to dominate the transport mechanism of HP's proposed structure. The value of w is proportional to the time integral of the voltage applied to the memristor and is normalized between 0 and 1 for the high resistive state and low resistive state, respectively. The current behavior of the HP memristor is described by (A.6).

$$I = w^n \beta \sinh(\alpha V) + \chi(exp(\gamma V) - 1) \tag{A.6}$$

In (A.6), the first term $\beta \sinh(\alpha V)$ is used to approximate the lowest resistive state of the device, and α and β are fitting parameters. The second term of (A.6), $\chi (exp(\gamma V)-1)$, approximates the rectifying behavior of the memristor with χ and γ as fitting parameters. The value for n suggests a nonlinear dependence of the vacancy drift velocity on the voltage applied to the memristive device. The value for n after fitting parameters ranges from 14 to 22, thereby suggesting applied voltage exhibits a highly nonlinear relationship with vacancy drift [12]. This highly nonlinear behavior has led to several drift models using effective ion drift values.

The dynamics of w which completes the description of the HP thin film device as a memristor is provided in [8] as

$$\dot{w} = f_{off} \sinh\left(\frac{i}{i_{off}}\right) \exp\left[-\exp\left(\frac{w - a_{off}}{w_c} - \frac{|i|}{b}\right) - \frac{w}{w_c}\right], \quad i > 0 \tag{A.7a}$$

$$\dot{w} = f_{on} \sinh\left(\frac{i}{i_{on}}\right) \exp\left[-\exp\left(\frac{w - a_{on}}{w_c} - \frac{|i|}{b}\right) - \frac{w}{w_c}\right], \quad i < 0$$

$$(A.7b)$$

As with (A.6), parameter fitting is used to describe w, so f_{off}, f_{on}, i_{off}, i_{on}, a_{off}, a_{on}, b, and w_c are the parameters to be set according to [15]. So with the dynamics of (A.7a) and (A.7b) combined with the current description (A.6), the complete model of the TiO_2 memristor describing the specific HP device is complete.

Metal oxides are not the only candidates for memristive devices. In [11], the memristor structure is composed of silver and silicon, specifically Ag/a-Si/p-Si. The insulating layer is the a-Si while the contacts are Ag and heavily doped p-type crystalline silicon. The memristor state change in this device is achieved by the drift of Ag ions towards the p-Si when voltage is applied to the device. The Ag ion drift into the a-Si layer causes traps that lower the effective resistance of the entire device as a whole. The ON to OFF resistance ratios of the a-Si memristor has been shown to range from 10^3 to 10^7. With realized memristors, the models provided are mostly parameters fitted to experimental results. This thesis takes a more generic approach based on the work presented in [14]. The next section describes the memristor model used for simulation.

Memristor Modeling

The memristor model used for simulation is based on the nonlinear drift model with window function F_p (A.9) as defined by [14] and [15]. The model is based on the HP TiO_2 device with the variables w and D identified in Figure A.2. The doped region width w is modulated according to (A.8) with the window function definition expressed in (A.9). For SPICE simulation the memristor model was implemented as a functional block in Verilog-A with parameter $p = 4$, memristor width $D = 10$ nm, and dopant mobility $\mu_D = 10^{-9} \text{ cm}^2/V \cdot s$.

$$\frac{dw}{dt} = \frac{\mu_D R_{ON}}{D} i(t) F \frac{w}{D} \tag{A.8}$$

$$F_p(x) = 1 - (2x - 1)^{2p} \tag{A.9}$$

The memristor's resistance is viewed in the 2D framework, whereby effective resistances of the oxygen deficient region (or doped region) and the effective resistance of the undoped region are weighted and added. This linear combination is described in (A.10), where R_{OFF} is the resistance of the undoped region and R_{ON} is the resistance of the doped region.

$$M(w) = \frac{w}{D}R_{ON} + \left(1 - \frac{w}{D}\right)R_{OFF} \qquad (A.10)$$

Joglekar and Wolf [14] performed two different derivations on the linear combination proposed. The first is dubbed the nonlinear drift model and is obtained by combining (A.8) and (A.9) with (A.10) using integer parameter $p > 1$. The second method is the linear drift model which is obtained by using an all pass window function which has the value of 1. The linear drift model provides the closed form analytic model in (A.11), while the nonlinear model must be solved numerically.

$$M_T = R_0\sqrt{1 - \frac{2 \cdot \eta \cdot \Delta R \cdot \phi(t)}{Q_o \cdot R_0^2}} \qquad (A.11)$$

The memristance values over time follow the definition of M_T in (A.11). In this definition, M_T is the total memristance, R_0 is the initial resistance of the memristor, η is related to applied bias (+1 for positive and −1 for negative), ΔR is the memristor's resistive range (difference between maximum resistance and minimum resistance), $\varphi(t)$ is the total flux through the device, and Q_0 is the charge required to pass through the memristor for dopant boundary to move a distance comparable to the device width. So $Q_0 = D^2/(\mu_D R_{ON})$, where D is device thickness and μ_D is dopant mobility, as previously discussed.

Modeling and setup applied to Memory Chapter: The memristor crossbar is an important element for ultra-dense digital memories. The crossbar structure has a device at each crosspoint, therefore possessing the quality of a very dense device population compared to CMOS. The crossbar is composed of nanowires connecting memristors in a pitch width smaller than that of CMOS. The crossbar also scales better than CMOS, thereby suggesting the process for building or fabricating this structure is different from the standard CMOS process. For memory simulation, the crosspoint devices have diode isolation of individual devices in accordance with [16]. The memristor is in series with a bi-directional diode model, representative of the MIM diode. In order to model worst case effects, P-N diode model is used for each direction of the bi-directional diode model, with each forward path presented in (A.12).

$$I_{Diode} = I_0(e(qVD/(nkT)) - 1) \qquad (A.12)$$

Overall, the simulation parameters for the diodes were: $I_0 = 2.2$ fA, $kT/q = 25.85$ mV, V_D is dependent on applied bias, and $n = 1.08$. A P-N diode model is used because it provides a weaker isolation than actual MIM diodes. Therefore, if the proposed adaptive method works with P-N diode configuration, then it will work better with actual MIM configuration that depends on tunneling currents and provides better isolation than P-N diodes. Nanowire modeling for simulation is a distributed pimodel, but for hand calculations, a lumped model will be used for simplicity. The numbers used for the crossbar are per unit length resistance in order to obtain fair results. From Snider and Williams [17], nanowire resistivity follows:

$$\rho/\rho_0 = 1 + 0.75 \times (1 - p)(\lambda/d) \tag{A.13}$$

Where ρ_0 is bulk resistivity, d is nanowire width, and λ is mean free path. The nanowire recorded values used for simulation were: 24 $\mu \cdot \Omega$cm for 4.5 nm thick Cu. Following a conservative estimate in the memory application of Chapter 5, the nanowire resistance was chosen to be 24 kΩ total. Using a nanowire capacitance of 2.0 pF $\cdot$ cm^{-1}, the nanowire modeling was made transient complete.

Modeling and setup applied to Neuromorphic Work: For the neuromorphic work, the memristor crossbar is not utilized, so individual memristor characteristics are more important. Since the overall model is based on the HP Lab's device, a detailed valuation of a separation of the analog memristor is pursued as opposed to the digital memristor. The memristor model of HP labs gives rise to a device whose resistance change is proportional to applied bias. If applied bias is relatively low for a certain time span, then the change in memristance is very small and can be neglected. This idea allows for the establishment of a device threshold, whereby the memristor's resistance is assumed to be unchanged when bias is below this threshold value. This memristor behavior is seen not just in HP's device but also in the a-Si memristor in [11]. The a-Si memristor shows conformity to the idea of a built-in threshold, thereby allowing the authors to use different voltage biases for read/write interpretation. This memristor can withstand low current without resistance change, and this quality is important for analog circuit design usage of the memristor.

The memristor behavior already described allowed for the creation of a threshold based SPICE model proportional to conductance change magnitude, Δ_C, that follows (A.14).

$$\Delta_C = -M \times \sqrt[3]{(V_{ab} - V_{thp})(-V_{ab} - V_{thn})} + V_{off} \qquad (A.14)$$

In the above relationship, M is an amplitude correcting factor, V_{ab} is the applied bias across the terminals of the memristor, V_{thp} and V_{thn} are both threshold voltages of the memristor with a positive and negative applied bias respectively. V_{off} corrects and maintains a zero change with no applied bias. Equation (A.14) works really well for a symmetric device, and the simulation done in this work uses a device with the same magnitude in threshold voltage for both the positive and negative directions. This threshold behavior, in conjunction with the linear-drift model presented in [14], is used to implement a memristor with threshold characteristics.

The memristor threshold model does not assume zero change below the applied threshold voltage. The change is minimal, but not negligible, to some above threshold voltage applications as shown in a normalized plot of Δ_C vs. V_{ab} in Figure A.3. In circuit design, depending on application, the voltage choices between read and write pulses will determine how the memristive device is used. The read pulse is chosen to not cause drastic change in memristance, while the write pulse is chosen to encourage higher levels of conductance change than the read pulse.

For hand design purposes, it is useful to determine appropriate pulse widths and approximate memristance changes, for the change in memristance for each pulse is very important. The exact role of the thresholding factor Δ_C needs to be quantified.

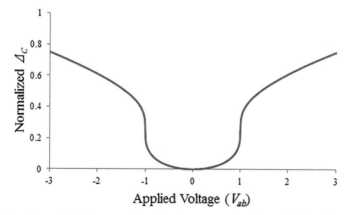

Figure A.3 Normalized Δ_C vs. V_{ab} showing proportional magnitude of conductance change as a function of applied bias. ± 1 V can be viewed as threshold voltages.

By taking the derivative of (A.11) with respect to $\varphi(t)$, the approximation of the change of memristance is:

$$\Delta M_T = \frac{-R_0 \cdot \eta \cdot \Delta R \cdot \phi(t)/(Q_0 R_0^2)}{\sqrt{1 - 2 \cdot \eta \cdot \Delta R \cdot \phi(t)/(Q_0 R_0^2)}} \cdot \Delta_C \qquad (A.15)$$

Equation (A.15) suggests that for successive small changes in $\Delta\varphi$ whereby $\varphi(t)$ is not affected significantly, then the change in memristance, ΔM_T, will respond with almost constant step changes. For analog memristor design applications, the designer is essentially taking advantage of this localized constant stepping for a range of $\varphi(t)$ values. The concept is represented in Figure A.4 by graphing (A.11) with respect to $\varphi(t)$.

The plot in Figure A.4 suggests an analog mode and a digital mode of operation for the memristor. The mode of operation is strongly linked to the concept of localized constant stepping range previously discussed. In Figure A.4, the decrease in memristance seems nearly linear at first and then exponentially increases. The nearly linear part of operation is where the memristor values should lie for the analog neural network functionality. In this region of operation, $\varphi(t) \leq 2.6$ Wb, the memristance decreases by about 2 MΩ to 3 MΩ in response to every 1 Wb change in $\varphi(t)$. This operating region is a design choice to allow for better flexibility in choosing voltage levels and pulse widths. Designs that desire higher changes with respect to chosen applied biases will most likely operate in the region closer to the digital device characteristics.

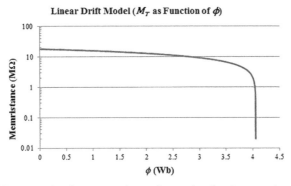

Figure A.4 M_T vs. φ showing two regions of operation for the memristor. In the slowly changing region, the magnitude of memristance change ranges from $\sim$2 MΩ to 3 MΩ for every 1 Wb flux change. The change in memristance increases drastically when φ is > $\sim$2.5 Wb. (Parameters used to simulate the analog memristor: $R_0 = 18$ MΩ, $Q_0 = 5 \times 10^{-7}$C, $\Delta R \approx 20 M\Omega$).

References

[1] Chua, L. O. (1971), Memristor – missing circuit element, *IEEE Transactions on Circuit Theory*, *CT18* (5), 507–519.

[2] Chua, L., and S. M. Kang (1976), Memristive devices and systems, *Proceedings of the IEEE*, *64*(2), 209–223, doi:10.1109/PROC.1976.10092.

[3] Strukov, D. B., G. S. Snider, D. R. Stewart, and R. S. Williams (2008), The missing memristor found, *Nature*, *453*(7191), 80–83.

[4] Scott, J., and L. Bozano (2007), Nonvolatile memory elements based on organic materials, *Advanced Materials*, *19*(11), 1452–1463.

[5] Waser, R., and M. Aono (2007), Nanoionics-based resistive switching memories, *Nature Materials*, *6*(11), 833–840.

[6] Pagnia, H., and N. Sotnik (1988), Bistable switching in electroformed metalinsulatormetal devices, *physica status solidi (a)*, *108*(1), 11–65, doi:10.1002/pssa.2211080102.

[7] Beck, A., J. G. Bednorz, C. Gerber, C. Rossel, and D. Widmer (2000), Reproducible switching effect in thin oxide films for memory applications, *Applied Physics Letters*, *77*(1), 139–141.

[8] Strukov, D. B., G. S. Snider, D. R. Stewart, and R. S. Williams (2008), The missing memristor found, *Nature*, 453(7191), 80–83.

[9] Strukov, D. B., J. L. Borghetti, and R. S. Williams (2009), Coupled ionic and electronic transport model of thin-film semiconductor memristive behavior, *Small*, 5(9), 1058–1063.

[10] Pickett, M. D., D. B. Strukov, J. L. Borghetti, J. J. Yang, G. S. Snider, D. R. Stewart, and R. S. Williams (2009), Switching dynamics in titanium dioxide memristive devices, *Journal of Applied Physics*, *106*(7), 074508, doi:10.1063/1.3236506.

[11] Jo, S. H., and W. Lu (2008), Cmos compatible nanoscale nonvolatile resistance, switching memory, *Nano Letters*, *8*(2), 392–397.

[12] Jo, S. H., T. Chang, I. Ebong, B. B. Bhadviya, P. Mazumder, and W. Lu (2010), Nanoscale memristor device as synapse in neuromorphic systems, *Nano Letters*, *10*(4), 1297–1301, doi:10.1021/nl904092h.

[13] Yang, J. J., M. D. Pickett, X. M. Li, D. A. A. Ohlberg, D. R. Stewart, and R. S. Williams (2008), Memristive switching mechanism for metal/oxide/metal nanodevices, *Nature Nanotechnology*, *3*(7), 429–433.

[14] Joglekar, Y. N., and S. J. Wolf (2009), The elusive memristor: properties of basic electrical circuits, *European Journal of Physics*, 30(4), 661–675.

[15] Biolek, Z., D. Biolek, and V. Biolkova (2009), Spice model of memristor with nonlinear dopant drift, *Radioengineering*, 18(2), 210–214.

[16] Rinerson, D., C. J. Chevallier, S. W. Longcor, W. Kinney, E. R. Ward, and S. K. Hsia (2005), Re-writable memory with non-linear memory element, patent Number: US 6 870–755.

[17] Snider, G. S., and R. S. Williams (2007), Nano/cmos architectures using a field-programmable nanowire interconnect, *Nanotechnology*, 18(3), 11.

Index

About the Authors

Pinaki Mazumder is currently a Professor with the Department of Electrical Engineering and Computer Science, University of Michigan (UM), Ann Arbor. He was for six years with industrial R&D centers that included AT&T Bell Laboratories, where in 1985, he started the CONES Project the first C modeling-based very large scale integration (VLSI) synthesis tool at Indias premier electronics company, Bharat Electronics, Ltd., India, where he had developed several high-speed and high-voltage analog integrated circuits intended for consumer electronics products. He is the author or coauthor of more than 200 technical papers and four books on various aspects of VLSI research works. His current research interests include current problems in nanoscale CMOS VLSI design, computer-aided design tools, and circuit designs for emerging technologies including quantum MOS and resonant tunneling devices, semiconductor memory systems, and physical synthesis of VLSI chips. Dr. Mazumder is a Fellow of the American Association for the Advancement of Science (2008). He was a recipient of the Digitals Incentives for Excellence Award, BF Goodrich National Collegiate Invention Award, and Defense Advanced Research Projects Agency Research Excellence Award.

Yalcin Yilmaz is currently a principal design engineer in the Tensilica microprocessor design team in Cadence Design Systems Inc. He has been working on developing micro-architecture and specification of microprocessor cores, multiprocessor sub-systems and their peripherals. His previous research experience includes the modeling, simulation, low power digital and analog circuit designs using CMOS and emerging technologies including resonant tunneling diodes and memristors. He holds a Ph.D in electrical engineering from the University of Michigan.

Idongesit Ebong is currently a patent agent at Leydig, Voit & Mayer, Ltd. with an emphasis in electrical, computer, and mechanical engineering. He is experienced in design of analog and digital circuits and circuits involving exotic device structures like memristors, resonant tunneling diodes, and tunneling transistors. He has prepared U.S. and International patent applications, for both small and large clients, in the following practice areas: machine learning, neural networks, telecommunications, optical systems, semiconductor devices, and industrial measurement and fabrication equipment. He holds a Ph.D in electrical engineering from the University of Michigan.

Woo Hyung Lee is currently an engineering manager in the CPU core development team in Intel corporation. He has been leading a design team that is responsible for design optimization for all the design quality criteria and design flow development to enhance the design tools. Prior to join intel, he served as a technical leader in Oracle and Apple corporation. He holds a Ph.D in electrical engineering from the University of Michigan.